Atomic Layer Growth and Processing

MATERIALS RESEARCH SOCIETY SYMPOSIUM PROCEEDINGS VOLUME 222

Atomic Layer Growth and Processing

Symposium held April 29-May 1, 1991, Anaheim, California, U.S.A.

EDITORS:

T.F. Kuech
University of Wisconsin, Madison, Wisconsin, U.S.A.

P.D. Dapkus
University of Southern California, Los Angeles, California, U.S.A.

Y. Aoyagi
Riken, Institute of Physical and Chemical Research, Wako-Shi, Japan

MRS MATERIALS RESEARCH SOCIETY
Pittsburgh, Pennsylvania

This work was supported in part by the Office of Naval Research under Grant Number ONR-N00014-91-J-1412.

Single article reprints from this publication are available through University Microfilms Inc., 300 North Zeeb Road, Ann Arbor, Michigan 48106

CODEN: MRSPDH

Published by:

Materials Research Society
9800 McKnight Road
Pittsburgh, Pennsylvania 15237
Telephone (412) 367-3003

Library of Congress Cataloging in Publication Data

Atomic layer growth and processing : symposium held April 29-May 1, 1991, Anaheim, California, U.S.A. / editors, T.F. Kuech, P.D. Dapkus, Y. Aoyagi

p. cm. — (Materials Research Society symposium proceedings : ISSN 0272-9172 ; v. 222)

Includes bibliographical references and index.

ISBN 1-55899-116-6

1. Compound semiconductors—Congresses. 2. Crystals—Growth—Congresses. 3. Epitaxy—Congresses. 4. High temperature superconductors—Congresses. I. Kuech, T.F. II. Dapkus, P.D. III. Aoyagi, Y. IV. Series: Materials Research Society symposium proceedings : v. 222.

TK7871.99.C65A86 1991 91-26832
621.381'52—dc20 CIP

Manufactured in the United States of America

Contents

*Invited Paper

*Invited Paper

*Invited Paper

*Invited Paper

Preface

The first symposium on Atomic Layer Growth and Processing was held during the Spring 1991 Materials Research Society Meeting, April 19 - May 1, Anaheim, California. This symposium consisted of 13 invited papers and 35 contributed talks, representing the major research activities throughout the world. The future demands of the semiconductor industry, requiring an ever finer control over the thickness and composition of the layers comprising modern device structures, has spurred activity into this technologically challenging field. The concept of Atomic Layer based processing is perhaps a primary road to achieving both the degree of control and processing uniformity for both etching and deposition. At present, this measure of control, affecting a single layer of atoms or molecules on surface per cycle of the process, is being extended to the whole range of basic processes in the semiconductor technology. Atomic layer growth, etching, and oxidation processes are being developed. While compound semiconductor deposition and epitaxy appears to be the main thrust of most of the research activity, this volume contains contributions developing the Atomic Layer Growth of Si, superconductors, metals, and oxides.

This volume consists of 46 papers which were submitted at the symposium covering the range of contributions. All the papers were peer reviewed and the authors were given the opportunity to respond the remarks of the referees. The organization of this volume is different from the order of presentation during the symposium. Part 1 consists of papers primarily concerned with basic studies of the underlying mechanisms of the self-limiting nature characteristic of Atomic Layer Processes. Atomic Layer Epitaxy of GaAs, using trimethyl gallium, is presently the most well understood chemical system, largely due to these and related studies. The application of new characterization techniques, sensitive to the nature of the growth surface, has become a major focus of the work in the field. The contributions to this area of activity form Part 2. The remainder of the papers are organized by the material system under study: III-V or II-VI semiconductors, group IV elemental semiconductors, high temperature superconductors and a variety of other materials.

This volume represents the best of our current understanding of the nature and application of Atomic Layer Processing. The contributions within this volume serve as excellent indicators of the direction of the field and variety of material processes and problems which may be addressed within the context of atomically controlled processing.

Thomas F. Kuech
P.D. Dapkus
Y. Aoyagi

July 1991

Acknowledgments

We would like to thank the invited speakers as well as the presenters of the contributed papers for their contributions to the wide variety of topics presented within the symposium. The organizers would also like to thank the session chairpersons for their help in running the three days of talks. The help of J. Redwing and J. Geisz in the processing of the various papers contained within this volume is gratefully acknowledged.

We would finally like to thank both the Office of Naval Research and EMCORE whose generous financial support of the symposium makes this volume possible.

Volume 201—Surface Chemistry and Beam-Solid Interactions, H. Atwater, F.A. Houle, D. Lowndes, 1991, ISBN: 1-55899-093-3

Volume 202—Evolution of Thin Film and Surface Microstructure, C.V. Thompson, J.Y. Tsao, D.J. Srolovitz, 1991, ISBN: 1-55899-094-1

Volume 203—Electronic Packaging Materials Science V, E.D. Lillie, P. Ho, R.J. Jaccodine, K. Jackson, 1991, ISBN: 1-55899-095-X

Volume 204—Chemical Perspectives of Microelectronic Materials II, L.V. Interrante, K.F. Jensen, L.H. Dubois, M.E. Gross, 1991 ISBN: 1-55899-096-8

Volume 205—Kinetics of Phase Transformations, M.O. Thompson, M. Aziz, G.B. Stephenson, D. Cherns, 1991, ISBN: 1-55899-097-6

Volume 206—Clusters amd Cluster-Assembled Materials, R.S. Averback, J. Bernholc, D.L. Nelson, 1991, ISBN: 1-55899-098-4

Volume 207—Mechanical Properties of Porous and Cellular Materials, K. Sieradzki, D. Green, L.J. Gibson, 1991, ISBN-1-55899-099-2

Volume 208—Advances in Surface and Thin Film Diffraction, T.C. Huang, P.I. Cohen, D.J. Eaglesham, 1991, ISBN: 1-55899-100-X

Volume 209—Defects in Materials, P.D. Bristowe, J.E. Epperson, J.E. Griffith, Z. Liliental-Weber, 1991, ISBN: 1-55899-101-8

Volume 210—Solid State Ionics II, G.-A. Nazri, D.F. Shriver, R.A. Huggins, M. Balkanski, 1991, ISBN: 1-55899-102-6

Volume 211—Fiber-Reinforced Cementitious Materials, S. Mindess, J.P. Skalny, 1991, ISBN: 1-55899-103-4

Volume 212—Scientific Basis for Nuclear Waste Management XIV, T. Abrajano, Jr., L.H. Johnson, 1991, ISBN: 1-55899-104-2

Volume 213—High-Temperature Ordered Intermetallic Alloys IV, L.A. Johnson, D.P. Pope, J.O. Stiegler, 1991, ISBN: 1-55899-105-0

Volume 214—Optical and Electrical Properties of Polymers, J.A. Emerson, J.M. Torkelson, 1991, ISBN: 1-55899-106-9

Volume 215—Structure, Relaxation and Physical Aging of Glassy Polymers, R.J. Roe, J.M. O'Reilly, J. Torkelson, 1991, ISBN: 1-55899-107-7

Volume 216—Long-Wavelength Semiconductor Devices, Materials and Processes, A. Katz, R.M. Biefeld, R.L. Gunshor, R.J. Malik, 1991, ISBN 1-55899-108-5

Volume 217—Advanced Tomographic Imaging Methods for the Analysis of Materials, J.L. Ackerman, W.A. Ellingson, 1991, ISBN: 1-55899-109-3

Volume 218—Materials Synthesis Based on Biological Processes, M. Alper, P.D. Calvert, R. Frankel, P.C. Rieke, D.A. Tirrell, 1991, ISBN: 1-55899-110-7

Volume 219—Amorphous Silicon Technology—1991, A. Madan, Y. Hamakawa, M. Thompson, P.C. Taylor, P.G. LeComber, 1991, ISBN: 1-55899-113-1

Volume 220—Silicon Molecular Beam Epitaxy, 1991, J.C. Bean, E.H.C. Parker, S. Iyer, Y. Shiraki, E. Kasper, K. Wang, 1991, ISBN: 1-55899-114-X

Volume 221—Heteroepitaxy of Dissimilar Materials, R.F.C. Farrow, J.P. Harbison, P.S. Peercy, A. Zangwill, 1991, ISBN: 1-55899-115-8

Volume 222—Atomic Layer Growth and Processing, Y. Aoyagi, P.D. Dapkus, T.F. Kuech, 1991, ISBN: 1-55899-116-6

Volume 223—Low Energy Ion Beam and Plasma Modification of Materials, J.M.E. Harper, K. Miyake, J.R. McNeil, S.M. Gorbatkin, 1991, ISBN: 1-55899-117-4

Volume 224—Rapid Thermal and Integrated Processing, M.L. Green, J.C. Gelpey, J. Wortman, R. Singh, 1991, ISBN: 1-55899-118-2

Volume 225—Materials Reliability Issues in Microelectronics, J.R. Lloyd, P.S Ho, C.T. Sah, F. Yost, 1991, ISBN: 1-55899-119-0

Volume 226—Mechanical Behavior of Materials and Structures in Microelectronics, E. Suhir, R.C. Cammarata, D.D.L. Chung, 1991, ISBN: 1-55899-120-4

Volume 227—High Temperature Polymers for Microelectronics, D.Y. Yoon, D.T. Grubb, I. Mita, 1991, ISBN: 1-55899-121-2

Volume 228—Materials for Optical Information Processing, C. Warde, J. Stamatoff, W. Wang, 1991, ISBN: 1-55899-122-0

Volume 229—Structure/Property Relationships for Metal/Metal Interfaces, A.D Romig, D.E. Fowler, P.D. Bristowe, 1991, ISBN: 1-55899-123-9

Volume 230—Phase Transformation Kinetics in Thin Films, M. Chen, M. Thompson, R. Schwarz, M. Libera, 1991, ISBN: 1-55899-124-7

Volume 231—Magnetic Thin Films, Multilayers and Surfaces, H. Hopster, S.S.P. Parkin, G. Prinz, J.-P. Renard, T. Shinjo, W. Zinn, 1991, ISBN: 1-55899-125-5

Volume 232—Magnetic Materials: Microstructure and Properties, T. Suzuki, Y. Sugita, B.M. Clemens, D.E. Laughlin, K. Ouchi, 1991, ISBN: 1-55899-126-3

Volume 233—Synthesis/Characterization and Novel Applications of Molecular Sieve Materials, R.L. Bedard, T. Bein, M.E. Davis, J. Garces, V.A. Maroni, G.D. Stucky, 1991, ISBN: 1-55899-127-1

Volume 234—Modern Perspectives on Thermoelectrics and Related Materials, D.D. Allred, G. Slack, C. Vining, 1991, ISBN: 1-55899-128-X

Prior Materials Research Society Symposium Proceedings available by contacting Materials Research Society.

PART I

Mechanistic Studies of Atomic Layer Processes

SURFACE CHEMISTRY AND MECHANISM OF ATOMIC LAYER GROWTH OF GaAs

MING L. YU, NICHOLAS I. BUCHAN*, RYUTARO SOUDA**, AND THOMAS F. KUECH***
IBM Research Division, T. J. Watson Research Center, Yorktown Heights, NY 10598

ABSTRACT

The success in attaining atomic layer epitaxy (ALE) of GaAs depends critically on the choice of the Ga precursor. Three systems were examined: trimethylgallium (TMGa) and diethylgallium chloride (DEGaCl) both of which give ALE, and triethylgallium (TEGa) which does not. We compared the surface reactions of these compounds on GaAs(100) and concluded that there was no evidence for reaction selectivity between Ga and As sites to cause ALE. Site blocking by the ligands on the Ga precursors alone also could not provide a self-limiting Ga deposition for ALE. We found evidence of a new mechanism by which self-limiting deposition of Ga resulted when the incoming Ga flux by the adsorption of Ga precursors was counter-balanced by an outgoing flux of Ga containing reaction product. For TMGa and DEGaCl with which ALE is successful, the products are CH_3Ga and GaCl, respectively. For TEGa, the corresponding compound C_2H_5Ga was not formed.

INTRODUCTION

Atomic layer epitaxy (ALE) is a process by which epitaxy growth can be controlled at the atomic level. ALE relies critically on the ability to deposit atomic or molecular species one atomic (molecular) layer at a time. By the proper choice of chemical precursors in chemical vapor deposition (CVD), the surface chemical reaction that leads to the material deposition can also be used to self-limit the deposition to the required amount. This concept has been successfully demonstrated in the ALE of GaAs. The most studied system is the combination of trimethylgallium (TMGa, $(CH_3)_3Ga$) and arsine (AsH_3) [1 – 5]. TMGa and arsine are usually introduced separately in an alternating sequence. TMGa decomposes on the As-rich GaAs surface to deposit Ga up to a self-limiting coverage. Then arsine reacts with the Ga to form a molecular layer of GaAs. The remarkable dependence of the ALE process on the chemistry is illustrated by replacing the CH_3 ligands by C_2H_5. The resulting triethylgallium (TEGa, $(C_2H_5)_3Ga$) can no longer provide self-limiting Ga deposition and ALE fails [5 – 7]. On the other hand, when one of the C_2H_5 ligands in TEGa is replaced by chlorine, the resulting diethylgallium chloride (DEGaCl, $(C_2H_5)_2GaCl$) again can give self-limiting Ga deposition. Recently, ALE of GaAs using diethylgallium chloride (DEGaCl) and arsine [8] or arsenic [9] combination has been demonstrated.

There have been several mechanisms proposed for the self-limiting deposition of Ga. The prevailing one is the site-blocking mechanism. Nishizawa and Kurabayashi [1], Watanabe et al. [3], and Creighton et al. [10] all proposed that the CH_3 ligands from the adsorbed TMGa molecules block adsorption sites on the GaAs surface either by preventing further reaction with the ambient gaseous TMGa or by causing excess TMGa to be desorbed. Another proposal is the site-selectivity model of Sakuma et al. [6]. They proposed that TMGa deposits Ga readily on As sites while the decomposition probability of TMGa on the Ga atoms is negligible. They also proposed that TEGa has no such selectivity and hence no self-limiting Ga deposition was observed.

Information on the chemistry of these Ga-alkyls on GaAs surfaces is becoming available through surface science experiments performed in ultrahigh vacuums [1, 11 – 16]. We have performed experiments to study the surface reactions of TMGa, TEGa, and DEGaCl to try to test the mechanisms proposed for ALE. In particular, the TEGa system where ALE fails is an excellent example with which other ALE successful systems can be compared. We found that both proposed mechanisms discussed above cannot adequately explain the ALE results. The self-limiting deposition of Ga depends instead on the balance between the incoming Ga flux and the outgoing Ga flux in the form of a reaction product.

EXPERIMENTAL

Our studies were all performed in ultrahigh vacuum systems with base pressures of about 2×10^{-10} torr. The experimental setup and procedures have been reported elsewhere [13,16]. Our samples were .001 Ωcm, n-type GaAs(100), $2° \rightarrow < 110 >$ wafers since they are most often used in the CVD of GaAs. The nominal starting surfaces were the c(2×8) As-rich (100) surfaces. We used X-ray photoemission (XPS) to identify the adsorbed species, ultraviolet photoemission (UPS) to monitor the valence orbitals, and high resolution electron energy loss spectroscopy (HREELS) [17] to study the vibrational modes of the organic ligands on the surfaces.

We also studied the reactions at growth temperatures by simulating the ALE process with pulsed molecular beams of the Ga alkyls. By dosing the GaAs surfaces with the molecular beam, and using a mass spectrometer to monitor the reaction products, the sticking coefficient of the Ga precursors and the branching of the reaction pathways can be studied during growth. When the mass spectrometer was operated in the high time-resolution mode, the desorption kinetics of the various reaction products can also be measured.

RESULTS AND DISCUSSIONS

Reactions of triethylgallium on GaAs(100)

The reaction of TEGa on GaAs(100) has been studied by Murrell et al. [14], and Donnelly and McCaulley [15] using temperature programmed desorption and other surface diagnostics. According to Murrell et al., the Ga deposition reaction proceeds by the evolution of ethene (C_2H_4), ethane (C_2H_6) and hydrogen:

$$(C_2H_5)_3Ga \rightarrow Ga + xC_2H_4\uparrow + (3-x)C_2H_6\uparrow + (x-1.5)H_2\uparrow \quad (1)$$

The ethene signal was about twenty five times stronger than the ethane signal. Donnelly and McCaulley detected instead the evolution of mostly C_2H_4 and some C_2H_5:

$$(C_2H_5)_3Ga \rightarrow Ga + xC_2H_4\uparrow + (3 - x)C_2H_5\uparrow + (x/2)H_2\uparrow \qquad (2)$$

In our molecular beam / mass spectrometry experiment, we observed that the C_2H_5 ligands were desorbed in the form of C_2H_4 and C_2H_5, supporting Connelly and McCaulley's observation. The $C_2H_6{}^+$ signal was only about one twentieth of the $C_2H_5{}^+$ signal indicating that the production of C_2H_6 was much smaller than that of C_2H_5. Interestingly, we also found that C_2H_4 and C_2H_5 desorbed with the same kinetics. Ultraviolet photoemission measurements indicated there were two bonding configurations for the C_2H_5 ligands. We found in the dilute coverage limit that the thermal removal of the C_2H_5 ligands could be modelled by the combination of two first-order desorption rates. These rates are listed in Table 1.

Table I. Rate of desorption of reaction products

TMGA	
CH_3:	$10^{(13.41 \pm 0.48)} \exp[-(37.9 \pm 1.6\ \mathrm{kcal/mole})/k_BT]s^{-1}$
CH_3:	$10^{(14.48 \pm 0.43)} \exp[-(45.0 \pm 1.4\ \mathrm{kcal/mole})/k_BT]s^{-1}$
TEGa	
C_2H_4/C_2H_5:	$10^{(8.13 \pm 0.74)} \exp[-(17.4 \pm 2.3\ \mathrm{kcal/mole})/k_BT]s^{-1}$
C_2H_4/C_2H_5:	$10^{(9.53 \pm 0.67)} \exp[-(23.9 \pm 2.1\ \mathrm{kcal/mole})/k_BT]s^{-1}$
DEGaCl	
C_2H_4/C_2H_5:	$10^{(11.1 \pm 0.7)} \exp[-(27.5 \pm 2.1\ \mathrm{kcal/mole})/k_BT]s^{-1}$
C_2H_4/C_2H_5:	$10^{(11.2 \pm 0.7)} \exp[-(30.3 \pm 2.2\ \mathrm{kcal/mole})/k_BT]s^{-1}$
GaCl:	$10^{(14.0 \pm 0.5)} \exp[-(38.9 \pm 1.6\ \mathrm{kcal/mole})/k_BT]s^{-1}$

Our sticking coefficient measurement indicated that the adsorbed C_2H_5 ligands caused the blocking of adsorption sites. We estimated the effective sticking coefficient by measuring the reflected $(C_2H_5)_2Ga^+$ signal [16]. The $(C_2H_5)_2Ga^+$ signal could come from the unreacted (reflected) TEGa and $(C_2H_5)_2Ga$, if produced. Both of which did not result in Ga deposition. No C_2H_5Ga was detected. Figure 1 depicts the changes of the effective sticking coefficient with TEGa dosage (in number of TEGa pulses) at various adsorption temperatures. The Ga coverage increased with the TEGa dosage. At room temperature, the initial sticking coefficient was close to unity and decreased only weakly with coverage, typical of molecular precursor mediated adsorption [18]. But as the saturation coverage was approached, the sticking coefficient dropped rapidly to zero, showing the effective blocking of the adsorption sites by the adsorbed C_2H_5 ligands. While site blocking was still observable at 250 and 350° C, the rate of thermal removal of the C_2H_5 ligands became comparable to the rate of arrival of the TEGa molecules. The steady state sticking

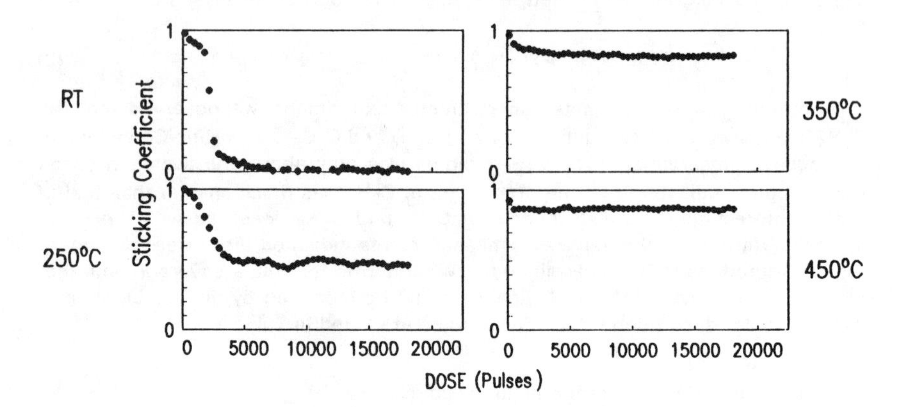

Fig. 1 Effective sticking coefficient of TEGa on GaAs(100) as a function of TEGa dosage at room temperature, 250°C, 350°C, and 450°C.

coefficient was no longer zero at higher temperatures. With this non-zero steady state sticking coefficient, Ga continued to accumulate on the surface and the deposition was hence not self-limiting.

That the sticking coefficient remained constant with dosage (hence Ga coverage) at high coverages shows that the adsorption of TEG has no selectivity between As and Ga sites. In fact, selectivity was also not observed for TMGa and DEGaCl (see below), though both could be used for ALE. Hence the issue is why site-blocking alone did not provide self-limiting Ga deposition at elevated temperatures. We shall discuss the implications of the TEGa results with a simple model. Let I_B be the rate of arrival of the Ga precursors on the surface, $[C_2H_5]$ the surface concentration of the C_2H_5 ligands, s_0 the sticking coefficient on the active sites. In this paper, the surface concentrations will all be normalized to the surface site-density and they will be in the unit of monolayers (ML). We assume for simplicity that C_2H_5 ligands were thermally desorbed (as C_2H_5 and C_2H_4/H_2) with a first order rate constant k. The rate equations are:

$$\frac{d[C_2H_5]}{dt} = 3I_B s - k[C_2H_5]$$
$$\frac{d[Ga]}{dt} = I_B s \tag{3}$$

where s, the sticking coefficient, is reduced due to site-blocking:

$$s = s_0\{1 - [C_2H_5]\} \tag{4}$$

At steady state, the rate of adsorption of C_2H_5 is balanced by the desorption rate. The steady state C_2H_5 coverage is hence given by:

$$[C_2H_5]_s = 3I_B s_0/(3I_B s_0 + k) \tag{5}$$

The C_2H_5 coverage is always less than unity. The rate of deposition of Ga, as given by eq. (3) is hence always finite:

$$\frac{d[Ga]}{dt} = \frac{kI_B s_0}{(3I_B s_0 + k)} \tag{6}$$

In the limit of large TEGa flux I_B, the deposition rate is k/3. For example, it is about 1.4 ML/s at 350°C if $k = k_2$ in Table 1. Hence the Ga deposition is not self-limiting. This is consistent with the failure of TEGa for ALE.

Reaction of diethylgallium chloride on GaAs(100)

With one of the C_2H_5 ligands in TEGa replaced by chlorine, the new molecule DEGaCl is again suitable for ALE. ALE was observed at about 400 to 600° C [8,9] growth temperatures, and at 5.9×10^{-3} torr [8] to above 3×10^{-2} torr [9] DEGaCl partial pressures. The reactor total pressure was about 100 torr to atmospheric pressure. The gas transport was in the hydrodynamic regime.

Our photoemission studies showed that the DEGaCl molecule adsorbed dissociatively on the surface. The initial sticking coefficient was close to unity and has little dependence on the Ga/As stoichiometry of the surface, showing the non-site specific adsorption as in TEGa. The adsorbed C_2H_5 ligands and Cl blocked adsorption sites as expected. At elevated temperatures we observed the desorption of C_2H_4, C_2H_5 and GaCl in our molecular beam experiment. Similar to TEGa, we found that C_2H_4 and C_2H_5 have practically identical desorption rates which are within a factor of two as those observed for TEG in the temperature range for film growth. Our surface studies indicated that the C_2H_5 ligands from TEGa and DEGaCl have similar adsorption properties. As the C_2H_5 ligands could not provide the mechanism for ALE, we shall focus our discussion on the behavior of GaCl. The overall reaction pathway is hence given by:

$$(C_2H_5)_2GaCl \rightarrow GaCl + xC_2H_4\uparrow + (2-x)C_2H_5\uparrow + (x/2)H_2\uparrow \tag{7}$$

Mori et al. [8] first proposed that DEGaCl decomposes to form a monolayer of GaCl on the GaAs surface. Chlorine can then be removed either by arsenic or arsine in the next step of the deposition cycle. Our surface science study was consistent with this picture [19]. The XPS data showed that Ga deposition was always associated with the presence of Cl, and GaCl was the only chlorine containing reaction product detected unambiguously in our molecular beam / mass spectrometry experiment. Figure 2 shows the Arrhenius plot of our measured GaCl desorption rate at dilute coverages.

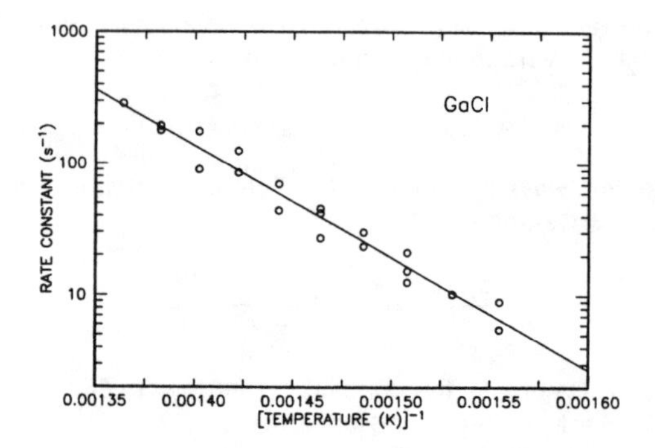

Fig. 2 Arrhenius plot of the GaCl desorption rate.

Mori et al. [8] proposed that GaCl was stable on the surface and its coverage followed the Langmuir-type adsorption isotherm to 1 ML. In view of the thermal desorption of GaCl, this model has to be modified. If k_{GaCl} is the desorption rate constant of GaCl, the rate equation is given by:

$$\frac{d[GaCl]}{dt} = I_B s - k_{GaCl}[GaCl] \quad \text{with} \quad s = s_0\{1 - [GaCl]\} \tag{8}$$

The GaCl deposition is hence self-limiting with the steady state concentration given by:

$$[GaCl]_s = I_B s_0/(I_B s_0 + k_{GaCl}) \tag{9}$$

In order for $[GaCl]_s$ to reach 0.95 ML at 450°C, since $k_{GaCl} \simeq 200\ s^{-1}$ according to Fig. 2 with $s_0 \simeq 1$, I_B has to exceed 3800 ML/s which corresponds to a partial pressure of DEGaCl of about 1.5×10^{-2} torr. This is in reasonable agreement with the high DEGaCl partial pressures used in ALE [8,9]. The self-limiting deposition resulted from the balance of the incoming GaCl flux and the desorbing GaCl flux. This derivation however is only qualitative since the hydrodynamic transport was not taken into account. In the next example, we shall extend this flux balance concept to the TMGa system.

Reaction of trimethylgallium with GaAs(100)

In significant contrast to DEGaCl, ALE with TMGa was observed even at TMGa pressures as small as 3×10^{-6} torr [5]. Experimentally, the deposited thickness only approximated one GaAs monolayer per ALE cycle. It ranged from below a monolayer to above a monolayer as temperature of deposition was increased [5]. In certain conditions, Ga droplets had been observed when the growth rate exceeded one monolayer per cycle [1].

The reaction of TMGa on GaAs(100) has been studied most extensively [11 – 13, 15,16]. The TMGa molecules were observed by photoemission to adsorb dissociatively on the surface with two bonding configurations for the CH_3 ligands [12]. Memmert and Yu [13] found that the hydrocarbons desorbed as CH_3 radicals with two first order desorption rates in the dilute coverage limit. Creighton [11] used temperature program desorption and reported coverage dependent desorption kinetics. The activation energies for the desorption of CH_3 are higher than those for C_2H_5 (Table 1). Qualitatively, the desorption of CH_3 is about an order of magnitude slower than that of C_2H_5 at equivalent temperatures.

The site blocking effect of the adsorbed CH_3 ligands on the adsorption of TMGa is clearly shown by the variations of the effective sticking coefficient with TMGa dosage as shown in Figure 3. At 50°C, the initial (low coverage) sticking coefficient was close to unity (curve a), and remained so until a critical coverage after which the sticking coefficient decreased rapidly to zero when all adsorption sites were occupied. This behavior was typical of molecular precursor mediated adsorption and was similar to that of TEGa. At 350°C, the thermal desorption of the molecular precursor was fast and the rate of CH_3 desorption became appreciable. As the coverage increased with dosage, the sticking coefficient dropped rapidly from the initial (low coverage) values of s_0 of about 0.9 to a CH_3 limited value of about 0.2 (curve b). The sticking coefficient s remained relatively constant as the Ga coverage increased with TMGa dosage. This behavior can be described in the same way as for TEGa. Assuming that CH_3 desorbed with a first order rate constant k with a steady state CH_3 coverage $[CH_3]_s$, the steady state sticking coefficient is given by:

$$s = s_0\{1 - [CH_3]_s\} \tag{10}$$

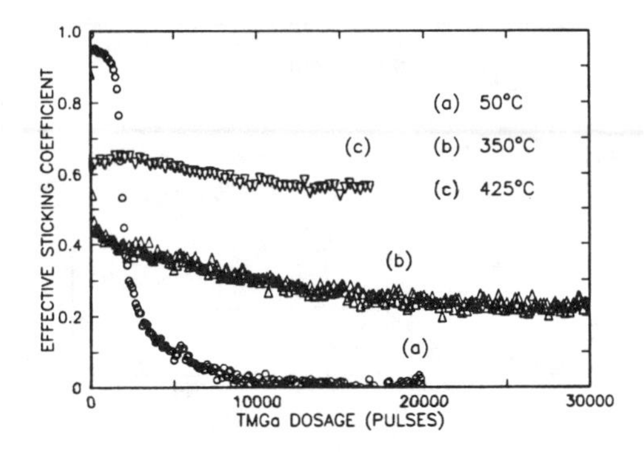

Fig. 3 Effective sticking coefficient of TMGa on GaAs(100) as a function of TMGa dosage at (a) 50°C, (b) 350°C, and (c) 425°C.

At 425°C, the CH_3 desorption rate was so high that the sticking coefficient was basically independent of dosage.

The nonzero steady state sticking coefficient at high coverages and elevated temperature again illustrates that the site blocking mechanism cannot explain the self-limiting deposition of Ga. An ALE experiment that clearly demonstrated this point was reported by Sakuma et al. [6]. In their experiment, two consecutive pulses of TMGa, applied with a controlled time delay up to 100 s in between, showed no dependence on the time delay in the ALE growth. If CH_3 site blocking was the mechanism, the desorption of the CH_3 ligands (from the first TMGa pulse) during the time delay would have affected the growth per cycle. None was observed. That the sticking coefficient remained relatively constant as the Ga accumulated also showed that TMGa has no selectivity between the Ga and As sites.

At temperatures above 350°C, we found a second important reaction pathway: the evolution of CH_3Ga. This production of CH_3Ga was also reported by Nishizawa and Kurabayashi [1] and Creighton [11], but its relationship to the ALE process was not recognized. Figure 4 depicts the CH_3Ga^+ and $CH_3{}^+$ signals as a function of TMGa dosage at 425°C from an As-rich c(2×8) GaAs(100) starting surface. The data was corrected for the cracking fragments from the reflected TMGa and residual gas (CH_4) background. The CH_3Ga^+ signal was small initially. At about 2000 pulses, it increased gradually and reached a steady state value at about 6000 pulses. Simultaneously, there was a noticeable decrease in the $CH_3{}^+$ signal in this dosage region as expected from the conservation of CH_3 ligands. The reaction pathway hence is:

$$(CH_3)_3Ga \rightarrow xGa + (1-x)CH_3Ga\uparrow + (2+x)CH_3\uparrow \quad (11)$$

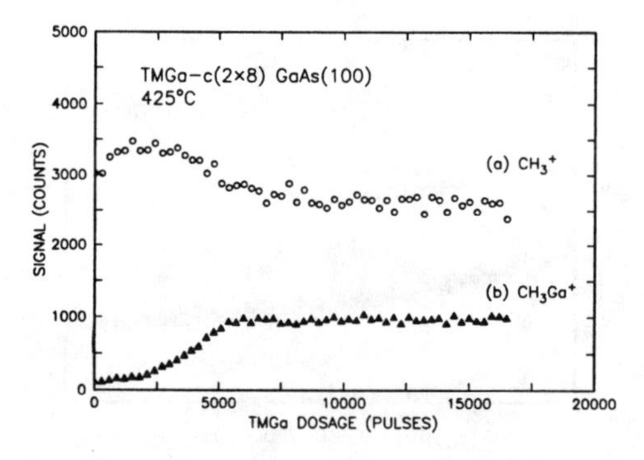

Fig. 4 $CH_3{}^+$ and CH_3Ga^+ signals as a function of TMGa dosage at 425°C. The starting surface was the As-rich c(2×8) GaAs(100).

The data clearly shows that the branching of the reaction to CH_3Ga desorption increased with the Ga coverages as the latter changed with the TMGa dosage. Time resolved measurements showed that the desorption rates of CH_3 and CH_3Ga were the same, indicating that the desorption of CH_3Ga was first order in $[CH_3]$ as expected.

Let us examine the implications of the experimental observation. The rate equations are as follows:

$$\frac{d[Ga]}{dt} = I_B s - \tilde{k}[CH_3]f(Ga)$$
$$\frac{d[CH_3]}{dt} = 3I_B s - k[CH_3] - \tilde{k}[CH_3]f(Ga) \qquad (12)$$
$$s = s_0\{1 - [CH_3]\}$$

Here f(Ga) is the functional dependence of the CH_3Ga desorption rate on the Ga coverage. $\tilde{k}$ is the desorption rate constant. At the steady state, $d[Ga]/dt = d[CH_3]/dt = 0$, and eq. (12) gives:

$$f(Ga) = k/2\tilde{k} \qquad (13)$$

We found from our experiment that we could approximate f(Ga) by:

$$f(Ga) = ([Ga] - [Ga]_0)^n \qquad (14)$$

where $[Ga]_0$ was close to the Ga coverage of the (1×6) surface and n was approximately 3. Details of these results will be presented in future publications. The steady state Ga coverage in this case is given by:

$$[Ga]_s = [Ga]_0 + (k/2\tilde{k})^{1/n} \qquad (15)$$

Hence the deposition of Ga is self-limiting and the coverage is temperature dependent through the $(k/2\tilde{k})$ term. This term reflects the competition between the desorption of CH_3 and the desorption of CH_3Ga. A higher value of k depletes the CH_3 coverage and gives a higher rate of Ga deposition by having a larger sticking coefficient (eq. (10)). On the other hand, a larger value of $\tilde{k}$ depletes the Ga coverage. The ALE growth data is consistent with $k/2\tilde{k}$ increasing with temperature. Only by the proper choice of temperature, can $[Ga]_s$ be adjusted to unity. This is consistent with the gradual increase in ALE growth rate per cycle with temperature until 1 ML per cycle was reached [1].

To explain the loss of ALE at higher temperatures, we examine the situation when [Ga] slightly exceeds unity. Since only the exposed top one monolayer of Ga atoms are involved in the surface chemistry, f(Ga) = f(1), the value at 1 ML. Using this value of f(Ga) and $d[CH_3]/dt = 0$ in eq.(12), the Ga deposition rate is:

$$\frac{d[Ga]}{dt} = \frac{I_B s_0[k - 2\tilde{k}f(1)]}{3I_B s_0 + k + \tilde{k}f(1)} \qquad (16)$$

The sign of d[Ga]/dt depends on the factor $[k - 2\tilde{k}f(1)]$. There are three possibilities depending on the temperature: (i) at low temperatures where $(k/2\tilde{k}) < f(1)$, d[Ga]/dt is negative and the Ga coverage would be reduced to the stable value given by eq.(15); (ii) at the proper temperature where $(k/2\tilde{k}) = f(1)$, d[Ga]/dt is zero and the coverage would not change; and, (iii) at high temperatures where $(k/2\tilde{k}) > f(1)$, d[Ga]/dt is positive and there will be continuous deposition which would lead to multilayers of Ga or Ga droplets. Again, the competition between the desorption of CH_3 and the desorption of CH_3Ga controls the Ga deposition.

The above discussion shows that the presence of a Ga coverage dependent CH_3Ga desorption flux can result in self-limiting Ga deposition for ALE. The derivation explains qualitatively why the ALE growth per cycle is temperature dependent, and how multilayer growth (droplet) can result.

SUMMARY

We have studied the surface reactions and the Ga deposition characteristics of three Ga precursors: TEGa, DEGaCl, and TMGa. While both DEGaCl and TMGa can be used for the ALE of GaAs, TEGa cannot. Our study of the sticking coefficients of these Ga precursors indicated no appreciable adsorption selectivity between Ga and As sites, hence site selectivity was excluded as the mechanism of ALE in these systems.

The study of the TEGa system indicated that the site blocking mechanism alone could not give self-limiting Ga deposition. The reason is simply that at elevated temperature, the surface was at a **dynamic** state chemically. Adsorption and thermal removal of C_2H_5 proceeded continuously, maintaining a steady rate of Ga deposition. The study of DEGaCl illustrated that an appropriate channel for the desorption of Ga (GaCl) could provide self-limiting deposition of Ga. The reaction of TMGa had a Ga desorption channel (CH_3Ga) which increased with Ga coverage. Our experimental data led us to propose that a dynamic equilibrium Ga coverage was obtained when the rate of removal of Ga by desorption of CH_3Ga balanced the Ga deposition rate. These three studies show that the successful ALE cases with DEGaCl and TMGa both can fit to a dynamic description where the balance between the incoming Ga flux and the outgoing Ga desorption flux is the mechanism of self-limiting deposition.

References

* Present address: IBM Zurich Research Laboratory, 8803 Rueschlikon, Switzerland
** Present address: Columbia University, New York, NY 10027
*** Present address: University of Wisconsin, Madison, WI 53706

1. J. Nishizawa and T. Kurabayasahi, J. Cryst. Growth **93**, 98 (1988).
2. M. A. Tischler and S. M. Bedair, J. Cryst. Growth **77**, 89 (1986).
3. A. Watanabe, T. Kamijoh, M. Hata, T. Isu, and Y. Katayama, Vacuum **41**, 965 (1990).

4. S. P. Denbaars, P. D. Dapkus, C. A. Beyler, A. Hariz and K. M. Dzurko, J. Cryst. Growth **93**, 195 (1988).
5. J. Nishizawa, T. Kurabayashi, H. Abe, and N. Sakurai, J. Electrochem. Soc. **134**, 945 (1987).
6. Y. Sakuma, M. Ozeki, N. Ohtsuka, and K. Kodoma, J. Appl. Phys. **68**, 5660 (1990).
7. A small processing window for ALE has been reported by H. Ohno, S. Ohtsuka, H. Ishii, Y. Matsubara, and H. Hasegawa, Appl. Phys. Lett. **54**, 2000 (1989).
8. K. Mori, M. Yoshida, A. Usui, and H. Terao, Appl. Phys. Lett. **52**, 27 (1988).
9. C. Sasaoka, M. Yoshida, and A. Usui, Jap. J. Appl. Phys. **27**, L490 (1988).
10. J. R. Creighton, K. R. Lykke, V. A. Shamanian, and B. D. Kay, Appl. Phys. Lett. **57**, 279 (1990).
11. J. R. Creighton, Surf. Sci. **234**, 287 (1990).
12. M. L. Yu, U. Memmert, and T. F. Kuech, Appl. Phys. Lett. **55**, 1011 (1989).
13. U. Memmert and M. L. Yu, Appl. Phys. Lett. **567**, 1883 (1990).
14. A. J. Murrell, A. T. S. Wee, D. H. Fairbrother, N. K. Singh, J. S. Foord, G. J. Davies, and D. A. Andrews, J. Appl. Phys. **68**, 4053 (1990).
15. V. M. Donnelly and J. A. McCaulley, Surf. Sci. **238**, 34 (1990).
16. M. L. Yu, U. Memmert, N. I. Buchan, and T. F. Kuech in Chemical Perspectives of Microelectronic Materials II, edited by L. V. Interrante, K. F. Jensen, L. H. Dubois, and M. E. Gross, (Mater. Res. Soc. Proc. **204**, Pittsburgh, PA 1991) pp. 37-46.
17. A. Närmann, R. Purtell, and M. L. Yu, Proceedings of this Symposium (to be published).
18. D. A. King and M. G. Wells, Surf. Sci. **29**, 454 (1972).
19. H. Ohno, S. Ohtsuka, H. Ishii, Y. Matsubara, and H. Hasegawa, Appl. Phys. Lett. **54**, 2000 (1989), reported that they were able to deposit a monolayer of Ga with DEGaCl at 300° C at a pressure of 1×10^{-5} torr but without the presence of Cl on the surface. However we failed to reproduce the result.

THE SURFACE CHEMISTRY OF GaAs ATOMIC LAYER EPITAXY

J. RANDALL CREIGHTON AND BARBARA A. BANSE
Sandia National Labs, Division 1126, Albuquerque, NM 87185-5800

ABSTRACT

In this paper we review three proposed mechanisms for GaAs ALE and review or present data in support or contradiction of these mechanisms. Surface chemistry results clearly demonstrate that TMGa irreversibly chemisorbs on the Ga-rich GaAs(100) surface. The reactive sticking coefficient (RSC) of TMGa on the adsorbate-free Ga-rich GaAs(100) surface was measured to be ~0.5, conclusively demonstrating that the "selective adsorption" mechanism of ALE is not valid. We describe kinetic evidence for methyl radical desorption in support of the "adsorbate inhibition" mechanism. The methyl radical desorption rates determined by temperature programmed desorption (TPD) demonstrate that desorption is at least a factor of ~10 faster from the As-rich c(2 X 8)/(2 X 4) surface than from the Ga-rich surface. It is this disparity in CH_3 desorption rates between the As-rich and Ga-rich surfaces that is largely responsible for GaAs ALE behavior. A gallium alkyl radical (e.g. MMGa) is also observed during TPD and molecular beam experiments, in partial support of the "flux balance" mechanism. Stoichiometry issues of ALE are also discussed. We have discovered that arsine exposures typical of atmospheric pressure and reduced pressure ALE lead to As coverages $\geq$ 1 ML, which provides the likely solution to the stoichiometry question regarding the arsine cycle.

INTRODUCTION

Atomic layer epitaxy (ALE) [1-6] has the potential for depositing compound semiconductors with precise thickness control and superb thickness uniformity. Ideally, ALE achieves monolayer scale growth control through the alternation of the group V and group III precursor with each step exhibiting self-limiting deposition, i.e. deposition ceases at 1 ML. However, the utility of GaAs ALE has been limited due to its relatively narrow operating window and also due to the high unintentional carbon doping levels that are typically produced. The narrow operating window is generally believed to be due to loss of self-limiting deposition during the gallium cycle. At high temperatures and/or large trimethylgallium (TMGa) exposures, excess gallium is deposited which leads to non-ideal ALE (>1ML/cycle). Therefore most discussions of GaAs ALE mechanisms deal primarily with the chemistry of the gallium cycle. There are several proposed mechanisms for GaAs ALE [1-3,6,7-16], and considerable disagreement as to which mechanism is correct. In this paper we review the merits and shortcomings of the proposed mechanisms. Both kinetic and stoichiometric (often overlooked) issues are discussed. Experimental surface chemistry results for (TMGa) and arsine decomposition are presented in support or contradiction of the proposed ALE mechanisms.

Before reviewing the proposed ALE mechanisms we comment on the role of gas-phase TMGa pyrolysis, which is often believed to be the major source of non-ideal ALE [5,12]. Indeed, Ozeki, et al. [5] have achieved excellent GaAs ALE results in a reduced pressure reactor designed to minimize gas-phase TMGa pyrolysis. While these results support the idea that gas-phase pyrolysis is the major cause of non-ideal ALE, we caution that the results are not conclusive since reduced reactor residence times will also remove potentially detrimental gas-phase species that were created via a surface reaction. Evidence that gas-phase pyrolysis is not a necessary (albeit perhaps sufficient) condition for excess gallium deposition is seen in the low-pressure ALE results of Nishizawa, et al. [2]. Their results clearly show >1ML/cycle ALE behavior and the formation of gallium droplets at temperatures $\geq$520°C and at pressures too low ($<10^{-4}$ Torr) for gas-phase pyrolysis to be significant. Obviously there are conditions where surface reactions alone can lead to excess gallium deposition. We believe it is still an open question as to whether gas-phase pyrolysis or surface reactions are the dominant cause of excess gallium deposition during atmospheric or reduced pressure ALE. The remainder of the paper deals with the surface chemistry aspects of ALE.

MECHANISMS OF ALE

Selective Adsorption

The key assumption of the selective adsorption (SA) mechanism of ALE is that TMGa converts the arsenic-terminated surface to a gallium-terminated surface whereupon TMGa adsorption ceases (see Fig 1A). The SA mechanism is appealing since one would expect TMGa to selectively chemisorb on arsenic sites via a Lewis acid-base (donor-acceptor) interaction, and there would be no arsenic sites on an ideally terminated (1ML) gallium-rich surface. This mechanism is often used to explain ALE results [5,7,8,11,13]. There is, however, no evidence that an ideally terminated gallium-rich surface exists. All theoretical [17,18] and experimental [18-23] evidence indicates that the gallium-rich surfaces have a significant number of gallium vacancies. The vacancies expose arsenic atoms in the second layer which could serve as sites for TMGa chemisorption. In fact, there are many experimental results demonstrating that TMGa chemisorbs and irreversibly decomposes on the gallium-rich surfaces [9,10,14,16] with surprisingly high rate (some results will be discussed later in this paper). The surface stoichiometry issue poses another problem for the SA mechanism, since it is difficult to see how ideal ALE (1ML/cycle) is achieved when there is not a complete monolayer of gallium in the gallium-rich surface.

There is one type of experiment used to support the the SA mechanism which we will comment on because we think the results are not conclusive. The GaAs surface is exposed to the group III alkyl species at temperatures typical of ALE conditions, whereafter the surface is inspected with a surface sensitive spectroscopy (such as XPS) to determine if excess gallium has been deposited. Normally no excess gallium is detected by such techniques. Results of this nature have been used by Dapkus, et al. [13], Kodama, et al. [7], and Yu, et al. [8] in support of the SA or SA-like mechanism. However, the conclusions drawn from these results are suspect because excess gallium does not "wet" the GaAs surface, but instead nucleates into three dimensional droplets [2] which can be difficult to detect. For instance, if 5 ML (1ML = 6.26×10^{14} atoms/cm^2) of gallium where distributed into hemispherical droplets 0.4 microns in diameter (see ref. 2), they would occupy <1% of the GaAs surface and thus be virtually indetectable by XPS or similar techniques. A second conclusion drawn from these types of experiments is that there are no adsorbates, e.g. CH_3, on the surface following the high temperature exposures [7,8,11,13]. The problem here is that the analysis is not performed in situ and it is difficult to properly "quench" the surface (i.e. turn off the gas and cool the sample) without altering its state. After terminating the gas exposure the GaAs substrate takes a finite time to cool, during which significant adsorbate desorption may occur. Determination of the methyl radical desorption rate constant [9,10,14,16] indicates that for the aforementioned studies [7,8,11,13] it is very likely that significant methyl group desorption occurred before the surface analysis was performed.

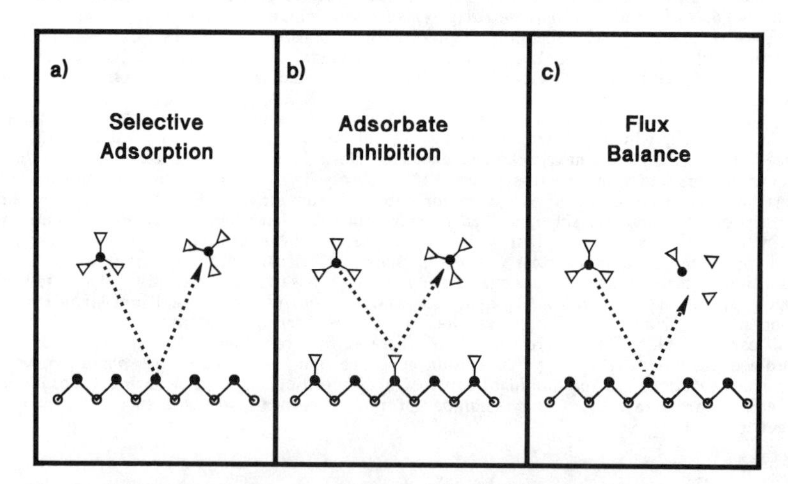

Fig. 1. Schematic representation of three proposed mechanisms for GaAs ALE. Solid circles = Ga atoms, open circles = As atoms, and traingles = CH_3 groups.

Adsorbate Inhibition

The key assumption of the adsorbate inhibition (AI) mechanism (sometimes referred to as site-blocking) of ALE is that TMGa converts the arsenic-terminated surface to a gallium-terminated surface covered with adsorbates, i.e. methyl groups (see Fig 1B). At this stage the adsorbates inhibit further decomposition of TMGa. The TMGa decomposition reaction rate is effectively limited by the rate of methyl group desorption. Most of the evidence in support of the AI mechanism is kinetic in nature, such as measurements of the methyl radical desorption rate constant on the gallium-rich surface [9,10,14,16]. Further kinetic details of the AI mechanism will be discussed later in this text. Some results have failed to observe methyl groups (or carbon) after TMGa exposures at typical ALE temperatures [7,8,11,13], but as discussed in the previous paragraph we believe the researchers did not account for the methyl group desorption rate (or residence time) and the finite time required to cool the surface before analysis. One problem with the AI mechanism is that the methyl residence time is too short to account for ideal ALE behavior at temperatures ≥500°C which has been reported by Ozeki et al [5]. This issue will be discussed later in the text. The AI mechanism does potentially explain the stoichiometry problem as the adsorbates may stabilize the gallium-rich surface with 1ML of gallium.

Flux Balance

The key feature of the flux balance (FB) mechanism (as proposed by Yu, et al. [14,15]) is that TMGa decomposes on the gallium-rich surface but a gallium-containing product, specifically monomethylgallium (MMGa), leaves the surface quantitatively so no net gallium deposition occurs (see Fig. 1C). Yu has made direct observations of MMGa desorbing from GaAs using a pulsed molecular beam technique [14,15]. While this proposed mechanism has merit, we believe conclusive evidence demonstrating that MMGa is produced quantitatively is needed. We also observe a gallium-alkyl radical (either MMGa or DMGa) during TPD [10] but it may not formed quantitatively (this work).

EXPERIMENTAL

Most of the relevant experimental details have been previously published [10]. Briefly, experiments were performed in a three-level UHV chamber equipped with a number of surface diagnostics. The upper level is isolatable by a gate valve and is used for "high" pressure exposures. The lower two levels are equipped with Auger spectroscopy, LEED, and a differentially pumped quadrupole mass spectrometer (QMS) for temperature programmed desorption (TPD) and molecular beam experiments. The heating rate used for TPD was typically 5 K sec^{-1}. An effusive molecular beam is designed to impinge on the GaAs surface while desorbed reactant and products are monitored line-of-sight by the QMS. This arrangement was used to study the temperature and flux dependence of the TMGa reaction rate on GaAs.

RESULTS

TMGa on Ga-rich (4 X 6) GaAs(100)

Previous work demonstrated that TMGa can irreversibly decompose on the Ga-rich (1 X 6) or "(4 X 6)" surfaces [9,10,14,16]. The main product seen during TPD experiments is methyl radical, which desorbs with maximum rate between 400-440°C [9,10]. This is demonstrated in Figure 2, curve a. The narrow FWHM (~30°C) of the peak indicates that there is in effect only one type of binding site for the methyl groups on the surface at this stage. A detailed kinetic analysis of methyl radical desorption indicates that the activation energy (E_a) and preexponential factor (ν) are coverage dependent [10]. A reasonable approximation of the kinetics can be made using a Redhead analysis [24], which yields E_a = 44.1 kcal/mole for an assumed $\nu = 10^{13}$ sec^{-1}. Using these parameters we can estimate the methyl group residence time (τ) at optimal ALE temperatures. At 450°C we estimate that τ = 2.1 sec, but since the methyl group coverage on the Ga-rich surface is ~1/4 ML [25], the "excess" gallium monolayer formation time is actually ~8 sec. These timescales are consistent with the results of DenBaars, et al. [3] who obtained their best ALE behavior (slightly greater than 1 ML/cycle) around 450°C for a TMGa pulse duration of 1 sec.

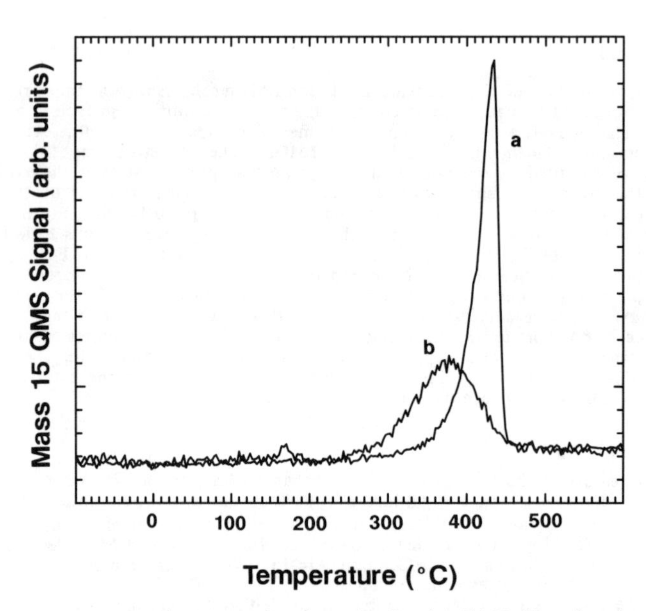

Fig. 2. Methyl radical desorption from the Ga-rich "(4 X 6)" surface (curve a) and from the As-rich c(2 X 8)/(2 X 4) GaAs(100) surface (curve b).

The other products observed during TPD are MMGa (or DMGa), which also desorbs ~430°C [10], and "excess" atomic gallium which desorbs between 500-600°C (see Figure 3). From the clean GaAs(100) surface the Ga^+ and As^+ TPD signals can be scaled so they are superimposable, indicating congruent evaporation, but following TMGa exposures the Ga^+ TPD signal always exceeds the As^+ at high temperatures. This latter observation is what we refer to as "excess" gallium desorption. Calibration of the desorption signals indicate that 70-90% of the gallium from irreversible TMGa dissociation desorbs as atomic Ga while ~10-20% desorbs as MMGa [25]. An interesting point is that if the surface is examined after heating (to ~450°C) just beyond the methyl group desorption temperature, the "excess" gallium is not detectable by AES or LEED, whereas extra gallium was detectable just before the onset of methyl group desorption [10]. We therefore believe the "excess" gallium resides in droplets after methyl radical desorption has occurred and can be detected at high temperatures by TPD. We would expect the droplets to evaporate first since the vapor pressure of metallic Ga exceeds the Ga vapor pressure over GaAs in this temperature range [26]. When the surface is examined after heating to temperatures just below methyl radical desorption, the excess gallium is detectable by AES and the surface exhibits a unique LEED pattern [10]. This result demonstrates that the methyl groups stabilize a gallium coverage that is higher than that obtained on the adsorbate-free gallium-rich (1 X 6) or "(4 X 6)" surface. This observation helps resolve part of the ALE stoichiometry issue.

Direct measurement of the TMGa + GaAs reaction rate can be made using the molecular beam scattering technique where the scattered/desorbed TMGa flux is monitored with a differentially-pumped QMS. The scattered/desorbed TMGa signal decreases when the TMGa surface reaction rate becomes appreciable, and this signal decrease (ratioed to the original signal level) is an absolute measurement of the reactive sticking coefficient (RSC) [25,27]. The RSC can exhibit both a temperature and incident flux dependence. Some preliminary results of the TMGa RSC vs. temperature on the Ga-rich surface are given in Fig. 4. The RSC for both fluxes studied reaches a limit of ~0.5 at high temperatures (>400°C). Since we know the TMGa arrival rate (flux) and the methyl radical desorption rate (from TPD), we know that the surface methyl coverage is essentially zero at temperatures where RSC~0.5. Therefore, when TMGa impinges on an adsorbate-free Ga-rich GaAs(100) surface at high temperatures it has a ~50% chance of irreversibly

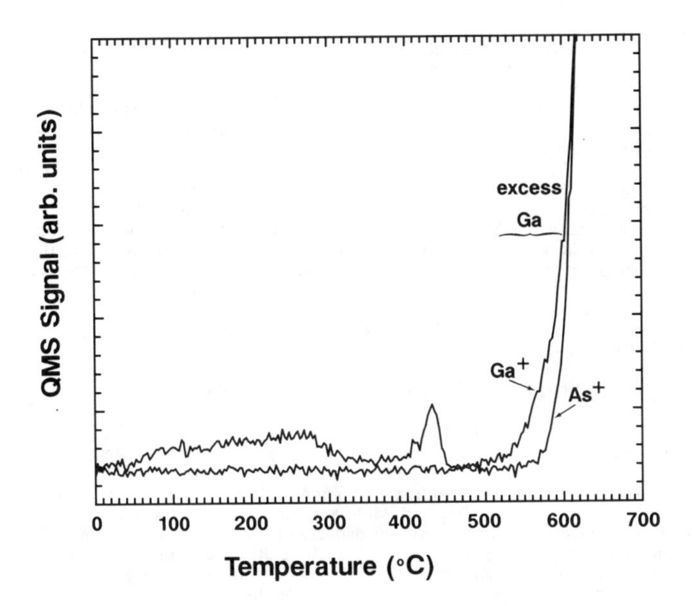

Fig. 3. TPD spectra from the TMGa-dosed Ga-rich "(4 X 6)" surface demonstrating excess atomic gallium desorption between 500-600°C. The lower temperature peaks in the Ga^+ signal are due to TMGa and MMGa (or DMGa).

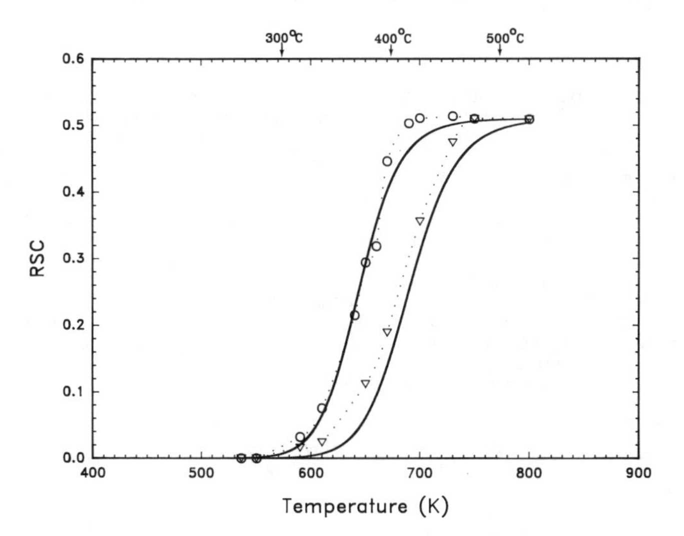

Fig. 4. The temperature and flux dependence of the reactive sticking coefficient (RSC) for TMGa impinging on Ga-rich "(4 X 6)". Open circles are for a TMGa flux of 10^{12} molecules cm^{-2} sec^{-1} and open triangles are for 10^{13} molecules cm^{-2} sec^{-1}. The dashed lines are guides for the eye while the solid lines are calculated from a simple adsorption-reaction mechanism (see text).

decomposing. These results conclusively demonstrate that the SA mechanism of ALE is not operable. Incidentally, after the RSC measurement at the higher TMGa flux (10^{13} molecules cm^{-2} sec^{-1}) we noticed that the GaAs surface was hazy. Examination of the surface under an optical microscope revealed the presence of droplets (presumably gallium), which once again demonstrates that there are conditions where excess gallium deposition occurs purely by a surface chemical process.

The solid lines if Fig 4. are the results from an extremely simple adsorption reaction model [25] that uses the approximate methyl radical desorption kinetics obtained from the TPD experiments (E_a=44.1 kcal/mole, $\nu=10^{13}$ sec^{-1}) and a classical first order Langmuirian sticking coefficient, i.e. $S(\Theta) = S_o(1-\Theta)$. The only "adjustable" parameter in the simple model is in fact S_o, which at high temperatures defines the RSC. Using S_o=0.51, this model reproduces the temperature and flux dependence of the experimental RSC values very well. Most of the deviations between model and experiment are due to the approximation of the methyl radical desorption kinetics.

It is instructive to write the TMGa reaction equation allowing for MMGa formation, as proposed in the FB ALE mechanism;

$$Ga(CH_3)_3(g) \rightarrow (X + 2)CH_3(g) + XGa(s) + (1-X)GaCH_3(g)$$

where X ($0 \leq X \leq 1$) defines the degree of gallium droplet formation vs. MMGa formation. For the FB mechanism to be significant, X has to be near-zero, i.e. most Ga leaves the surface as MMGa. While the molecular beam technique can measure desorbing products such as methyl radicals and gallium alkyl radicals, it cannot directly monitor the formation of gallium droplets. If the rate of CH_3 and MMGa evolution could be accurately measured, then it would be possible to infer the gallium deposition rate from mass balance. However, quantitative determination of the product yield is difficult at best, but reasonable estimates can be made by comparing the signals to the scattered TMGa signal. We normally detect a $MMGa^+$ signal (in agreement with Yu et al [14,15]) in the 400-500°C range. Our initial attempts to calibrate the $MMGa^+$ signal indicate that for many conditions **X** is near unity (MMGa desorption insignificant) while for other conditions **X** is smaller and may be near zero (MMGa desorption significant). It is also sometimes difficult to reproduce the magnitude of **X** for nominally identical conditions (i.e. temperature and flux). The source of this variation is currently being investigated. We have also measured the methyl radical signal and found it to be proportional to the reaction rate over the temperature range studied.

TMGa on As-rich c(2 X 8)/(2 X 4) GaAs(100)

We have also performed TMGa TPD and molecular beam experiments on the As-rich c(2 X 8)/(2 X 4) GaAs(100) surface. Theoretical and experimental results indicate that this surface is terminated with 3/4 ML of As [18,22,23]. This surface structure is normally produced during MBE and by moderate arsine exposures (500-10,000 Langmuirs, $1L=10^{-6}$ Torr-sec) at temperatures >250°C [8,29]. This arsine exposure range is typical of the exposures used in Nishizawa's low pressure ALE [2]. A methyl radical TPD trace following low-temperature TMGa exposure to this As-rich surface is shown in Fig. 2, curve b. Note that the peak temperature is significantly lower (~380°C) than that seen from the Ga-rich surface (curve a). This indicates that CH_3 desorption is significantly faster from the As-rich surface. Although there is only one TPD peak from the As-rich surface, its FWHM is too broad to be consistent with one type of binding site. We can nevertheless estimate an "average" rate constant for CH_3 desorption using the Redhead analysis [24]. This technique yields E_a = 40.4 kcal/mole assuming $\nu = 10^{13}$ sec^{-1}. We note that these values will somewhat underpredict the CH_3 desorption rate, especially at low temperatures. When we compare the rate constant for CH_3 desorption from the Ga-rich and As-rich surface over the typical ALE temperature range we find that there is a factor of ~10 difference, i.e. $k_{As}(CH_3) \sim 10 \times k_{Ga}(CH_3)$. This disparity in the CH_3 desorption rate is a major factor contributing to successful GaAs ALE. At high fluxes, conversion from the As-rich to the Ga-rich surface is rate limited by CH_3 desorption, and this process occurs at least ~10 times faster than gallium deposition (also CH_3 desorption rate limited) on the Ga-rich surface.

Review of ALE-Related Kinetics

Due to the significant variation in ALE results and considerable disagreement as to the mechanism of ALE, we have found it enlightening to review the current kinetic studies relevant to ALE. The data basically falls into two categories; direct kinetic measurements of CH_3 desorption at low pressure conditions, and measurements of surface conversion or growth rate during ALE or ALE-like conditions. We have compiled three data sets of the first type and two of the latter, all of which are plotted in Arhennius fashion in Fig. 5. The "good" news is that similar activation energies (38 to 45 kcal/mole) are measured, but the "bad" news is that there is a 3 order-of-magnitude variation in the absolute rates. The data includes our TPD results from the Ga-rich [9,10] and As-rich surface (this work) and the TPD results from the Ga-rich surface of Donnelly, et al. [16]. The results for the Ga-rich surfaces are in relatively good agreement. The "slow" component of CH_3 desorption from the pulsed molecular beam results of Yu, et al. [14] compares favorably with our As-rich rate, but the "fast" component is a decade higher. The high flux kinetics from the reflectance-difference (RD) data of Aspnes, et al. [28] (E_a = 39 kcal/mole, ν = 1.2 X 10^{13} sec^{-1}) and the ALE kinetics of Ozeki, et al. [5] (E_a = 42 kcal/mole, ν = 6 X 10^{11} sec^{-1}) are also displayed in Fig. 5. The RD experiments monitored the reaction of TMGa with an As-rich surface and are in reasonable agreement with our As-rich results. The ALE kinetic data of Ozeki, et al. [5] is also a monitor of the reactivity of TMGa with the As-rich surface, and there is a large discrepancy between this data and both the As-rich data of Aspnes, et al. [28] and our As-rich data. Some of the variation seen in the data in Fig. 5 is due to differences in reactivity of the As-rich and Ga-rich surface, and some variation may be due to subtle differences in the experimental techniques. However, we strongly suspect the remainder of the discrepancy is due to temperature measurement errors. If this is true, then the temperatures for the ALE data of Ozeki, et al. [5] appears to be at least 50°C too high with respect to the other results. In other words, their temperature scale must be shifted down at least 50°C in order to bring their data in agreement with the other techniques. Our interest in pointing out this discrepancy in rates is partly because the AI mechanism of ALE is incompatible with the higher temperature (≥500°C) ALE results of Ozeki, et al. [5]. The methyl radical desorption rate is too fast at these higher temperatures and excess gallium deposition would occur. If Ozeki's temperature scale is shifted down ≥50°C, then their result are compatible with the AI mechanism as well as other ALE kinetic data.

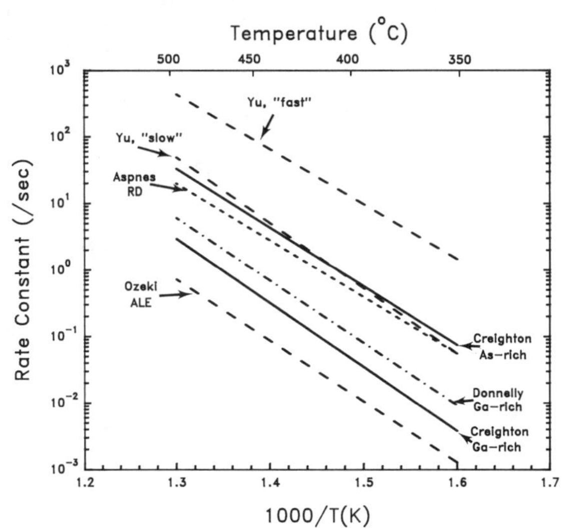

Fig. 5. Summary of literature data for methyl radical desorption kinetics and ALE growth/reaction rates. See text for references.

On a general note, we suggest that researchers fully explain their temperature calibration procedure at the early phase of publication when they are reporting kinetic results. If this is done, other workers can at least be aware of the possible differences between the calibration procedures and comment on the likely sources of error.

Stoichiometry of the As-rich Surfaces

Most discussions of ALE mechanisms deal with kinetic issues while the questions regarding stoichiometry have often been overlooked. For instance, if the adsorbate-free Ga-rich and As-rich surface reconstructions are not ideally terminated (θ(As or Ga) < 1ML) then how is ideal ALE (1ML/cycle) achieved? We have shown that methyl groups stabilize a higher coverage gallium-rich surface (presumably at 1ML). This could resolve stoichiometry issue for the gallium cycle, but the As-rich c(2 X 8)/(2 X 4) surface still is terminated with only 3/4 ML of arsenic. Most experimental results indicate that is takes 500-10,000 L of arsine to convert the Ga-rich surface to the As-rich c(2 X 8)/(2 X 4) surface [8,29]. We have for some time suspected that higher arsine exposures may lead to a higher arsenic coverage that is stabilized by adsorbates such as hydrogen. This is in fact the case and it is demonstrated in Fig. 6 by TPD results following high arsine exposures at 275°C. Curve (a) is the As^+ signal from the Ga-rich surface and serves as a benchmark. Curve (b) is for a 4,500 L dose which populates an arsenic state which desorbs with a peak at 570°C. This excess arsenic corresponds to the c(2 X 8)/(2 X 4) As-rich surface as evidenced by LEED. Curve (c) and (d) correspond to exposures of 2.3 X 10^4 and 1.9 X 10^6 L. For comparison, a typical arsine exposure during an atmospheric pressure or reduced pressure ALE cycle would be 10^5-10^7 L. Note that these high exposures generate two new arsenic states at 490 and 420°C. Using the c(2 X 8)/(2 X 4) state as a reference point, the

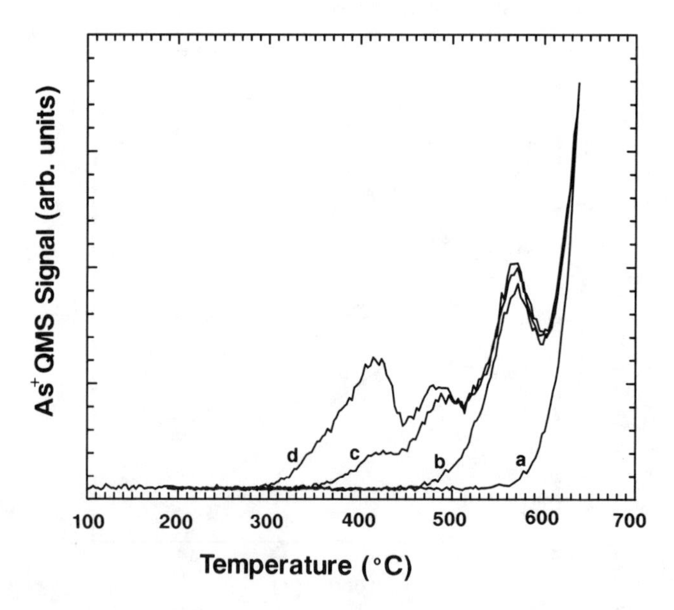

Fig. 6. TPD spectra of excess arsenic desorption monitored by As^+ (m/e = 75) following arsine exposures at 275°C. Arsine exposures are; a) 0 L, b) 4,500 L, c) 2.3 X 10^4 L, and d) 1.9 X 10^6 L.

490°C state corresponds to ~1 ML of arsenic while the 420°C state is clearly > 1ML. Preliminary evidence indicates that the 420°C state exhibits a (2 X 1) + $\frac{1}{2}nX^*$ LEED pattern, and that this excess arsenic desorbs as the tetramer (As_4) rather than the dimer (As_2) as it does in the higher temperature states. Significant hydrogen desorption also occurs in conjunction with the 420°C state. Further details will be reported at a later date [29]. As mentioned above, TPD shows that this "super" As-rich surface has > 1 ML of arsenic on it. This conclusion has been reached independently by Aspnes [30] by noting the similarity in the RD spectrum of the arsine-dosed surface and the c(4 x 4) MBE prepared surface. The c(4 X 4) surface has 1.25-1.75 ML of arsenic and consists of arsenic dimers on top of a complete ML of arsenic [23,31,32].

We suggest that the intermediate arsenic desorption state (~490°C) which corresponds to ~1 ML of total arsenic coverage is the answer to the stoichiometry question posed earlier. If this complete monolayer of As were available during ALE, then it is possible to achieve 1 ML deposition during the arsine cycle. It is also necessary to invoke the assumption that the arsenic in excess of 1 ML (the 420°C state) does not participate in the ALE growth. This could occur for chemical reactivity reasons or because of the shorter residence time this state has on the surface. A simple estimate of the residence time indicates that a significant amount of this state will desorb during a 1 second purge time at 450°C. Future work is needed to characterize the chemistry and kinetics of the 420°C and 490°C As states.

The low-pressure ALE results of Nishizawa, et al. [25] can also be reconciled by these results. Their arsine exposures (10^3-10^4 L) would not have completely saturated the 490°C state (1 ML As), which could explain why there ALE results typically saturated at 0.8-0.9 ML/cycle. In other words, their arsine cycle did not lead to ≥ 1 ML of arsenic so the optimal growth per cycle was < 1 ML.

SUMMARY

We have reviewed three proposed mechanisms for GaAs ALE and reviewed or presented data in support or contradiction these mechanisms. Both kinetic and stoichiometry issues were discussed. TPD results clearly demonstrate that TMGa irreversibly chemisorbs on the Ga-rich GaAs(100) surface. Measurement of the reactive sticking coefficient (RSC) of TMGa on the adsorbate-free Ga-rich GaAs(100) surface yields a value of ~0.5, conclusively demonstrating that the SA mechanism of ALE is invalid. We also briefly describe the pitfalls in the conclusions drawn from the previous research used in support of the SA mechanism. We offer kinetic evidence for methyl radical desorption in support of the AI mechanism. Methyl radical desorption kinetics from the Ga-rich "(4 X 6)" surface are consistent with ALE results at ~450°C. The methyl radical desorption rate is at least a factor of ~10 faster from the As-rich c(2 X 8)/(2 X 4) surface. It is this disparity in CH_3 desorption rates between the As-rich and Ga-rich surfaces that is largely responsible for GaAs ALE behavior. A gallium alkyl radical (e.g. MMGa) is also observed during TPD and molecular beam experiments, in partial support of the FB mechanism proposed by Yu, et al. [14,15]. However, conditions that lead to quantitative formation of MMGa, if they do exist, need further elucidation.

Ideal ALE (1 ML/cycle) seems inconsistent with the known Ga-rich and As-rich surface reconstructions which are not ideally terminated. We have found that there are reasonable solutions to this apparent dilemma. In the case of the gallium ALE cycle, the adsorbed CH_3 groups stabilize a higher Ga coverage (apparently at 1 ML), thus solving part of the stoichiometry issue. Since the As-rich c(2 X 8)/(2 X 4) surface has Θ_{As} = 3/4 ML, it is unlikely that this surface can lead to ideal ALE. In fact, we have discovered that arsine exposures typical of atmospheric pressure and reduced pressure (10< P <760 Torr) ALE lead to As coverages ≥ 1 ML. A similar observation has been made by Aspnes [30] using RD spectroscopy. Our TPD experiments demonstrate that As_2 desorbs from the Θ_{As} = 3/4 ML surface ~570°C, while for Θ_{As} ~ 1 ML the extra ~1/4 ML of arsenic desorbs around 490°C. This observation provides the likely solution to the stoichiometry question regarding the arsine cycle.

ACKNOWLEDGEMENTS

The authors thank Kevin Killeen and Mike Coltrin for stimulating scientific discussions and thank Gary Karpen for technical support. This work was performed at Sandia National Laboratories supported by the US Department of Energy under contract #DE-AC04-76DP000789.

REFERENCES

1. C.H.L. Goodman and M.V. Pessa, J. Appl. Phys. 60, R65 (1986).
2. (a) J. Nishizawa, T. Kurabayashi, H. Abe and A. Nozoe, Surface Sci. 185, 249 (1987). (b) J. Nishizawa, T. Kurabayashi, H. Abe and N. Sakurai, J. Electrochem. Soc. 134, 945 (1987).
3. S.P. DenBaars, P.D. Dapkus, C.A. Beyler, A. Hariz and K.M.Dzurko, J. Cryst. Growth 93,195 (1988).
4. M.A. Tischler and S.M. Bedair, Appl. Phys. Lett. 48, 1681 (1986).
5. M. Ozeki, K. Mochizuki, N. Ohtsuka and K. Kodama, Appl. Phys. Lett. 53, 1509 (1988).
6. G.B. Stringfellow, Organometallic Vapor-Phase Epitaxy, (Academic Press, San Diego, 1989), pp. 363-367.
7. K. Kodama, M. Ozeki, K. Mochizuki and N. Ohtsuka, Appl. Phys. Lett. 54, 656 (1989).
8. M.L. Yu, U. Memmert and T.F. Kuech, Appl. Phys. Lett. 55, 1011 (1989).
9. J.R. Creighton, K.R. Lykke, V.A. Shamamian, B.D. Kay, Appl. Phys. Lett. 57, 279 (1990).
10. J.R. Creighton, Surface Sci. 234, 287 (1990).
11. H. Ishii, H. Ohno, K. Matsuzaki and H. Hasegawa, J. Crystal Growth 95, 132 (1989).
12. P.D. Dapkus, S.P. DenBaars, Q. Chen, W.G. Jeong and B.Y. Maa, Prog. Crystal. Growth and Charact. 19, 137 (1989).
13. P.D. Dapkus, B.Y. Maa, Q. Chen, W.G. Jeong and S.P. DenBaars, J. Crystal Growth 107, 73 (1991).
14. M.L. Yu, U. Memmert, N.I. Buchan and T.F. Kuech, in Chemical Perspectives of Microelectronic Materials II, edited by L.V. Interrante, K.F. Jensen, L.H. Dubois and M.E. Gross (Mater. Res. Soc. Proc. 204, Pittsburgh, PA 1991) pp. 37-46.
15. M.L. Yu, N.I. Buchan, R. Souda and T.F. Kuech, presented at the 1991 MRS Spring Meeting, Anaheim, CA, 1991 (unpublished).
16. V.M. Donnelly, J.A. McCaulley and R.J. Shul, in Chemical Perspectives of Microelectronic Materials II, edited by L.V. Interrante, K.F. Jensen, L.H. Dubois and M.E. Gross (Mater. Res. Soc. Proc. 204, Pittsburgh, PA 1991) pp. 15-23.
17. D.J. Chadi, J. Vac. Sci. Technol. A5, 834 (1997).
18. P.K. Larsen and D.J. Chadi, Phys. Rev. B 37, 8282 (1988).
19. P. Drathen, W. Ranke and K. Jacobi, Surface Sci. 77, L162 (1978).
20. J. Massies, P. Etienne, F. Dezaly and N.T. Linh, Surface Sci. 99, 121 (1980).
21. D.J. Frankel, C. Yu, J.P. Harbison and H.H. Farrell, J. Vac. Sci. Technol. B5, 1113 (1987).
22. M.D. Pashley, K.W. Haberern, W. Friday, J.M. Woodall and P.D. Kirchner, Phys. Rev. Lett. 60, 2176 (1988).
23. D.K. Beigelsen, R.D. Bringans, J.E. Northrup and L.-E. Swartz, Phys. Rev. B 41, 5701 (1990).
24. P.A. Redhead, Vacuum 12, 203 (1962).
25. B.A. Banse and J.R. Creighton, to be published. We have found that previous CH_3 coverage determinations (ref. 10) were too large by a factor of ~3.
26. C.T. Foxon, J.A. Harvey and B.A. Joyce, J. Phys. Chem. Solids 34, 1693 (1973).
27. D.A. King and M.G. Wells, Surface Sci. 29, 454 (1972).
28. D.E. Aspnes, E. Colas, A.A. Studna, R. Bhat, M.A. Koza and V.G. Keramidas, Phys. Rev. Lett. 61, 2782, (1988).
29. B.A. Banse and J.R. Creighton, to be published.
30. D.E. Aspnes, presented at the 1991 MRS Spring Meeting, Anaheim, CA, 1991 (unpublished).
31. P.K. Larsen, J.H. Neave, J.F. van der Veen, P.J. Dobson, B.A. Joyce, Phys. Rev. B 27, 4966 (1983).
32. M. Sauvage-Simkin, R. Pinchaux, J. Massies, P. Calverie, N. Jedrecy, J. Bonnet and I.K. Robinson, Phys. Rev. Lett. 62, 563 (1989).

THE MECHANISMS AND KINETICS OF SURFACE REACTIONS OF TRIMETHYLGALLIUM ON GaAs (001) SURFACES AND ITS RELEVANCE TO ATOMIC LAYER EPITAXY

B.Y. MAA and P.D. DAPKUS

Department of Electrical Engineering and Center for Photonic Technology, University of Southern California, Los Angeles, California 90089-0483

Abstract

X-ray photoelectron spectroscopy (XPS) and reflection high energy electron diffraction (RHEED) have been applied to study the stable adsorbed Ga species and surface structures after GaAs (001) 2 x 4 As-rich surfaces are exposed to TMGa. These studies show that Ga atoms are the final adsorbed species, that Ga deposition is saturated at one atomic layer at temperatures between 360 and 530 °C and that the surface converts from a 2 x 4 to a 4 x 6 reconstruction after TMGa adsorption. To understand the surface reaction kinetics involved, reflectance-difference spectroscopy (RDS), an *in situ* real-time optical technique developed by Aspnes *et al.* , is applied to investigate TMGa adsorption on (001) GaAs surfaces. The kinetics of the surface reactions and reconstructions have been characterized over the temperature range from 400 to 500 °C using RDS. The transient RDS behaviors are interpreted by the application of a model that involves selective adsorption and reaction of TMGa at surface As sites and at Ga vacancies on Ga-rich reconstructed surfaces. Based upon these interpretations, rates of reaction and by product desorption are determined that suggest optimal strategies for ALE growth of GaAs.

1. Introduction

Atomic Layer Epitaxy (ALE) of III-V compound semiconductors using organometallic precursors holds great promise for crystal growth of layers with uniform thickness and control at the atomic level [1,2]. ALE is a process which proceeds by separately exposing the semiconductor surface with group III and V reactants. Conditions are chosen which result in the formation of a single monolayer of adsorbate containing each element during the alternate exposures. By careful choice of the reactants, reaction pathways can be found that result in the formation of the desired compound via surface reactions. In this way, saturated surface reactions are utilized to deposit exactly one monolayer of the compound over the entire surface on each cycle of exposures. This results in a conformal monolayer by monolayer epitaxial growth habit.

This paper will focus on studies of surface reactions of trimethylgallium (TMGa) on (001) GaAs and its relevance to atomic layer epitaxy of GaAs using TMGa. To meet this goal, a surface science approach is taken. X-ray photoelectron spectroscopy (XPS) and reflection high energy electron diffraction (RHEED) are applied to characterize the reacted surfaces after TMGa adsorbs on GaAs (001) 2 x 4 As-rich surfaces. To understand the surface reaction kinetics involved, reflectance-difference spectroscopy (RDS), is applied to investigate TMGa adsorption on (001) GaAs. A model consistent with mass spectroscopy study is proposed to explain the self-limiting mechanism during TMGa adsorption. And the kinetics study provides us optimal strategies for ALE growth of GaAs.

2. Experiments

All experiments were performed in a UHV system consisting of an analysis chamber and a reaction chamber as shown in Fig. 1 [3-5]. The

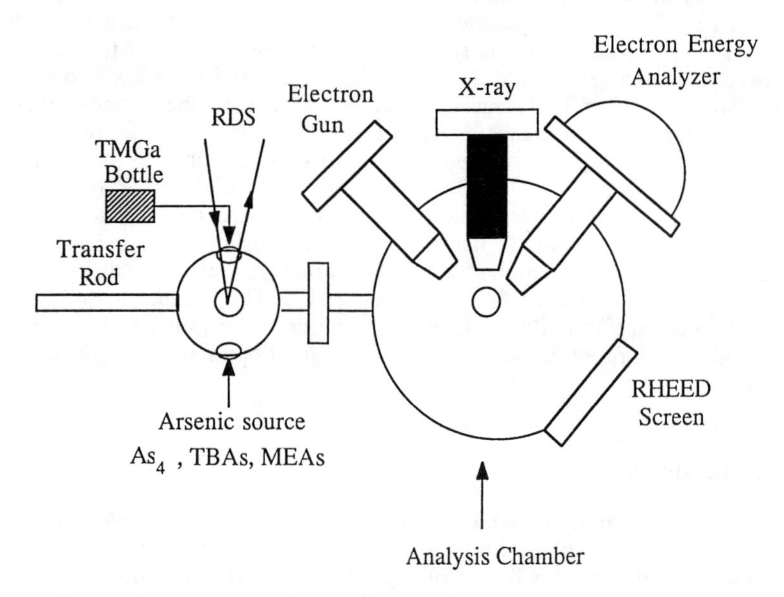

Fig. 1 Schematic diagram of a UHV system used for studies of surface reactions of trimethylgallium and various arsenic sources.

analysis chamber is a modified PHI ESCA 5100 system equipped with XPS, ion gun, RHEED gun, and a phosphor-coated view port. The reaction chamber provides for exposure to arsenic and TMGa. Arsenic metal, tertiarybutylarsine (TBAs), or monoethylarsine (MEAs) is used to create As-stabilized surfaces. A collimated beam of TMGa is introduced into the chamber via a UHV precision leak valve. The RDS setup is similar to that described by Aspnes *et al.* [6] but is simplified through the use of a single wavelength He-Ne laser (633 nm) as the light source [7]. Aspnes and coworkers have shown that at this energy (1.96 eV), the major contribution to the RDS signal from a GaAs (001) surface is due to the presence of Ga dimers. The RDS signal at 1.96 eV then acts as a monitor of the surface Ga dimer population. The substrate is radiatively heated and its temperature is monitored by an infrared pyrometer calibrated against the melting point of InSb at ~525° C. A base pressure of 4 x 10^{-10} Torr in the reaction chamber is obtained by a 170 L/s turbomolecular pump.

The samples, 10 x 12 mm^2 n-type (~2 x 10^{18} cm^{-3} Si-doped) GaAs (001) substrates, were degreased with solvents, etched by HCl and 5:1:1 (H_2SO_4:H_2O:H_2O_2) solution, and rinsed in flowing DI water. The cleaned sample was indium soldered on a Mo sample holder. The sample was heated in the reaction chamber to ~ 620° C with an As_4 overpressure of 10^{-6} Torr for about 20 minutes then transferred back to the analysis chamber. RHEED patterns and XPS spectra of Ga $2p_{3/2}$, As 3d, and C 1s were taken for this starting surface. The Al K_α x-ray line was used for XPS to avoid overlap between C 1s and Ga LMM Auger peaks. The photoelectron exit angle between surface and the axis of detector was normally 45°. To enhance surface sensitivity, the C 1s spectra were recorded with 30° exit angle. The details of the acquisition of the XPS spectrum were mentioned before and are not repeated here [3,4].

After sample treatment a clean surface was prepared showing no evidence of carbon as indicated by the XPS C 1s spectrum. A well defined RHEED 2 x 4 As-stabilized surface was also observed. All experiments involving TMGa exposure and reactions described here were carried out with this starting surface reconstruction. We have found that it is stable, once formed, from room temperature to the temperature used in these experiments. The sample was then moved to the reaction chamber and exposed to TMGa. Samples can be transferred between two chambers for XPS and RHEED investigations or for *in-situ* real time TMGa and TBAs adsorption experiments by RDS. Substrate temperatures were varied between 320 and 550° C.

3. Results

3.1 On the Observations by XPS and RHEED

Fig. 2 shows XPS intensity ratio of Ga $2p_{3/2}$ to As 3d at various substrate temperatures and TMGa exposure levels. At T_s=320° C with exposure time of either 10 or 15 seconds and TMGa pressures of P_{TMGa} = 2 x 10^{-6} ~ 5 x 10^{-6} Torr no change in the Ga intensity can be observed. The surface still exhibits a RHEED 2 x 4 pattern after TMGa exposures. As the substrate temperature is increased to higher than 370° C, we observe both a gradual increase of Ga $2p_{3/2}$ and decrease of As 3d intensities with increasing exposures to TMGa and finally their ratio reaches a saturation level. At T_s=370° C, the Ga intensity saturates after 60 L (1 L=1 Langmuir=10^{-6} Torr-sec) of TMGa is supplied. At this level RHEED shows a vague 1 x 1 pattern. After we heat this surface

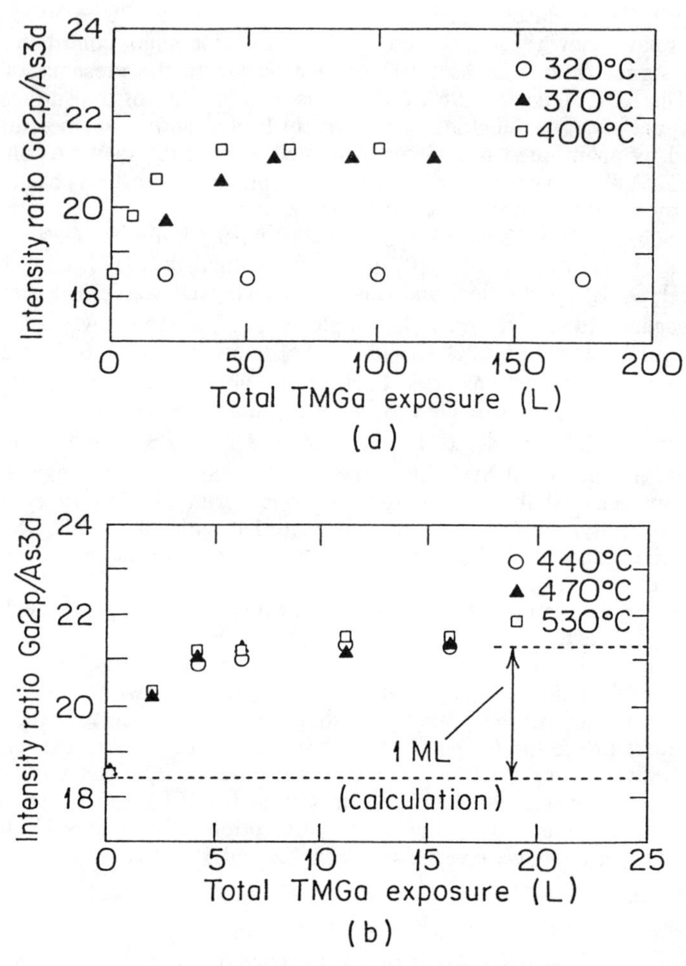

Fig. 2 Intensity ratio of Ga $2p_{3/2}$ to As 3d vs. total exposure of TMGa at (a) 320° C, 370° C, and 400° C (b) 440° C, 470° C, and 530° C. The starting surface is a 2 x 4 As-stabilized surface. Note the different exposure levels in (a) and (b). The arrow in (b) shows the calculated result for one ML Ga deposition which results in a final RHEED 4 x 2 structure.

structure to 500° C without any additional TMGa, a 3 x 1 RHEED pattern is observed. When the substrate temperature is above 440° C during TMGa exposure, the Ga intensity saturates rapidly at ~ 6 L exposure of TMGa. In this temperature regime we observe a 3 x 1 RHEED pattern at lower exposure levels and a 4 x X (X=1 or 2) pattern in the saturation region.

The XPS C 1s spectra before and after TMGa exposures at several substrate temperatures are shown in Fig. 3. The C 1s level on the starting 2 x 4 As-stabilized surface is typical of an MBE prepared sample. Throughout all TMGa adsorption processes, with increasing TMGa exposure and Ga surface coverages, no observable change in the carbon level can be detected within the sensitivity of XPS. For comparison, we also perform room temperature TMGa adsorption on a GaAs surface and record its C1s spectrum in Fig. 3(e). Under these conditions, the amount of carbon is estimated to be ~0.2 ML.

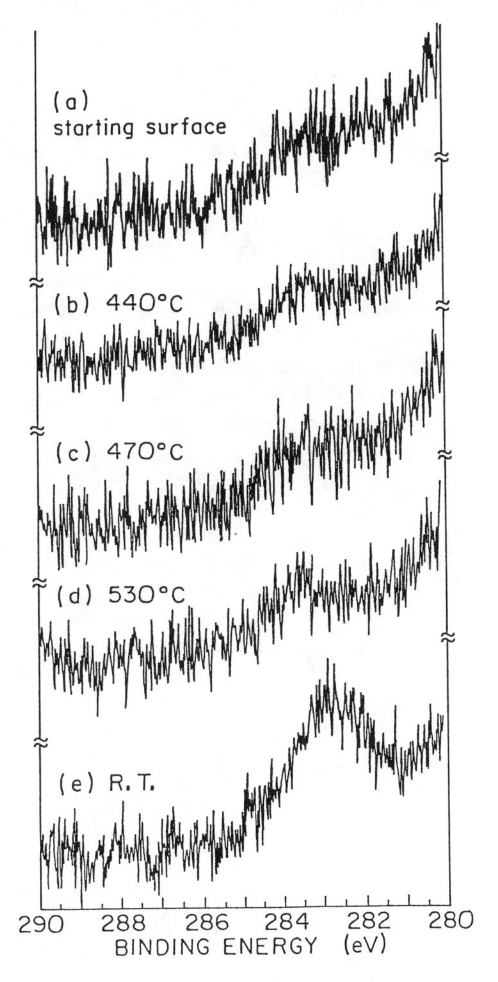

Fig. 3 XPS C 1s spectra (a) starting clean surface, and in the saturation regime of TMGa adsorption at (b) 440° C (c) 470° C (d) 530° C. (e) is the spectrum of room temperature adsorption of TMGa.

To gain knowledge about the ML deposition of Ga atoms, we derive for GaAs (001) the following expressions for the intensity ratio of the Ga 2p to As 3d core electrons from the exponential absorption law

$$\frac{I_{Ga2p}}{I_{As3d}} = S\frac{[\theta_{As}\sum_{i=0}^{\infty}\exp(-\frac{(2i+1)a}{\lambda_{Ga}}) + (1-\theta_{As})\sum_{i=0}^{\infty}\exp(-\frac{2ia}{\lambda_{Ga}})]}{[\theta_{As}\sum_{i=0}^{\infty}\exp(-\frac{2ia}{\lambda_{As}}) + (1-\theta_{As})\sum_{i=0}^{\infty}\exp(-\frac{(2i+1)a}{\lambda_{As}})]}$$

for the As-stabilized surface, and

$$\frac{I_{Ga2p}}{I_{As3d}} = S\frac{[\theta_{Ga}\sum_{i=0}^{\infty}\exp(-\frac{2ia}{\lambda_{Ga}}) + (1-\theta_{Ga})\sum_{i=0}^{\infty}\exp(-\frac{(2i+1)a}{\lambda_{Ga}})]}{[(1-\theta_{Ga})\sum_{i=0}^{\infty}\exp(-\frac{2ia}{\lambda_{As}}) + \theta_{Ga}\sum_{i=0}^{\infty}\exp(-\frac{(2i+1)a}{\lambda_{As}})]}$$

for the Ga-stabilized surface. Here S is the sensitivity factor ratio of Ga $2p_{3/2}$ to As 3d photoelectrons, 2a is one ML thickness along (001) GaAs growth direction, λ is the inelastic mean free path of the corresponding photoelectron at an exit angle of 45°, and θ is the surface coverage. $\lambda_{Ga2p3/2}$ = 10 sin45° Å and λ_{As3d} = 25 sin45° Å are used for these calculations. θ_{As} and θ_{Ga} are both taken to be 0.75 according to a recent observation by Pashley *et al.* [8]. The calculated change of this ratio in transition from a 2 x 4 As-stabilized to a 4 x X Ga-stabilized surface is shown by the arrow in Fig. 2(b).

Based on the results from XPS and RHEED we conclude that Ga atoms are the stable adsorbed species after GaAs (001) As-stabilized surfaces are exposed to TMGa in the temperature range between 370 and 530° C. To investigate further if there is any appreciable reaction between TMGa and GaAs (001) Ga-stabilized surfaces and the self-limiting mechanism does exist during TMGa exposure, we performed experiments of TMGa adsorption on thin GaAs (001) layers [4]. A thin layer of GaAs on thick AlAs was grown by MOCVD. The AlAs layer, which was thicker than 1000 Å, was used to completely suppress the Ga signal from the underlying GaAs substrate during the XPS measurements. A 50 Å layer of GaAs was grown on AlAs and the sample was transferred into UHV system. After heat treatment in the reaction chamber at 600° C to remove the oxide layer, the sample was sputtered with Ar^+ beam such that a ~ 10 Å thick GaAs overlayer was obtained. Before exposure to TMGa, the sample was annealed under As_4 overpressure at 510° C. The XPS spectra of Ga $2p_{3/2}$, As $2p_{3/2}$, Al 2p, and C 1s were taken for this starting surface. RHEED patterns were recorded after the XPS measurements. The sample was moved to the reaction chamber and exposed to TMGa. It was then transferred back to the analysis chamber to take RHEED pattern and XPS spectra of Ga $2p_{3/2}$, As $2p_{3/2}$, Al 2p, and C 1s again. The sample was thus transferred back and forth between the reaction chamber and the analysis chamber with successive TMGa exposures.

The starting surfaces are characteristic of clean and As-stabilized surfaces. A 2 x 4 RHEED pattern is shown in Fig. 4(a). By exposure to TMGa at 500° C we form Ga-stabilized surfaces that exhibit a 4 x 6 RHEED pattern as shown in Fig. 4(b). No carbon can be detected within the sensitivity of XPS. Typically a dosage of TMGa of 5 L is supplied for

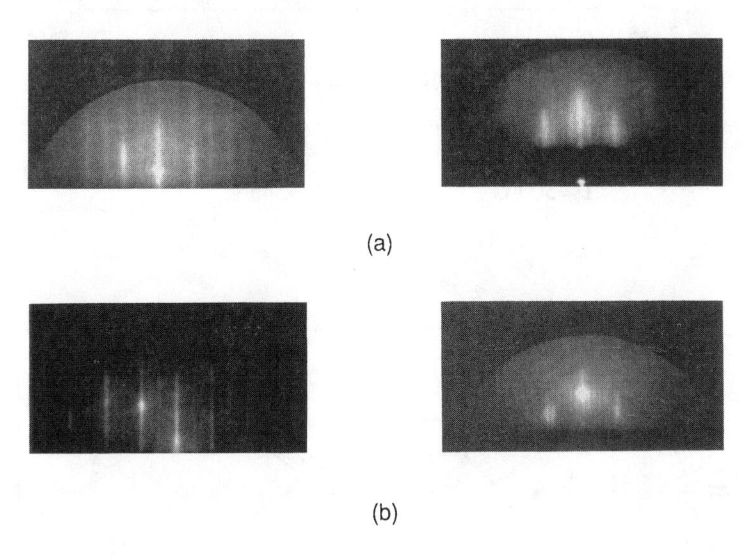

Fig. 4 RHEED patterns of GaAs/AlAs samples: (a) 2 x 4 for the starting As-stabilized surfaces, (b) 4 x 6 for Ga-stabilized surfaces after TMGa adsorption at 500 °C. Photos on the left are along [110] and those on the right [$\bar{1}$10].

such a surface conversion. To investigate TMGa adsorption on Ga-stabilized surfaces we expose this Ga rich surface to 3 ~ 5 L of TMGa at T_s=545° C four to six times with two minutes break in between. Then the sample is annealed under arsenic ambient. This process is repeated a few times and is schematically shown in Fig. 5(a). Results of the changes of XPS intensity ratio of Ga 2p to As 2p are shown in Fig. 5(b) for two samples. Note that there is a definite increase in the intensity ratio after exposure to TMGa/As cycle (points D and G) compared to the starting sample. This reflects an increase in the thickness of the GaAs overlayer.

The GaAs overlayer thickness, t, could be obtained from the measured XPS intensity ratios through

$$\frac{(\text{Ga/As) in GaAs/AlAs structure}}{\text{(Ga/As) in bulk GaAs}} = 1 - e^{-t/\lambda(\text{Ga})}$$

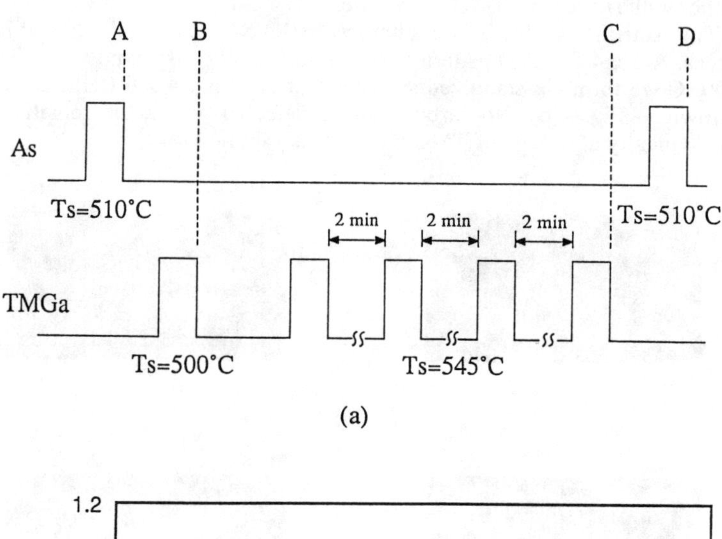

(a)

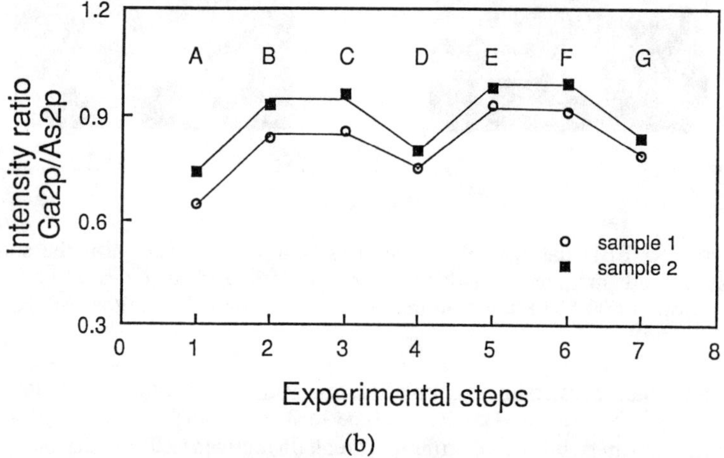

(b)

Fig. 5 (a) A is the starting As-stabilized surface. A Ga-stabilized surface is first formed at stage B and then several pulses of TMGa are directed to this surface with 2 min break in between. This is the stage of C. Stage D is after As_4 annealing at 510° C. Similarly for E and F under TMGa exposures and G after As_4 annealing as shown in (b). (b) XPS intensity ratio of Ga 2p to As 2p during different stages of experiments (A to G) as described in (a). The additional TMGa injections at stage C is four consecutive pulses of 3 L (3 x 10^{-7} Torr x 10 s) for sample 1 and 6 consecutive pulses of 5 L (3 x 10^{-7} Torr x 16 s) for sample 2.

This equation is derived by assuming a uniform excitation and an exponential absorption of emitted electrons with λ known as the mean free path. λ and cross section for As 2p photoelectrons in AlAs and GaAs are also assumed identical. Table 1 lists the calculated GaAs overlayer thicknesses for the measured XPS intensity ratios. In spite of the fact that Ga-stabilized surfaces receive multiple exposures of TMGa, almost one monolayer deposition of GaAs (2.83 Å) is obtained between t_D and t_A or t_E and t_B. It should be noted that at these temperatures atomic Ga is relatively mobile on the surface even under an As_4 overpressure [9]. Thus the formation of a single monolayer of GaAs after repeated TMGa exposures precludes the formation of appreciable excess Ga on the surface. Were such Ga present in significant quantities, surface migration during As_4 exposure could lead to the formation of more than one monolayer of GaAs. Recent MBE results have shown that as much as 3 monolayers of Ga resident on the surface of GaAs will redistribute to grow GaAs during As_4 exposure. These results show that, on time scale and at exposure levels relevant to ALE, the TMGa reaction on Ga-stabilized surfaces does not produce appreciable excess Ga atoms. They further support the concept that the differential chemisorption and decomposition rates of TMGa on surface As and Ga atoms is one of the factors that contribute to the self-limiting mechanism of ALE of GaAs using TMGa.

Table 1 Calculated GaAs overlayer thicknesses for the measured XPS intensity ratios shown in Fig. 5(b)

Thickness (Å) / samples	t_A	t_B	t_D	t_E	t_G
sample 1	8.5	9.6	11.4	12.3	13.9
sample 2	11.2	12.5	13.9	15.3	16.2

3.2 On the Results by RDS

RDS is applied to investigate the reaction mechanisms and kinetics of TMGa adsorption on (001) GaAs surfaces [5]. RDS, a surface sensitive technique [6], is particularly suitable for such a study because of the relatively good understanding of surface reconstructions of GaAs (001) surfaces and a surface chemistry which is closely related to such surface structures [10,11]. In the following experiments TBAs is used as an arsenic source, which has been shown as a promising arsine alternative source that under appropriate exposure conditions produces a 2 x 4 stabilized surface with no appreciable carbon contamination.

Fig. 6 shows the transient behavior of $\Delta R/R$ ($\Delta R = R_{110} - R_{\bar{1}10}$) for three TMGa exposure times at a substrate temperature of 500 °C and a TMGa injection level of 1.2 x 10^{-5} Torr. When the TMGa flux is initiated, the RDS signal starts to rise. Although TMGa flow occurs for only 1 s, the RDS signal continues to increase until a final saturation level is reached at the end of 3 s as shown in Fig. 6(a). When the TMGa exposure time is increased to 3 s and 5 s as shown in Fig. 6(b) and 6(c), a distinct change in the RDS signal is observed. The initial rise shows a pronounced kink for the longer exposures once the RDS signal reaches a certain level. After the termination of the TMGa flux, the RDS signal continues to rise until it reaches the same level as that for the 1 s exposure. These surface configurations are stable with time based upon our observation of prolonged stability of the RDS signal. The XPS spectrum of these saturated surfaces show no trace of carbon and the surfaces exhibit RHEED 4 x 6 patterns. When TBAs flow is directed on these saturated surfaces, the RDS signal falls off rapidly to the starting level. RHEED 2 x 4 patterns are observed on these surfaces and no trace of carbon could be found within the sensitivity of XPS.

Fig. 7 shows the RDS transient when RHEED 4 x 6 surfaces are exposed to TMGa at 500 °C. At a TMGa injection level of 1.2 x 10^{-5} Torr and an exposure time of 6 s, the RDS signal decreases rapidly to some lower level at which the signal saturates. Once the TMGa supply is stopped, the RDS signal returns to the previous level.

Similar RDS behavior is also observed at other substrate temperatures. At 450 °C, a 2 s TMGa exposure leads to a maximum saturation level after an additional 7 s as shown in Fig. 6(d) and a total of 25 s is required to reach this level for a 4 s TMGa exposure at 430 °C as shown in Fig. 6(e).

Fig. 8 shows the RDS transients when RHEED 4 x 6 surfaces are exposed to TMGa at 500, 450, and 430 °C. At a TMGa injection level of 1.2 x 10^{-5} Torr and an exposure time of 6 s, the RDS signal decreases rapidly at all three temperatures to some lower level at which the signal saturates. Once the TMGa supply is stopped, the RDS signal for samples exposed at various temperatures returns to the previous level with dramatically different time constants, increasing with decreasing temperature. These recovery time constants are similar to those observed after TMGa termination in the transition from an As-stabilized to a Ga-stabilized surface; suggesting a common mechanism.

Recent studies on (001) GaAs surfaces identify the existence of surface Ga-Ga dimers and As-As dimers as the main contributors to the surface anisotropy observed in RDS. The wavelengh of the red He-Ne laser used in our RDS study is sensitive to the existence of Ga-Ga dimer bond [12]. In Fig.

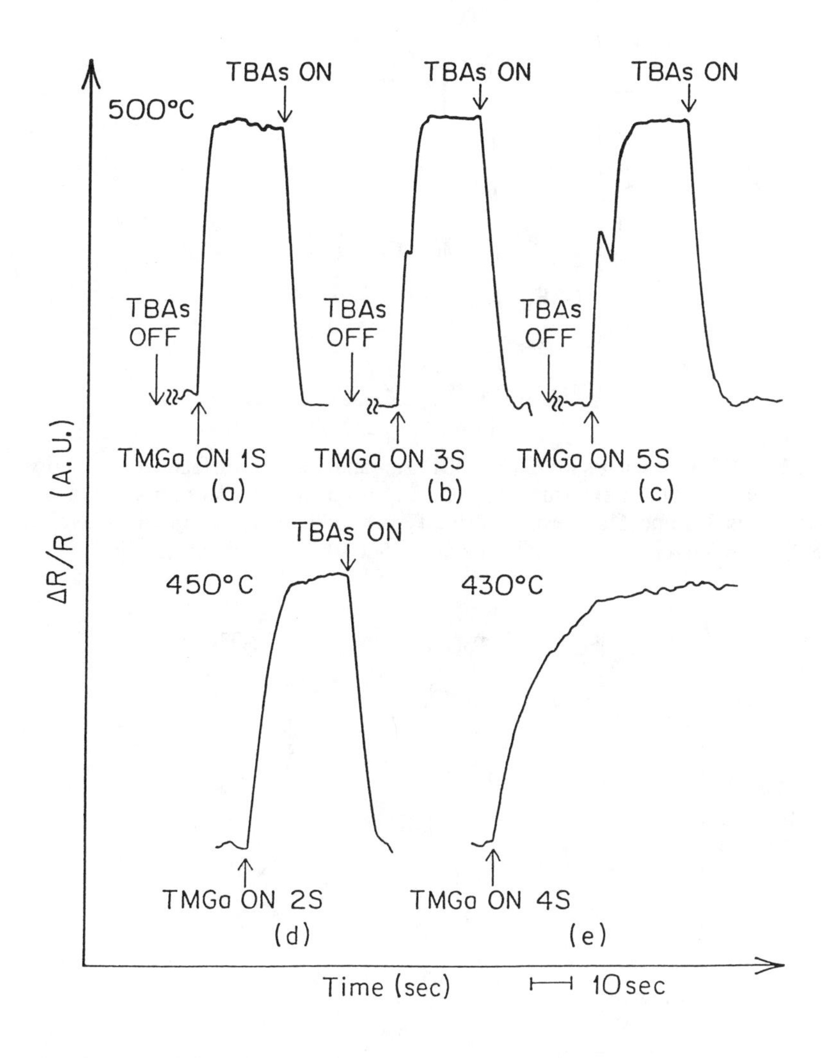

Fig. 6 RDS transients on (001) GaAs for initiation of TMGa and TBAs exposures. (a) 1 s, (b) 3 s, and (c) 5 s TMGa exposure at 500 °C. (d) 2 s TMGa exposure at 450 °C. (e) 4 s TMGa exposure at 430 °C. The exposure level for TMGa and TBAs is 1.2×10^{-5} and 1.5×10^{-5} Torr respectively.

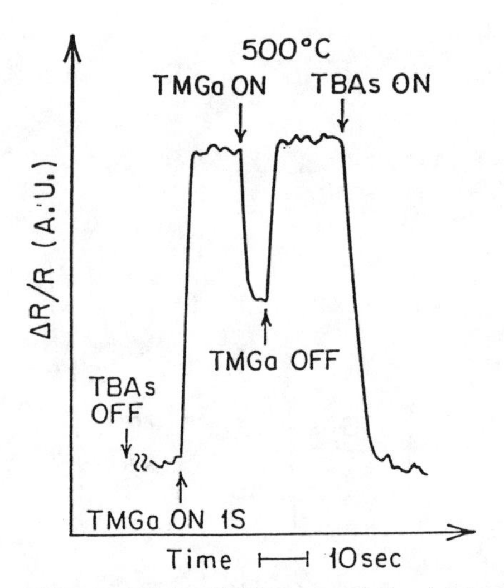

Fig. 7 RDS transients on GaAs (001) As- and Ga-rich surfaces for initiation of TMGa and TBAs exposures. TMGa exposure time on As-rich and Ga-rich surfaces is 1 s and 6 s , respectively. T_s = 500 °C and the injection level of TMGa and TBAs is 1.2 x 10^{-5} Torr and 1.5 x 10^{-5} Torr, respectively.

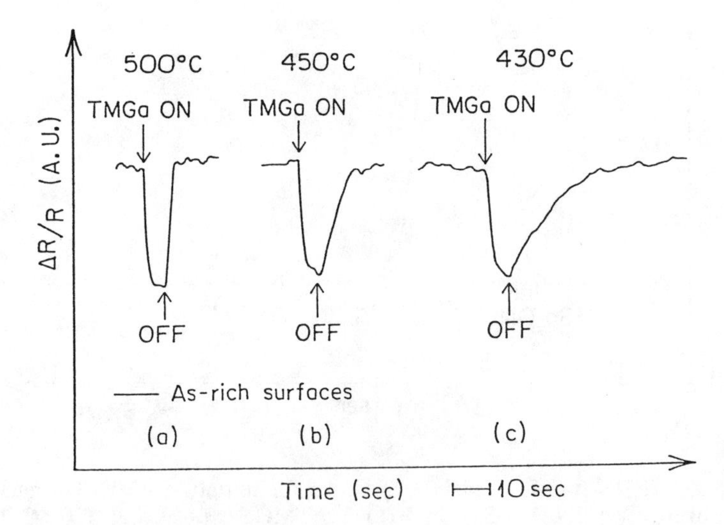

Fig. 8 RDS transients on RHEED 4 x 6 GaAs (001) surfaces upon introduction of TMGa flux. 6 s TMGa exposure at (a) 500 °C, (b) 450 °C, and (c) 430 °C. The TMGa injection level is 1.2 x 10^{-5} Torr. Also shown is the RD signal level for As-rich surfaces.

6(a) the increasing RDS signal when 2 x 4 As-rich surfaces are exposed to TMGa indicates that Ga-Ga dimers are formed on the GaAs surfaces by exposure to TMGa. On the other hand, the continued increase in the RDS signal after the termination of TMGa supply suggests that the formation of Ga-Ga dimers also takes place on the surface after the removal of TMGa supply. Recently, Memmert *et al.* proposed a model for TMGa decomposition on GaAs surfaces [13]. The TMGa molecule reacts with surface As atoms and dissociates into $GaCH_3$ and two CH_3 methyl radicals on the surface. The CH_3 methyl radicals desorb in two channels: one for the desorption of CH_3 from the adsorbed $GaCH_3$, the other for the desorption of CH_3 from GaAs (001) surface. Desorption of CH_3 from the adsorbed $GaCH_3$ results in formation of gallium atoms which, judging from the increasing RDS signal observed in this work, tend to form Ga-Ga dimer bonds. The two released CH_3 methyl radicals associated with TMGa dissociation could be attached to surface As or Ga atoms [13,14]. Attachment of CH_3 methyl radicals to nearby As atoms hinders further reactions between these As atoms and impinging TMGa molecules temporarily. Not until the desorption of CH_3 methyl radicals from As atoms will these As atoms become active again toward TMGa dissociation. The data of Memmert *et al.* suggest that the desorption rate for this process is fast with a time constant of 20 ms at 500 °C [13]. On the other hand, there exists the possibility for CH_3 methyl radicals to be attached to nearby newly formed gallium dimers, which requires that gallium dimer bonds be broken. Recovery of Ga-Ga dimers is anticipated once CH_3 methyl radicals desorb from gallium atoms. Nevertheless, reactions involving broken gallium dimer bonds would introduce a decrease in the signal level of RDS which is only sensitive to the existence of Ga-Ga dimer bonds.

When 2 x 4 As-rich surfaces are exposed to TMGa molecules, these reactions proceed simultaneously. In the initial stage, formation of gallium dimers dominates the other processes to show an increase in RDS signal. Although a steady increase in RDS signal is observed, the RDS signal level is in fact reflecting only those gallium atoms possessing dimer bonding configurations. Those deposited gallium atoms which are bonded to CH_3 can not be detected by RDS. The number of deposited gallium atoms resulting from TMGa decomposition on the reactive As sites will not increase indefinitely and in fact is self-limited to one atomic layer [15,4], a matter to be discussed in more detail later. If the TMGa flow is terminated near the time when one atomic layer gallium atoms are deposited as shown in Fig. 6(a), those CH_3 radicals which are attached to gallium atoms desorb and Ga-Ga dimers appear on the surface. The RDS signal therefore, continues to increase with the desorption of residual CH_3 radicals until all Ga-Ga dimers are recovered.

If the TMGa supply is not terminated in time and extended TMGa exposure occurs, the RDS signal decreases and results in the kinks as shown in Fig. 6(c). A Ga-rich GaAs (001) surface is not terminated with a complete atomic layer of gallium dimers but instead contains a number of gallium vacancies that expose underlying As atoms to vacuum [16]. Such exposed As atoms are active toward TMGa dissociation with formation of two CH_3 methyl radicals and $GaCH_3$ [14]. Hence, we suggest that extended TMGa exposure on Ga-rich surfaces brings about the adsorption of undesired CH_3 methyl radicals that are attached to nearby gallium dimers and break Ga-Ga dimer bonds. A decrease in the RDS signal is observed through this decrease in the Ga-Ga dimer population. Once the TMGa supply is terminated, the remaining attached CH_3 radicals desorb from the surface to restore the Ga-Ga dimers. The RDS signal increases until all gallium dimers are recovered. Similar and consistent behavior is observed when reconstructed 4 x 6 Ga-rich surfaces are exposed to TMGa. As shown in Fig. 7, a rapid initial drop of the RDS signal was observed which indicates both a relatively fast reaction between TMGa and active As atoms and the breaking of existing gallium dimers by the attachment of CH_3 methyl radicals. After the TMGa flux is terminated, the RDS signal rises again to the previous level.

The surface lifetime of CH_3 methyl radicals attached to As and Ga atoms is mainly determined by the substrate temperature. At lower temperatures, e.g. T_s= 430 °C, deactivated As atoms due to CH_3 attachment stay inactive to impinging TMGa molecules until the desorption of long-lived CH_3 radicals. Such long-lived CH_3 radicals and slower TMGa reaction rates require a longer TMGa exposure to achieve one atomic layer gallium deposition. However, even with all available gallium sites filled and TMGa supply terminated, time is required for those CH_3 radicals attached to gallium atoms to desorb. A dramatic difference in the total amount of time to recover the maximum Ga-Ga dimer coverage was noticed at substrate temperatures 500, 450, and 430 °C in Fig. 8. A reaction mechanism is attempted to account for these RDS behaviors and the resultant kinetics simulation shows that desorption of CH_3 methyl radical from Ga atoms dominates the process in the RDS signal recovery when TMGa supply is terminated [17]. The best fitted time constant for desorption of CH_3 methyl radicals is 1 s, 4.5 s, and 10 s at substrate temperature 500, 450, and 430 °C respectively. These values are longer than those reported by Memmert *et al.* In Memmert's experiments, however, the incident TMGa flux was 10^{-4} monolayer coverage per pulse. In the present experiments, a much higher TMGa flux is used and a coverage dependent CH_3 desorption process as observed by Creighton [18] takes place.

Gallium droplets were not found by Nishizawa *et al.* during atomic layer growth at temperatures below 500 °C even with a prolonged TMGa exposure [19]. We believe that the gallium vacancies associated with a Ga-rich surface reconstruction enhance the possibility of desorption of the $GaCH_3$ species. Although gallium-vacancy induced As atoms are active toward TMGa dissociation, we suggest that $GaCH_3$ desorbs from the As site in order to maintain a stable electronic state with a minimum free energy [20]. This is consistent with the observation by Yu *et al.* that an additional channel of desorption for $GaCH_3$ is observed when TMGa flux and/or exposure time is increased [21]. Therefore, several mechanisms contribute to the self-limiting growth in ALE using TMGa. Surface As atoms are the only active species for TMGa dissociation and they exist on both 2 x 4 As-rich and 4 x 6 Ga-rich surfaces. As the surface becomes more populated with gallium atoms, however, reconstruction of the Ga-rich surface plays a further decisive role to help achieve self-limiting gallium deposition by driving off the methylgallium species occupying unfavorable Ga sites.

4. Conclusions

XPS, RHEED, and RDS are applied to study surface reactions of TMGa on (001) GaAs surfaces. The stable TMGa-exposed surfaces show no trace of carbon within the sensitivity of XPS and the surfaces exhibit 4 x 6 RHEED pattern above 440 °C characteristic of Ga-rich surfaces. One atomic layer deposition of Ga atoms is obtained even though the Ga-rich surfaces are exposed to additional TMGa molecules. The use of RDS provides us invaluable real-time access to investigate TMGa adsorption process. A model consistent with various kinetics studies is established to explain the distinct behavior observed in RDS during TMGa exposure. The self-limiting mechanism which occurs in TMGa exposure cycle is believed to result from both selective adsorption and reaction of TMGa at As atoms and Ga vacancy associated Ga-rich surface reconstruction. It is also shown that optimal growth conditions of TMGa-based ALE can be achieved through RDS monitoring.

5. Acknowledgements

The authors would like to acknowledge the support of the Office of Naval Research and Solar Energy Research Institute for portions of this work and to thank D. E. Aspnes for advice given during the early stages of our development of the RDS technique.

References

1. S.P. DenBaars, C.A. Beyler, A, Hariz, and P.D. Dapkus, Appl.Phys.Lett., 51, 1530 (1987)
2. K. Mori, M. Yoshida, A. Usui, and H. Terao, Appl. Phys. Lett. 52, 27 (1988)
3. B. Y. Maa, and P. D. Dapkus, J. Electron. Mater. **19**, 289 (1990)
4. B. Y. Maa, and P. D. Dapkus, J. Crystal Growth **105**, 213 (1990)
5. B. Y. Maa, and P. D. Dapkus, to be published in Appl. Phys. Lett **58**, 20 May 1991
6. D. E. Aspnes, J. P. Harbison, A. A. Studna, and L. T. Florez, J. Vac. Sci. Technol. **A6**, 1327(1988)
7. G. Paulsson, K. Deppert, S. Jeppesen, J. Jonsson, L. Samuelson, and P. Schmidt, J. Crystal Growth **105**, 312 (1990)
8. M.D. Pashley, K.W. Haberern, and W. Friday, Phys. Rev. Lett. **60**, 2176 (1988)
9. T. H. Chiu, Appl. Phys. Lett. **55**, 1244 (1989)

10. D. E. Aspnes, J. P. Harbison, A. A. Studna, L. T. Florez, and M. K. Kelly, J. Vac. Sci. Technol. **B 6**, 1127 (1988)
11. D. E. Aspnes, R. Bhat, E. Colas, V. G. Keramidas, M. A. Koza, and A. A. Studna, J. Vac. Sci. Technol. **A 7**, 711 (1989)
12. D. E. Aspnes, Y. C. Chang, A. A. Studna, L. T. Florez, H. H. Farrell, and J. P. Harbison, Phys. Rev. Lett. **64**, 192 (1990)
13. U. Memmert, and M. Yu, Appl. Phys. Lett. **56**, 1883 (1990)
14. J. R. Creighton, K. R. Lykke, V. A. Shamamian, and B. D. Kay, Appl. Phys. Lett. **57**, 279 (1990)
15. J. Nishizawa, H. Abe, and T. Kurabayashi, J. Electrochem. Soc. **132**, 1197 (1985)
16. G. X. Qian, R. M. Martin, and D, J. Chadi, J. Vac. Sci. Technol. **B5**, 1482 (1987)
17. B. Y. Maa, PhD. Thesis, University of Southern California (1991)
18. J. R. Creighton, surface science **234**, 287 (1990)
19. J. Nishizawa, and T. Kurabayashi, J. Crystal Growth **93**, 98 (1988)
20. H. H. Farrell, J. P. Harbison, and L. D. Peterson, J. Vac. Sci. Technol. **B 5**, 1482 (1987)
21. M. L. Yu, invited talk in Material Research Society Spring Meeting, Anaheim, CA, 1991

HREELS STUDY OF THE ADSORPTION OF ORGANOMETALLICS ON GaAs(001) SURFACES

A. NÄRMANN, R. J. PURTELL AND M. L. YU
IBM Research Division, T. J. Watson Research Center, Yorktown Heights, NY

ABSTRACT

Trimethylgallium (TMGa), Triethylgallium (TEGa) and Diethylgalliumchloride (DEGaCl) are used in combination with arsine (AsH3) in atomic layer epitaxy of GaAs. We simulated this process in a UHV system by dosing a c(2x8) or (1x6) reconstructed GaAs(001) surface with arsine and organometallics at different temperatures. After the surface reconstruction was verfied by LEED, the sample was transferred in situ to the HREELS (high resolution electron energy loss spectroscopy) chamber. We present new studies on the adsorption and pyrolysis of organometallics on GaAs(001) surfaces.

As a result we find that TMGa adsorbs molecularly on an As-rich cooled surface but decomposes upon adsorption on a Ga-stabilized (1x6) surface. These two adsorption states are identified by the geometrical position of the Ga-methyl compound on the surface (Ga-C bond parallel or normal to the surface). The removal of the adsorbed species occurs on an As-rich surface at temperatures higher than 450°C and on the Ga-stabilized surface at 350°C.

For TEGa and DEGaCl we found similar results.

INTRODUCTION

In HREELS the long ranged Coulomb field of an electron impinging on a surface interacts with dipoles on and in the surface. By exciting vibrational modes of e.g. adsorbates, the electron loses a certain amount of energy which can then be detected. Thus an energy loss spectrum of backscattered electrons provides information about which adsorbates there are at the surface. Geometrical information may be extracted as well.

Another mechanism of excitation is by electron impact. The dipole and impact mechanism can be distinguished via their angular dependence, the dipole scattering leading to a strong signal in specular reflection direction whereas the impact mechanism produces a broad signal with respect to scattering angle. For details see e.g. Ref [1]. Due to selection rules, dipole scattering is sensitive to vibtrations which cause a change of the dipole moment perpendicular to the surface, whereas for impact scattering no such selection rule applies.

Most of earlier HREELS work concerning the adsorption of organometallics on surfaces was done on Si, see Refs [2-6]. Their results can be shortly summarized as follows: TMGa is adsorbed molecularly below 300 K and decomposes upon heating by a reaction of a methyl group (CH_3) in the TMGa with a hydrogen on another methyl group thus yielding methane (CH_4) and a CH_2 fragment. The methane desorbs, the CH_2 is bound to a gallium. Then the next methyl group reacts with the CH_2 fragment to give methane and a CH compound. Further heating causes the hydrogen to desorb, leaving the carbon atom back on the Si surface.

Our results indicate that the decomposition and desorption mechanism on a GaAs surface is different from that on a Si surface.

EXPERIMENTAL

The system in which all the experiments were performed consists of four connected UHV-chambers. A glove-box is connected to the first chamber ($<10^{-8}$ mbar) where all the sample preparation and dosing was performed. From there the sample was transferred to the LEED chamber ($<5 \times 10^{-10}$ mbar) where the surface reconstruction was checked. To the third chamber ($<2 \times 10^{-10}$ mbar) an XPS-system is attached from which the sample was finally brought into the HREELS chamber ($<2 \times 10^{-11}$ mbar).

Our samples were cut out of Si-doped $10^{-3}\Omega$cm n-GaAs(001) wafers with the surface normal 2° off the <110> direction. The electron density was $n=3 \times 10^{18}$cm^{-3}. The samples were degreased and etched by a standard treatment. After transfer to vacuum, the samples were outgased and sputtered before the procedure used to obtain the c(2×8) reconstructed surface was applied. The temperature of the sample was determined by a chromel-alumel (type K) thermocouple pair which touched the back of the center of the sample. The temperature readings obtained this way were off at high temperatures. In another system[7] where the heating was performed in the same way, the temperature readings were gauged with an IR-pyrometer. In the following we give the corrected temperatures.

To obtain the As-rich c(2×8) surface the sample was heated for 10 min at 650°C in arsine ($p=6\times10^{-5}$ mbar) and subsequently annealed in vacuum at 450°C for another 10 min. Annealing at 560°C instead gave the Ga-stabilized (1×6) reconstructed surface. In this way we obtained reproducible LEED patterns.

The dosing with metallorganics onto either surface was performed through a leak valve attached to the preparation chamber. During dosing the sample surface faced the inlet which was about 4 cm away. Usually a dosage of about 200L was applied (40 seconds exposure at a background pressure of 5×10^{-6} mbar). Once the dosing was finished (i.e. the leak valve closed), the sample was transferred within about 20 seconds to the XPS-chamber.

The HREEL spectrometer (ELS 22 from Leybold) produced electrons of 5 eV incident energy which impinged onto the surface at an angle of 30° and were detected in specular direction. The scans were taken in steps of 1 meV with a dwell time of 20 msec. The spectra shown are an average of 500 to 5000 of those scans. The spectrometer resolution usually was about 8 meV.

During data aquisition the sample was cooled down to about -80°C with liquid nitrogen.

RESULTS AND DISCUSSION

In Fig. 1 we show a HREEL-spectrum of a clean GaAs(001)c(2×8) as well as spectra obtained after dosing cooled GaAs(001)(1×6) and c(2×8) surfaces with 200L TMGa. Six peaks appear at 35, 70, 88, 146, 176 and 365 meV energy loss.

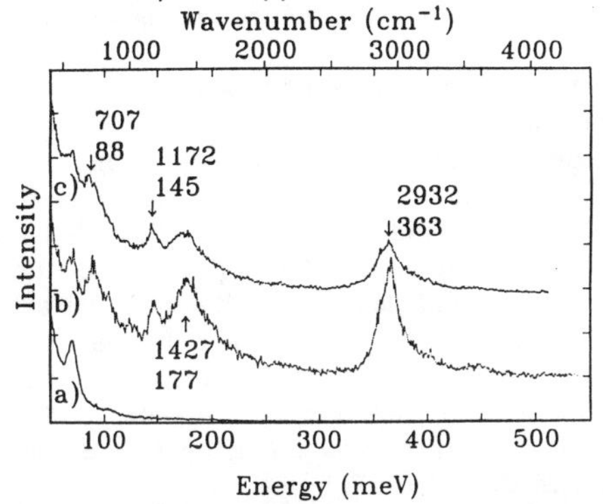

FIG. 1 Spectra of clean c(2×8) surface (a) and after dosing a c(2×8) (b) and (1×6) (c) surface with about 200 L TMGa. The elastic peak and phonon mode are not shown.

The first two modes are due to the GaAs-phonon and its first harmonic. The elastic peak is skewed to the high-energy-loss side. This effect is produced by the GaC_3 deformation modes, especially the out-of-plane deformation. The line at 88 meV can be assigned to the CH_3 rocking vibration. At 146 meV we find the CH_3 symmetric deformation (umbrella mode), at 176 meV the CH_3 anti-symmetric defor- mation and at 365 meV the C-H stretching modes. Due to the resolution of the spectrometer we cannot assign the C-H stretching modes as originating from the CH_3 anti-symmetric or CH_3 symmetric stretching mode. The skewness of the C-H strech peak can be readily explained by electrons which suffered two successive losses: one to a C-H stretch mode and another one to a phonon mode. The small peak at about 450 meV comes from co-adsorbed water which was verified by leaving a clean cooled c(2×8) sample a long time (25h) in vacuum. The HREEL-spectrum of this sample then showed water and substrate-hydrogen vibrational lines.

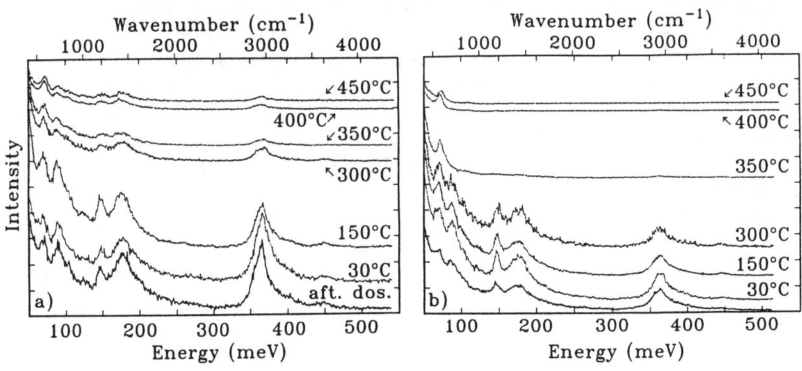

FIG. 2 Temperature series taken after dosing a cooled a) c(2×8) and b) (1×6) surface.

We present in Fig. 2 the series of measurements after annealing to different temperatures and then cooling back again taken from a c(2×8) and a (1×6) surface.

The first striking difference between the two temperature series is that on a (1×6) surface there are no lines after annealing to 350°C, whereas on a c(2×8) surface there are still some modes after heating to 450°C. This indicates that the bonding of TMGa and its decomposition products to an As-rich surface is stronger than to a Ga-stabilized one.

We did not find substrate-hydrogen lines on either surface at any annealing stage, implying that CH_3 did not decompose. If there is no hydrogen on the surface, adsorbed methyl groups cannot transform to methane like they do on a Si surface. This is consistent with TPD studies[8,9] and molecular beam experiments[7] where methyl groups were found to be the desorbing species. Next, the relative intensities of the CH_3 rocking mode, the umbrella mode and the CH_3 anti-symmetric deformation modes deserve some attention. In the following we attempt to extract the geometrical configuration of the adsorbates and their fragments from the HREEL spectra and from considerations concerning the change of the dipole moment in z-direction. The rocking and the anti-symmetric deformation modes produce large changes in the z-dipole if the Ga-CH_3 compound is lying flatly on the surface and should in that case give rise to strong HREELS peaks in the specular direction. This is what we observe on the c(2×8) surface. On the other hand, if the Ga-CH_3 group is standing up, with the Ga-C bond normal to the surface, one would expect only a small contribution from the rocking and anti-symmetric deformation mode, but a larger contribution from the umbrella mode which now produces a strong change of the z-dipole. On the (1×6) surface the relative intensity of the umbrella mode is much stronger than on the As-rich c(2×8) surface.

All the above indicates that after adsorption on a cooled GaAs(001) surface, the TMGa molecules are hardly decomposed on the c(2×8), lying flatly on the surface. But on the (1×6) surface an appreciable amount is already decomposed on adsorption. This difference is even more obvious after annealing to room-temperature, when we can expect to have only a monolayer of TMGa on the surface.

A possible explanation is that on a Ga-stabilized surface the methyl groups of the TMGa molecules are offered lots of sites on the surface where they might also bind to a Ga atom, hence seeing a similar chemical environment. If the energetic difference between a Ga-methyl bond within a TMGa molecule and on the surface were small, this would induce enhanced decomposition an a Ga-stabilized surface compared to an As-rich one.

Upon annealing the condensed film, all the vibrational lines are strongly attenuated except the phonon mode and its harmonic. But additionally, on both surfaces, the relative intensity of the umbrella mode increases, indicating an increasing fraction of CH_3 groups standing up on the surface, produced by TMGa decomposition.

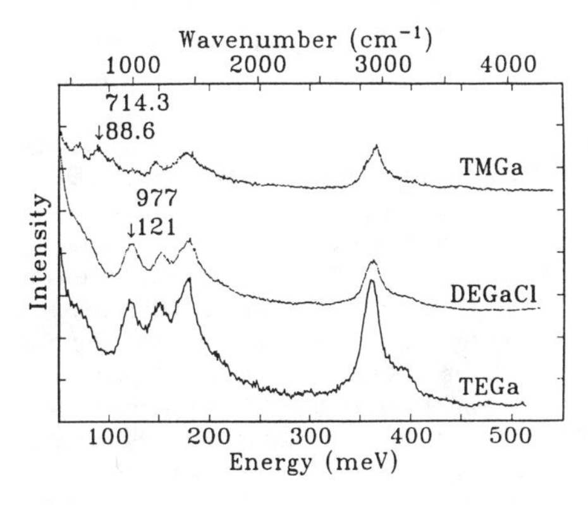

FIG. 3 Comparison of different organometallics dosed onto a cooled GaAs(001)c(2×8) surface.

Fig. 3 shows the spectra obtained after dosing TEGa and DEGaCl on a GaAs surface. In a study on Ga deposition from TEGa on GaAs Murrell at al. also used some results obtained by HREELS [10]. The assignment is very similar to the TMGa case. The CH_3 rocking mode now shifts to about 120 meV because within the ethyl group (C_2H_5) the methyl subgroup is bonded to a C atom instead to a Ga atom as in the TMGa case. This fact also explains the slight shift in the other modes. The low-energy-loss background is much higher for DEGaCl because in this energy region there are a lot of contributions coming from Ga-Cl vibrational modes. However the different bending and stretching modes are not resolved in the spectra. The interpretation is further complicated by an overlap of some vibrational lines from within the ethyl group. The peak at 120 meV can be assigned to the CH_2 twist mode, the C-C stretch and to the CH_3 rock. The line at 150 meV consists of contributions coming from the CH_2 wagging mode and the CH_3 symmetric deformation. The 180 meV mode can be assigned both to the CH_2 scissor vibration as well as to the CH_3 anti-symmetric deformation.

Next we are trying to extract geometrical information out of the HREEL spectra as before. If an ethyl group is lying flatly on a surface (i.e. the C-C bond parallel to the surface), the CH_3 anti-symmetric deformation as well as the CH_2 scissor type deformation mode produce changes in the dipole moment in z-direction (i.e. give rise to a stronger HREELS peak at 180 meV). To first order, this change is much smaller if the ethyl group is not lying parallel to the surface but e.g. with the Ga-C bond normal to the surface. On the other hand, the C-C stretch only produces a change in the z-dipole if it is not parallel to the surface. The CH_3 rocking mode can be expected to give a strong contribution in both cases, since the C-C bond is not normal to the surface in either case. Finally the CH_2 twisting mode's contribution is small to first order.

Taking all the above into account, one might say that after dosing on a cooled c(2×8) surface the TEGa and DEGaCl molecules are lying flatly on the surface (like TMGa). This is supported by the relatively strong 180 meV mode. Upon heating, the peak at 120 meV (C-C stretch) increases with respect to the other ethyl peaks, indicating a breakup of the molecules on the surface and a re-arrangement of the ethyl groups which now are no longer parallel to the surface.

Fig. 4 shows the temperature series taken after TEGa dosing on a cooled GaAs(001)c(2×8) surface. After annealing to room temperature the peak at 120 meV

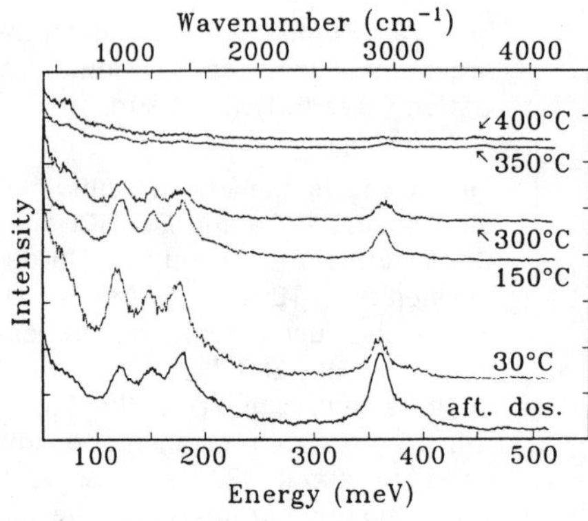

FIG. 4 Temperature series after dosing TEGa on a cooled c(2×8) surface.

is the strongest of the ethyl peaks, as expected from the above. Upon heating all peaks are attenuated until nothing can be seen after heating to 350°C. The temperature series for DEGaCl looks very similar, except for the low-energy modes and for the fact that all peaks have vanished upon heating to 250°C. More details will be published in forthcoming papers.

CONCLUSION

We studied the adsorption and pyrolysis of TMGa, TEGa and DEGaCl on GaAs(001)c(2×8) and (1×6) surfaces with HREELS. Our data indicates that the molecules are adsorbed molecularly onto a cooled surface. Upon heating these molecules decompose into methyl groups and a Ga atom (TMGa) or ethyl groups and a Ga atom (TEGa and DEGaCl). In the case of DEGaCl Cl is an additional decomposition product. In contrast to previous HREELS studies on Si, we did not find any indication for hydrogen originating from decomposition. The HREEL spectra were interpreted in terms of changes in the dipole in z-direction (normal to surface) of the molecules and their fragments. Upon heating to sufficiently high temperature, no adsorbates were left on the surface. The temperature to achieve this increases in the order DEGaCl, TEGa and TMGa.

References

1. H. Ibach and D. L. Mills, Electron Energy Loss Spectroscopy,(Academic Press, New York, 1982).
2. A. Förster and H. Lüth, J. Vac. Sci. Technol., B7,720(1989).
3. H. Lüth, J. Vac. Sci. Technol.,A7,696(1989).
4. F. Lee et al., Mat. Res. Soc. Symp. Proc.,131,339(1989).
5. F. Lee et al., Surf. Sci.,216,173(1989).
6. F. Lee, T. R. Gow, and R. I. Masel,J. Electrochem. Soc., 136,2640(1989).
7. U. Memmert and M. L. Yu, Appl. Phys. Lett.,56,1883(1990).
8. V. M. Donnelly and J. A. McCaulley, Surf. Sci.,238, 34(1990).
9. J. R. Creighton et al., Appl. Phys. Lett.,57,279(1990).
10. A. J. Murrell et al.,J. Appl. Phys.,68,4053(1990).

A STUDY OF HYDROGEN ATOM ADSORPTION ON GALLIUM ARSENIDE (100) BY MULTIPLE INTERNAL REFLECTION INFRARED SPECTROSCOPY

PAUL E. GEE AND ROBERT F. HICKS
Department of Chemical Engineering, University of California, Los Angeles, CA 90024-1592

ABSTRACT

We have studied the adsorption of hydrogen atoms on GaAs (100) by multiple internal reflection infrared spectroscopy. The crystal was etched in 1:1:10 H_3PO_4/H_2O_2-/H_2O solution and in 1:1 HCl/H_2O solution, then annealed to 580°C in the vacuum chamber. Hydrogen adsorption was carried out at -90 and 45°C. At both temperatures, a monolayer forms giving rise to infrared bands for arsenic hydride and gallium hydride at 2105 and 1860 cm^{-1}, respectively. The arsenic hydride vibration is polarized parallel to the surface, whereas the gallium hydride vibration is polarized normal to the surface. By monitoring the changes in the intensity of the infrared absorption bands with time during exposure to H atoms and during heating, the kinetics of hydrogen adsorption and desorption can be measured. At -90°C, the H atom sticking probability follows the Langmuir model, $S/S_o = (1 - \theta_H)$. Upon heating the crystal, the arsenic hydride rapidly decomposes near 120°C, while the gallium hydride slowly decomposes between 150 and 400°C.

INTRODUCTION

Infrared spectroscopy of adsorbed hydrogen can provide details of the chemistry of compound semiconductor surfaces which is not obtainable by other techniques [1,2]. Hydrogen atoms occupy the available adsorption sites on the semiconductor surface. The infrared spectrum reveals the distribution of hydrogen between each element, because the hydride stretching vibrations, eg., Ga-H and As-H, appear at different frequencies. Whether or not mono-, di-, or trihydride bonds are formed can be determined by monitoring the changes in the infrared bands of adsorbed hydrogen upon dilution with deuterium [3]. The number of hydride bonds formed is a measure of the number of dangling bonds on a surface atom. When multiple internal reflection is used to collect the infrared spectrum, one can determine the polarization dependence of the vibrations. This provides insight into the orientation of the bonds relative to the crystal lattice. By combining the polarized IR spectra with theoretical cluster calculations, it is possible to determine the hybridization of the semiconductor surface atoms [4,5].

The kinetics of hydrogen chemisorption on semiconductor surfaces can also be studied by infrared spectroscopy. The hydrogen desorption kinetics is an important issue in the organometallic vapor-phase epitaxy (OMVPE) of gallium arsenide. It has been proposed that the growth of gallium arsenide below 500°C is controlled by a surface reaction between adsorbed trimethylgallium and arsine [6-11]. The kinetics and mechanism of trimethylgallium decomposition on GaAs (100) has been determined [12,13]. However, the kinetics and mechanism of arsine decomposition on GaAs (100) remains a mystery [14]. Exposing gallium arsenide to hydrogen atoms will create adsorbed arsenic hydrides. The arsenic hydride decomposition kinetics can then be measured by temperature programmed desorption [15], or by infrared spectroscopy, as is shown here.

This paper summarizes our recent investigation of hydrogen atom adsorption on GaAs (100) by multiple internal reflection infrared spectroscopy. The gallium arsenide surface is prepared by wet-chemical etching and annealing in vacuum. Polarized infrared spectra reveal that the gallium and arsenic atoms are highly ordered on this surface. The temperatures for decomposition of the adsorbed arsenic and gallium hydrides are shown. We also demonstrate how the kinetics of hydrogen chemisorption can be determined by IR spectroscopy.

EXPERIMENTAL

The experiments are conducted in an ultrahigh vacuum chamber (base pressure $1x10^{-10}$ Torr) equipped with a PHI hemispherical analyzer for x-ray photoelectron spectroscopy (XPS), a Digilab FTS-40 IR spectrometer with Infrared Associates MCT detector ($D^* = 2x10^{10}$), and a Balzers QMG 112 mass spectrometer. The sample consists of a single crystal of semi-insulating gallium arsenide. It is cut into a trapezoid 40 mm long, 10 mm wide, 0.64 mm thick, with 45° bevels at each end. The long axis of the trapezoid is coincident with either the [110] or [-110] crystal directions. The sample face is 2.1° off axis in the [100] direction. The infrared beam internally reflects 40 times off each face. The infrared spectra are collected at 8 cm^{-1} resolution, coadding 512 scans at 10 Hz scan speed.

The XPS spectra are collected with the analyzer axis at 45° to the sample normal and at a 70° angle from the x-ray source. A pass energy of 35.75 eV is used, which yields a silver $3d_{5/2}$ FWHM of 1.18 eV. Atomic concentrations are determined by fitting the peaks to a Gaussian-Lorentzian curve, then dividing the areas under the curves by the atomic sensitivity factors given in the PHI handbook [16]. The observed binding energies are: Ga 3d, 18.5 eV; As 3d, 40.7 eV; C 1s, 284.6 eV; and O 1s, 531.6 eV.

The GaAs wafer is degreased in boiling acetone and methanol, then etched in 1:1:10 $H_3PO_4/H_2O_2/H_2O$ solution and in 1:1 HCl/H_2O solution, following the procedure of Lu et al. [17]. After etching, the sample is removed from the distilled water, dried under flowing nitrogen, and immediately mounted on the sample holder in the chamber loadlock. The sample is transferred to the main chamber, flashed to 580°C for 3 min to remove adsorbed oxygen and carbon, and cooled to the temperature for hydrogen adsorption. Hydrogen atoms are adsorbed onto the gallium arsenide by leaking $5x10^{-7}$ Torr of hydrogen into the chamber, while passing 0.6 mA through the mass spectrometer filament located 4 cm from the sample surface. Infrared and XPS spectra are taken before, during and after the hydrogen exposures.

RESULTS AND DISCUSSION

Table 1 summarizes the XPS analysis of the gallium arsenide at different stages of etching, annealing, and hydrogen adsorption. The hydrogen is adsorbed at -90°C. After the etch, the surface contains 31.3% carbon and 8.8% oxygen. Annealing to 580°C in vacuum reduces the carbon to 5.3% and removes all the oxygen. Over 6 cycles of hydrogen adsorption and annealing (2 weeks in the chamber), the carbon increases to 7.0%. Throughout the cycles of adsorption and annealing, the Ga/As 3d area ratio stays constant at 0.73. Bringans and Bachrach [18] observed that the LEED pattern obtained after H atom exposure is the same regardless of whether the surface is As-rich c(4x4) or Ga-rich (4x6). After annealing, the hydrogen exposed surface always reverts to a c(2x8) structure. They observed Ga/As 3d area ratios of 0.71 during H exposure and 0.72 after annealing. These values are in close agreement with ours.

Table I
Changes in GaAs (100) Surface Composition
During Hydrogen Exposure and Annealing

Treatment	Atomic Concentration (%)				Ga/As 3d Area Ratio
	C	O	Ga	As	
After Etch	31.3	8.8	31.2	28.7	0.64
After Anneal	5.3	0.0	52.6	42.1	0.73
With Adsorbed H	4.3	0.0	53.5	42.2	0.74
After H Desorbed	5.3	0.0	52.7	42.0	0.74
6 Cycles[1]	7.0	0.0	51.0	42.0	0.71

[1]Cycle = annealing to 580°C and H adsorption at -90°C.

Shown in Fig. 1 are infrared spectra of H atoms adsorbed on GaAs (100) as a function of exposure to hydrogen gas. The surface sites saturate after 1800 L of H_2. We estimate this equals about 1 L of H atoms. The arsenic hydride IR bands are located at 2105 and 2020 cm^{-1} [2,19,20]. At low coverages, the main peak at 2105 cm^{-1} appears to separate into two peaks 20 cm^{-1} apart. The gallium hydride IR band is at 1860 cm^{-1} [19,20]. It is comprised of two, or possibly three, bands separated by 20 cm^{-1}. On exposure to hydrogen, the gallium hydride sites corresponding to the low frequency peak fill first, followed by the sites corresponding to the middle and high frequency peaks.

The s- and p-polarized infrared spectra of a monolayer of hydrogen on GaAs (100) is presented in Fig. 2. The arsenic hydride peak at 2105 cm^{-1} exhibits strong s polarization, while the gallium hydride peak at 1860 cm^{-1} exhibits strong p polarization. This remarkable result indicates that there is a high degree of local order on the chemically etched and annealed GaAs (100) surface. The p electric vectors are normal to the surface and down the long crystal axis (ie., towards the beveled edges). The weak p-polarized intensity of the arsenic hydride suggests that this bond is nearly

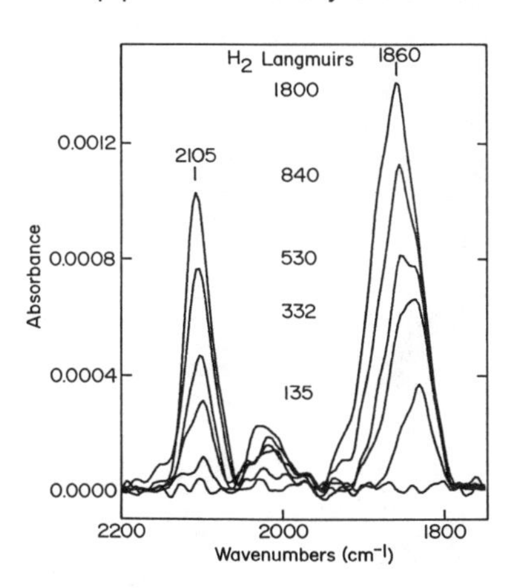

Fig. 1. Infrared spectra of H atoms adsorbed on GaAs (100) at -90°C as a function of dosage (0.001 abs. units = 5.75×10^{-5} $\Delta R/R$ per reflection).

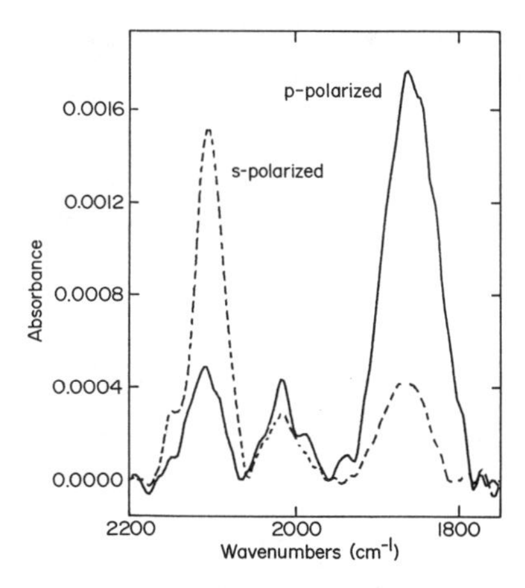

Fig. 2. Comparison of the s- and p-polarized infrared spectra for saturation coverage of hydrogen on GaAs (100) at -90°C.

parallel to the surface and points across the crystal (ie., parallel to the beveled edges). The weak s-polarized intensity of the gallium hydride suggests that this bond is approximately normal to the surface.

The polarization effects can be rationalized in terms of valence bond theory [4,5]. Arsenic has a valence configuration of $4s^2 4p^3$. In arsine, the hydrogen bonds to the three p orbitals at angles near 90°, leaving a lone pair of electrons in the s orbital. On the GaAs (100) surface, the As atom has two bonds down to Ga atoms. A hydrogen bond at a 90° angle from the As-Ga bonds is parallel to the surface. This As-H bond will be strongly s-polarized provided the long crystal axis of the GaAs trapezoid is in the [110] direction. Gallium, on the other hand, has a valence configuration of $4s^2 4p^1$. It prefers sp^2 hybridization, forming trigonal molecules with 120° bond angles. On GaAs (100), a Ga-H bond approximately 120° from the two Ga-As bonds is normal to the surface. This Ga-H bond will be strongly p-polarized. Further experiments need to be done to confirm this valence bond description of the H:GaAs system. In particular, one must carry out adsorption experiments on GaAs (100) surfaces which are characterized by LEED and STM. Also, the IR spectra need to be compared to theoretical cluster calculations as described in [1].

The kinetics of adsorption of hydrogen atoms can be examined by following the change in the area of the infrared bands with time during dosing. Assuming that the infrared absorbance area is proportional to coverage, the fractional coverage, θ_H, equals the IR peak area at time (t) divided by the IR peak area at saturation. For a constant flux of H atoms, the relative sticking probability is

$$S/S_o = \frac{d\theta_H}{dt}(t) / \frac{d\theta_H}{dt}(t=0) \tag{1}$$

where the change in coverage with time is determined from the change in the IR peak area with time. Shown in Fig. 3 is a plot of the relative sticking probability versus coverage for hydrogen adsorption on gallium arsenide at -90°C. The plot is linear

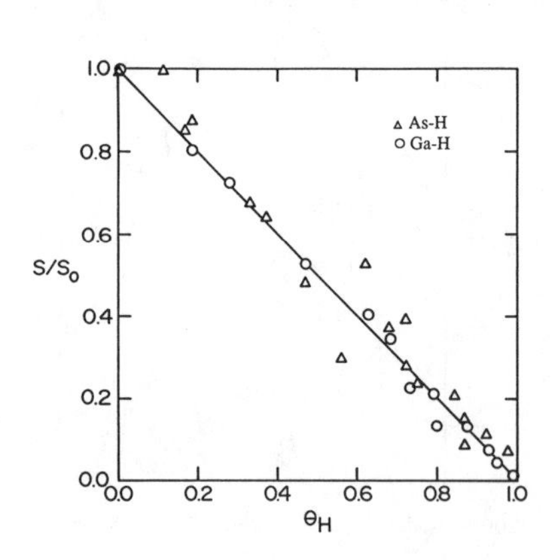

Fig. 3. The dependence of the relative sticking probability on the fractional coverage of hydrogen at -90°C.

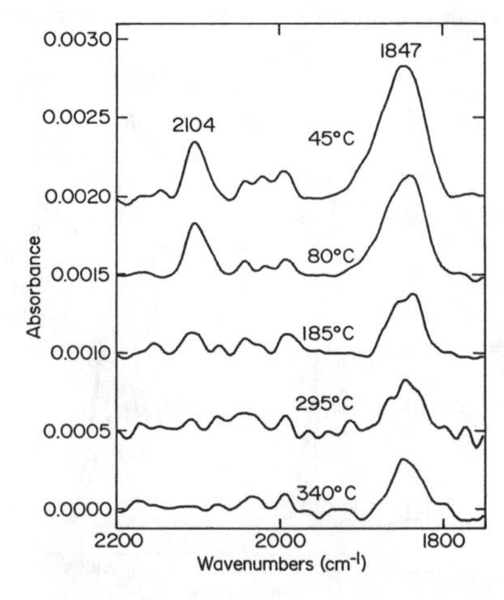

Fig. 4. Change in the infrared spectra of H atoms adsorbed on GaAs (100) during heating (0.001 abs. units = 5.75×10^{-5} $\Delta R/R$ per reflection).

indicating Langmuir adsorption, ie., $S/S_o = (1 - \theta_H)$.

The change in the infrared spectrum of adsorbed hydrogen on heating the gallium arsenide crystal is shown in Fig. 4. In this experiment, the hydrogen is adsorbed at 45°C. Both the As-H and Ga-H species are stable at 45°C. The infrared band intensities do not change significantly over 1 h at this temperature. On heating, hydrogen desorbs from arsenic at some temperature between 80 and 190°C. In a separate experiment, it was found that the arsenic hydride decomposes rapidly at 120°C. Conversely, hydrogen desorbs slowly from gallium over a temperature range extending from 80 to 340°C, and is not yet complete at 340°C.

These results are consistent with Creighton's study of deuterium desorption from GaAs (100) [15]. He found that following adsorption of D atoms at 17°C, deuterated arsine desorbs at 80°C, while deuterium molecules desorb in a broad peak centered at 210°C. The deuterium molecule desorption is ascribed to recombination of D atoms bonded to surface gallium. This assignment is confirmed by the infrared spectra presented in Fig. 4. We are now conducting experiments to determine the kinetics of hydrogen desorption from the As and Ga sites.

CONCLUSIONS

On chemically etched and annealed GaAs (100), hydrogen atoms adsorb onto arsenic and gallium atoms, and exhibits As-H and Ga-H stretching vibrations at 2105 and 1860 cm^{-1}, respectively. Polarized infrared spectra reveal that the As-H bonds are oriented parallel to the surface and across the long crystal axis, whereas the Ga-H bonds are oriented normal to the surface. At -90°C, the dependence of the hydrogen sticking probability on coverage is linear: $S/S_o = (1 - \theta_H)$. Infrared spectra taken during heating the crystal indicate that the As-H bond is weaker than the Ga-H bond. Arsenic hydride rapidly decomposes below 200°C. Gallium hydride slowly decomposes at temperatures ranging from 150 to 400°C.

REFERENCES

1. Y.J. Chabal, Surf. Sci. Rep. 8, 211 (1988).
2. D.M. Joseph, R.F. Hicks, L.P. Sadwick, and K.L. Wang, Surf. Sci. 204, L721 (1988).
3. V.A. Burrows, Y.J. Chabal, G.S. Higashi, K. Raghavachari, and S.B. Christman, Appl. Phys. Lett. 53, 998 (1988).
4. J.J. Barton, W.A. Goddard, and T.C. McGill, J. Vac. Sci. Technol. 16, 1178 (1979).
5. C.A. Swarts, W.A. Goddard, and T.C. McGill, J. Vac. Sci. Technol. 17, 982 (1980).
6. D.H. Reep and S.K. Ghandhi, J. Electrochem. Soc. 130, 675 (1983).
7. P.D. Dapkus, S.P. DenBaars, Q. Chen, W.G. Jeong, and B.Y. Maa, Prog. Crystal Growth 19, 137 (1989).

8. M. Tirtowidjojo and R. Pollard, Mat. Res. Soc. Symp. Proc. 131, 109 (1989).

9. T.J. Mountziaris and K.F. Jensen, Mat. Res. Soc. Symp. Proc. 131, 117 (1989).

10. G.B. Stringfellow, Organometallic Vapor-Phase Epitaxy, Theory and Practice, (Academic Press, New York, 1989).

11. C.A. Larsen, S.H. Li, N.I. Buchan, G.B. Stringfellow, and D.W. Brown, J. Crytal Growth 102, 126 (1990).

12. J.R. Creighton, Surf. Sci. 234, 287 (1990).

13. U. Memmert and M.L. Yu, Appl. Phys. Lett. 56, 1883 (1990).

14. K. Tamaru, J. Phys. Chem. 59, 777 (1955).

15. J.R. Creighton, J. Vac. Sci. Technol. A 8, 3984 (1990).

16. C.D. Wagner, W.M. Riggs, L.E. Davis, J.F. Moulder, and G.E. Muilenberg, Handbook of X-ray Photoelectron Spectroscopy, (Perkin-Elmer Corp., Eden Prairie, MN, 1979), p. 188.

17. Z.H. Lu, C. Lagarde, E. Sacher, J.F. Currie, and A. Yelon, J. Vac. Sci. Technol. A 7, 646 (1989).

18. R.D. Bringans and R.Z. Bachrach, Solid State Commun. 45, 83 (1983).

19. L.H. DuBois and G.P. Schwartz, Phys. Rev. B 26, 794 (1982).

20. H. Luth and R. Matz, Phys. Rev. Lett. 46, 1652 (1981).

Ideal Crystal Growth from Kink Sites and Fractional-Layer Growth on GaAs Vicinal Substrate by MOCVD

Takashi Fukui and Hisao Saito
NTT Basic Research Laboratories
3-9-11 Midori-cho, Musashino-shi, Tokyo 180, Japan

ABSTRACT

$(AlAs)_{1/2}(GaAs)_{1/2}$ fractional-layer superlattices (FLSs) are grown on (001) vicinal substrates by metalorganic chemical vapor deposition. Various kinds of GaAs substrates are used. When the substrate is misoriented to [$\bar{1}$10] direction by 1.92° and [110] by 0.10°, uniform superlattice periods in a large surface area are observed with a bright field transmission electron microscope (TEM). The results suggest ideal crystal growth from kink sites during MOCVD growth, and the distances between the kink sites are equal.

On a substrate misoriented to [110] by 1.90°, the superlattice periods exhibit an undulation. This shows that kink flow mode growth is not dominant in the [110] direction. On a substrate misoriented to [010] by 2.0°, no superlattice periods were observed. From the above results, we discuss the growth mechanisms.

Polarization dependent photoluminescence and optical absorption spectra of FLS were also observed. Electron wave interference devices with lateral periodic potential were fabricated.

I. INTRODUCTION

III-V semiconductor growth on slightly misoriented substrates from low index surfaces has been actively studied in recent year, because of interest in fundamental research of growth kinetics and application to the fractional-layer superlattices (FLSs)[1-3]. Most cases are AlAs/GaAs growth on (001) vicinal GaAs substrate, using metalorganic chemical vapor deposition (MOCVD), or migration enhanced epitaxy (MEE).

A traditional growth model called the Kossel model[4], has occasionally been used to explain the crystal growth on vicinal surfaces which include monolayer step sites and kink sites. During crystal growth on vicinal substrates, surface migrating atoms preferentially attach to kink sites which are the most stable lattice sites. P. M. Petroff suggested that if ideal crystal growth proceeds from kink sites following Kossel model, AlAs/GaAs sequential half-monolayer growth on vicinal substrates should result in new superlattices having a periodicity parallel to the surface, i.e. lateral superlattices. His proposal, however, failed to mention that the terrace width must be equal.

In an earlier study, we fabricated $(AlAs)_{1/2}(GaAs)_{1/2}$ FLSs on GaAs (001) vicinal substrates by metalorganic chemical vapor deposition (MOCVD), and observed step flow growth with an transmission electron microscope (TEM)[1,5]. Furthermore, we determined the equal terrace width, and clarified the terrace width ordering mechanisms. In this review article, we report on the step-flow growth and on the fractional-layer superlattices grown on GaAs vicinal substrates by MOCVD. When the substrates are (001) misoriented in the $[\bar{1}10]$ direction, uniform superlattice periods are observed with a bright field TEM. On the other hand, when the substrates are (001) misoriented in the [110] and [010] directions weak and no superlattice contrasts are observed, respectively. From these results, we discuss the growth mechanisms.

II. Experiments

A horizontal reactor system was used. The working pressure of 76 Torr was automatically controlled to be within ±0.1 Torr. Triethylaluminum(TEAl), triethylgallium(TEGa) and AsH_3 were used as source gases. Purity in the gas line is a very important factor to achieve step flow growth. If the dew point is worse than -90°C, the crystal quality of an FLS degrades. The dew point of the gas line is always held at about -100°C. Although FLS was grown with 1-3 second growth interruptions, the quality of crystal was the same as that of FLSs without interruptions. This indicates that the source gases are switched quickly from TEAl to TEGa, and TEGa to TEAl within 1 sec. Therefore, most FLSs were grown without growth interruptions between each half monolayer growth of AlAs and GaAs. Growth rate stability is another important factor to fabricate FLSs. The growth rate fluctuation during crystal growth and that between run-to-run are both less than 0.5 %. The layer thicknesses of m and n in (AlAs)m(GaAs)n were determined by X-ray diffraction. The m+n values were adjusted exactly to 1.00 by using the relationship between the flow rates of organometallic source gases and the layer thicknesses.

The substrates were (001) GaAs misoriented about 2 degrees. The misorientation directions were $[\bar{1}10]$, [110] and [010]. The mean distance between each monolayer step was 8nm.

III. Results and discussions

The electron diffraction pattern and bright field TEM images of FLSs were observed from the (001) direction. Figures 1(a), (b) and (c) show TEM plane view images of FLSs on (001) GaAs substrates. For sample B, clear contrasts of AlAs and GaAs are observed. In addition, there are no fluctuations of the superlattice distance in the sub-micron-wide area. The period of the superlattice is 8.45 nm, which corresponds to a 1.92° misorientation, and there are 21 atomic rows of Al and Ga included between steps.

The FLS is expected to tilt by a small deviation in the m+n value from 1.0. The TEM sample thickness is assumed to be more than 300 nm. Although the sample is very thick compared with the superlattice periods (8nm), Al and Ga contrast in the surface view can be distinguished. This shows that the superlattice period is not tilted, but is almost vertical. In this sample, the deviation δ (=1-(m+n)) is estimated to be less than 0.5%.

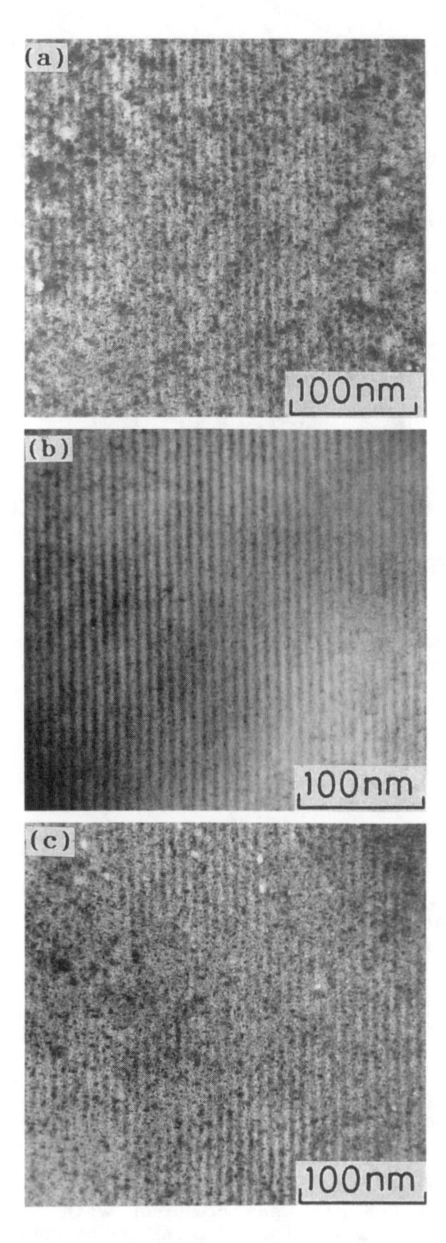

Fig.1. TEM bright field surface view images of FLSs. The substrates are (a) exactly misoriented to [110] direction by 2.0° (Sample A), (b) to [$\bar{1}$10] by 1.92° and to [110] by 0.10° (Sample B), and (c) to [$\bar{1}$10] by 1.90° and to [110] by 0.20° (Sample C). The dark region is GaAs, and the bright region is AlAs.

By comparing diffraction image and bright field image, the crystallographic direction of the superlattice could be exactly determined. The results are schematically shown in Fig.2. The superlattice direction is 2.8° (±0.5°) misoriented from [$\bar{1}$10] to [110]. This indicates that the substrate direction is misoriented to [$\bar{1}$10] by 1.92°, and to [110] by 0.10°. Therefore, the step front on the vicinal surface includes many kink sites. A single kink site is 0.40 nm wide on the (001) plane. Therefore, the mean distance between kink sites is estimated to be 8 nm.

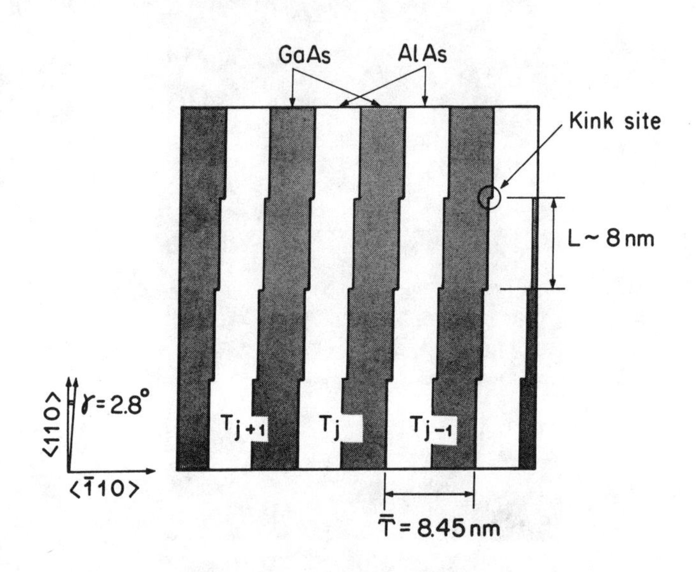

Fig.2. Schematic surface view observed by TEM of the FLS of sample B. The GaAs/AlAs interfaces ideally include equally spaced kink sites.

We have already pointed out that the distances between steps are equal in FLSs grown by MOCVD. Fig. 1 (b) shows further evidence that the distances between kinks are also equal. If the kink distribution within a step has significant fluctuations, the superlattice image should undulate. The equally spaced kinks could be explained by assuming a simple dynamic process, as was mentioned in a previous paper[5].

In TEM photographs of samples A and C, the contrasts of AlAs and GaAs of FLS are weak compared with that of sample B. The weak contrasts suggest that kink flow mode growth is not always dominant in these sample types. For sample A there are no regular kink sites because of strict misorientation in the [110] direction. Therefore, surface migrating Al or Ga atoms do not always adsorb kink sites. On the other hand, for sample C, there should be regular kink sites whose distances are 4nm apart. However, the kink flow mode growth here seems to be weak. Consequently, it is thought that an appropriate kink distance, i.e. about 8nm, probably exists for kink flow mode growth.

X-ray superlattice satellite intensities are slightly weak for both samples A and C, which means that the compositional contrasts between AlAs and GaAs are weak compared with sample B.

Next, FLS was grown on (001) GaAs misoriented in the [110] direction. Fig.3 shows a TEM plane view image of FLS on (001) GaAs misoriented to [110] by 1.9° and to [$\bar{1}$10] by 0.1° (Sample D). Weak AlAs/GaAs contrast with undulation was observed. X-ray superlattice satellite intensity was also weak, about half that of sample B.

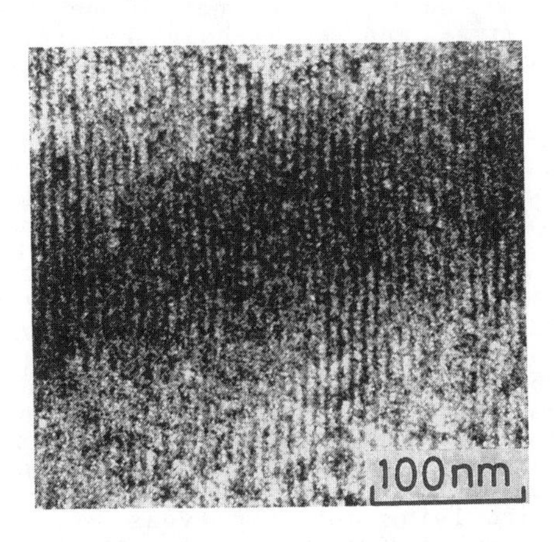

Fig.3 TEM plane view image of FLS on (001) GaAs misoriented to [110] by 1.9° and to [$\bar{1}$10] by 0.1° (Sample D).

These step flow growth dependences on misorientation can be explained using the local atomic structures of GaAs near the monolayer steps. According to the MOCVD growth model proposed for GaAs vicinal surfaces [1,6], growth proceeds along the step front for [$\bar{1}$10] misoriented substrates. On the other hand, for [110] misoriented surfaces, growth proceeds in the direction perpendicular to the step front. Undulation of superlattice image and weak superlattice satellite intensity in sample D probably result from the roughening effects of step fronts.

Step flow mode growth and FLS were also obtained by migration-enhanced epitaxy[2,3], and a surface view of a GaAs vicinal surface was recently observed with a scanning electron microscope[7]. However, the step front image, in which clear 2x4 reconstruction was observed depicted roughness and undulation. Furthermore, the distances between steps were unequal. These results suggest that, compared with growth by MOCVD, growth from kink sites is imperfect and the terrace ordering mechanisms are insufficient. The growth mode difference probably reflects the microscopic surface characteristics, such as the surface reconstruction, and depends on whether growth takes place in a hydrogen atmosphere or in a vacuum.

IV. Optical and electrical properties

Polarization-dependent optical absorption and photoluminescence of FLS show anisotropy at the band edge caused by the carrier confinement effects in the parallel direction of superlattices[8,9]. The observed separation in the band edge wavelengths corresponds to the energy difference between the heavy- and light-hole related transitions. However, the observed band edge of FLS was higher than the calculation, and close to that of $Al_{0.5}Ga_{0.5}As$ alloy. The result suggests that significant alloying occurs at the AlAs/GaAs interface. Qualitative analysis at the interface and the reason for the alloying are now under study.

Modulation-doped $Al_{0.3}Ga_{0.7}As$/GaAs heterostructure electron wave interference devices with FLS were fabricated[10]. The device structures were similar to those of high electron mobility transistors (HEMT). The epitaxial layers were 200nm-thick undoped GaAs, undoped FLS, Si-doped $Al_{0.3}Ga_{0.7}As$ and Si-doped n^+-GaAs cap layer, in which quasi-2 dimensional electron gas was accumulated at the FLS/GaAs interface. The FLS periods were 16nm, 12nm, and 8nm. These devices displayed drain current oscillation due to electron wave interference at 4.2K. From an analysis of the drain current oscillation, the peaks of the structure function agreed with the multiples of the FLS periods. The FLS structure is promising for new types of low-dimensional electron and optical devices.

V. Summary

We fabricated $(AlAs)_{1/2}(GaAs)_{1/2}$ FLS on a (001) vicinal GaAs substrate by MOCVD. A high-resolution TEM image depicted uniform superlattice periods of the FLS on the $[\bar{1}10]$ misoriented substrate. The monolayer step sites and kink sites are ordered and equally spaced. These results suggest that ideal crystal growth from kink sites occurs using the MOCVD model, which corresponds to the Kossel model. On the other hand, FLSs on [110] misoriented substrates show undulated superlattice periods. The difference in these superlattice periods is probably caused by the local atomic structures near the monolayer steps.

FLS optical and electrical properties and its device applications were also presented. The lateral-controlled MOCVD method appears to be a very promising technique for fabricating lateral corrugations less than 10nm on a semiconductor surface.

References

[1] T.Fukui and H.Saito, J. Vac. Sci.& Technol. B4, 1373(1988).
[2] J.M.Gaines, P.M.Petroff, H.Kroemer, R.J.Simes, R.S.Geels, and J.H.English, J.Vac.Sci.& Technol. B4, 1378(1988).
[3] H.Yamaguchi and Y.Horikoshi, Jpn. J. Appl. Phys. 28, L1456(1989).
[4] W.Kossel, Nach. Ges. Wiss. Gottingen, 135(1927).
[5] T.Fukui and H.Saito, Jpn. J. Appl. Phys.29, L731(1990).
[6] Y.Horikoshi, Oyo Butsuri 59, 27(1990)(in Japanese)
[7] M.D.Pashley, K.W.Haberern, and J.M. Gaines, Appl. Phys. Lett.58, 406(1991).
[8] H.Ando, T.Fukui, and H.Saito, Extended Abstracts of the 22nd Conf. on Solid State Devices and Materials, Sendai, Japan, B-4-7(1990).
[9] M.Kasu, H.Ando, H.Saito, and T.Fukui, Appl. Phys. Lett., to be published.
[10] K.Tsubaki, T.Honda, H.Saito, and T.Fukui, Appl. Phys. Lett. 58, 376(1991)

PART II

Characterization of Atomic Layer Processes

REAL-TIME OPTICAL DIAGNOSTICS FOR MEASURING AND CONTROLLING EPITAXIAL GROWTH

D. E. ASPNES,* R. BHAT,* E. COLAS,* L. T. FLOREZ,* S. GREGORY,* J. P. HARBISON,* I. KAMIYA,** W. E. QUINN,* S. A. SCHWARZ,* H. TANAKA,*** and M. WASSERMEIER****
*Bellcore, Red Bank, NJ 07701-7040 USA
**University of Illinois Urbana-Champaign, Urbana, IL 61801 USA
***Fujitsu Laboratories Ltd., Atsugi 243-01, JAPAN
****University of California, Santa Barbara, CA 93106

ABSTRACT

A variety of optical methods are now available for studying surface processes and for monitoring layer thicknesses and compositions during semiconductor crystal growth by molecular beam epitaxy (MBE), organometallic chemical vapor deposition (OMCVD), and related techniques. New capabilities for surface analysis are being provided by developing techniques such as reflectance-difference spectroscopy (RDS), which use intrinsic symmetries to suppress ordinarily dominant bulk contributions. Bulk and microstructural properties such as compositions and layer thicknesses can be determined by techniques such as spectroellipsometry (SE), which return information integrated over the penetration depth of light. Recent advances include the application of reflectance to monitor dynamic surface processes, RDS to characterize (001) GaAs surfaces in OMCVD environments, and SE to control growth of $Al_xGa_{1-x}As$ materials and structures.

INTRODUCTION

Technology continues to place new demands on crystal growth capabilities as samples become increasingly complex and constraints on compositions, interface abruptness, and layer thicknesses become more severe. At the same time yields must be maintained at high levels to minimize costs.

As a result, a substantial effort is now underway to monitor and control growth of semiconductor crystals in real time. Much of this effort is centered on optical probes, because they are nondestructive, noninvasive, and can be used in any transparent ambient. However, optical probes have two major disadvantages: their spectral range is limited, and their surface sensitivity is low. Fortunately, the 1.5 to 6.0 eV range accessible to quartz-optics systems contains the major bonding-antibonding transitions for most materials encountered in semiconductor growth, which alleviates spectral-range limitations for these applications. The surface-sensitivity problem is less easily dismissed; optical absorption coefficients rarely exceed 10^6 cm^{-1} (10^{-4} cm^{-1} in the infrared), so penetration depths are rarely less than 100 Å. Consequently, the surface seldom contributes more than 1% to the total optical signal.

However, within the last few years a number of techniques have been developed to address the surface-sensitivity problem as well. These include reflectance-difference spectroscopy (RDS, also known as reflectance-anisotropy spectroscopy, RAS), second-harmonic generation (SHG), and laser light scattering (LLS), all of which take advantage of intrinsic symmetries to suppress the ordinarily dominant bulk contribution to the overall optical signal. We now have it *both* ways, with the newer optical probes possessing surface sensitivity and specificity and the classical probes such as spectroellipsometry (SE) returning information integrated over the penetration depth of light. Because composition and layer thicknesses are both bulk quantities, for control purposes the latter capability is actually the more important. The first optical system to control semiconductor crystal growth in real time used a spectroellipsometer as the feedback element.

As several reviews of the present topic are already available [1], this work will be restricted to a discussion of optical techniques that have already demonstrated a practical capability for monitoring and control. Our closed-loop growth-control system will also be discussed.

OPTICAL APPROACHES

Optical probes can be classified as surface- or bulk-sensitive according to whether or not symmetry is used to suppress the ordinarily dominant bulk contribution to the total optical signal. The bulk-sensitive optical probes of interest for real-time applications are spectroreflectometry (SR) and SE. SR and SE deal with power (intensity) and polarization state, respectively. They bear much the same relationship as a wattmeter does to an impedance bridge, and provide a trade-off between experimental simplicity and diagnostic capability. SR and its transmission analog, spectrotransmittivity (ST) are more naturally suited for configurations where interference effects are important, e.g., for transparent materials and characteristic thicknesses of the order of the wavelength of light. Not surprisingly, SR and ST are both used extensively to monitor and control the deposition of optical thin films [2]. For semiconductor applications where layers are thin and materials absorbing, SE is the preferred choice. Most operating spectroellipsometers use the rotating-analyzer (-polarizer) configuration, and commercial systems are available. As SE is computation-intensive, the practicality of a system depends as much on the quality of its software as its hardware.

Both SE and SR data are affected by surface conditions, but surface information can only be accessed indirectly by monitoring changes that occur when the condition of the surface is changed. This is not adequate for detailed analysis because it is usually impossible to determine to what extent the initial and final surface conditions contribute to the change. Nevertheless, as Horikoshi, Kobayashi, and co-workers have shown [3,4], single-wavelength reflectometry, which they call surface photoabsorption (SPA), is a powerful technique for monitoring cyclic processes such as the deposition of alternating monolayers (ML) of material, as encountered in atomic layer epitaxy (ALE). Here, precision, not accuracy, is the primary consideration, with the issue being simply to establish when a ML of material has been deposited.

Surface-specific probes such as RDS, SHG, and LLS rely on the surface having a different (usually lower) symmetry than that of the bulk. In order of increasing experimental complexity, those currently under development include LLS, reflectance as mentioned above, RDS, and SHG. LLS trades on the property that specular surfaces do not scatter light. In its most rudimentary form LLS requires only a laser source; morphological degradations of growth surfaces are easily perceived by eye. The addition of a detector makes the technique quantitative, as exemplified by several recent experiments [5,6]. The same combination is also all that is needed for reflectance monitoring.

In RDS, the difference between the near-normal-incidence reflectances of light polarized along the two principal axes of the sample in the plane of the surface is determined experimentally. Because cubic materials are nominally optically isotropic, the bulk contribution essentially cancels in reflectance leaving that from the surface. Not all surfaces exhibit linear-optical anisotropy, but fortunately, those that do include the technologically important (001) surfaces of III-V and II-VI semiconductors. Because RDS, like ellipsometry, deals with polarization states, the necessary instrumentation is more complex than that needed for LLS or reflectance monitoring. All currently operating systems are based on some form of photoelastic modulation, and commercial versions are available. Reflectance-difference (RD) spectrometers can be made compact – a necessary condition for applications involving crystal growth where monitoring equipment must fit into existing configurations. Our present RDS system mounts on a $12 \times 16\ in^2$ baseplate.

In SHG, the difference between bulk and surface symmetries allows the bulk and surface components of the second harmonic generated by a laser beam incident on a III-V or II-VI semiconductor to be separated by suitable polarization dependences [7]. For Group IV semiconductors the situation is trivial because bulk SHG does not occur. SHG [7-9] and related techniques such as coherent anti-Stokes Raman scattering (CARS) [10] and sum-frequency generation [9] require large equipment investments, and for that reason their possibilities have not been as thoroughly explored.

Surface- and bulk-sensitive probes are not competitive but complementary. Natural applications of the former are to dynamic situations, to analyze growth processes, and to ensure that growth is occurring under proper conditions. The latter are naturally suited to assessing thicknesses and compositions and to controlling growth in real time.

The theory and practice of SE, RDS, SHG, and LLS have been discussed in the literature, and further details are given in refs. 11-12; 12-14; 7-10; and 6; respectively. All optical techniques require unobstructed optical access to the sample. Approaches that involve polarization, such as SE and RDS, require in addition windows that are essentially strain-free. We recently developed a fused-quartz window assembly that mounts on a standard 2 3/4 in vacuum flange and reduces strain effects to negligible levels for most applications [15]. As a bonus, the window face can be heated externally to remove arsenic deposits without having to vent the station. The effect of moderate window strains can also be eliminated by compensation, either by applying an external stress directly to the window or preferably to a quartz plate mounted ahead of the window.

APPLICATIONS AND EXAMPLES

Monitoring dynamic surface processes: SR

Horikoshi, Kobayashi, and co-workers [3,4] recently used p-polarized reflectance with HeNe, Ar^+, and HeCd laser sources to monitor (001) GaAs surfaces during crystal growth in the migration-enhanced epitaxy (MEE) mode of molecular beam epitaxy (MBE) and by organometallic chemical vapor deposition (OMCVD). As mentioned above, the approach requires minimal equipment and offers unparalleled simplicity. Typical data are shown in Fig. 1 [3]. The inset compares the reflection-high-energy-electron-diffraction (RHEED) response over a complete MEE growth cycle with the reflectance response obtained for p-polarized 325 nm (HeCd laser) light incident near the Brewster angle. Of the laser lines investigated, the maximum relative reflectance response, nearly 3%, was found at this energy. The effects of opening and closing the Ga and As shutters are clearly seen, with the reflectance increasing with increasing Ga coverage. The main part of the figure shows the evolution of the relative reflectance response with increasing Ga coverage. A tendency of the signal to saturate at 1 ML coverage of Ga is also evident. In contrast, the s-polarized reflectance did not change. These results were interpreted in terms of transitions involving lone-pair electron orbitals of As-As surface dimers to the Ga-As antibonding orbitals, although detailed calculations remain to be done. Measurements of the p-polarized response as a function of sample azimuth angle would allow both isotropic and anisotropic parts of the signal to be determined.

Monitoring static and dynamic surface processes: RDS

RDS has excellent surface sensitivity, is specific to the surface (more accurately to species that have already reacted with the surface), can perform surface spectroscopy with the surface under steady-state conditions and in any transparent ambient, and can be used with existing growth systems without requiring their modification. Not surprisingly, since its first application to MBE growth four years ago [16] the use of RDS has increased rapidly [17,18]. A complete summary of RDS is now beyond the scope of the present work, but recent reviews are available [13]. In uhv applications RDS is complementary to RHEED, which is primarily sensitive to long-range order. This is important for kinetic studies, because surfaces tend to change from one reconstruction to another over relatively narrow phase boundaries even though surface coverage may be varying continuously. To date most RD studies have concentrated on this aspect.

The other dimension of RDS, its spectral capability, is just beginning to be explored. Spectral responses offer the possibility of positively identifying surface species, although this is not trivial. In simple cases, as for example (001) GaAs, reference data can be obtained from known reconstructions prepared by MBE and verified by RHEED. The more general identification problem remains to be solved. Fortunately, the theory of electronic structure has now progressed to the point where surface optical properties can be calculated reliably [19,20]. Theoretical contributions can be expected to play a large part in data analysis in the future.

The above points can be illustrated by recent work done at Bellcore. Figure 2 shows a series of RD spectra obtained at different temperatures for a (001) GaAs surface under MBE

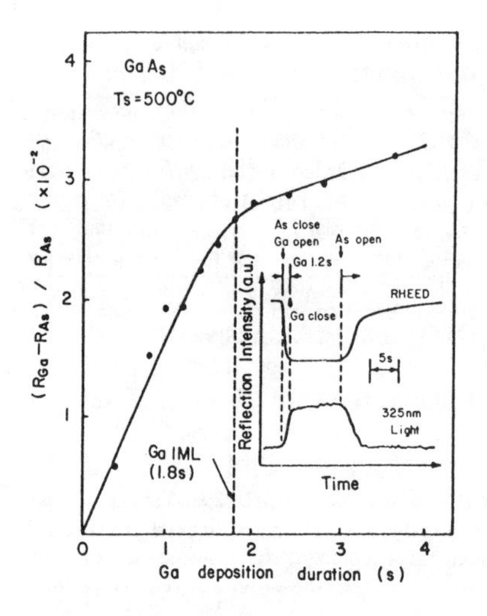

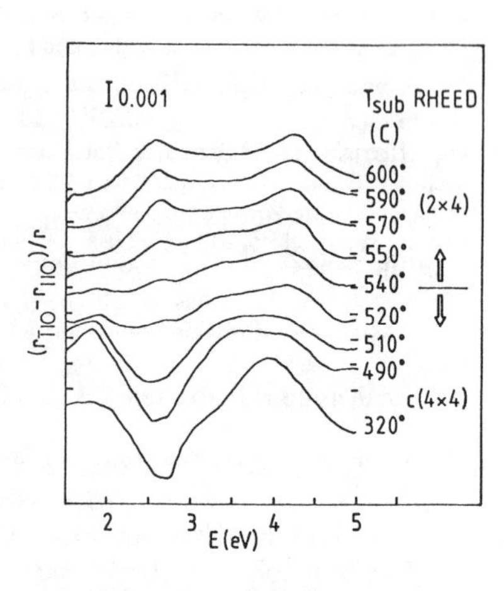

Fig. 1. Relative change in p-polarized reflectance at 3.81 eV with increasing Ga coverage for MEE growth of GaAs at 500 °C. The inset compares typical optical and RHEED responses over a complete growth cycle (after ref. 3).

Fig. 2. RD spectra of As-terminated (001) GaAs surfaces as a function of temperature at constant As_4 beam equivalent pressure of 2.4 × 10^{-5} Torr. The RHEED pattern changes between c(4× 4) and (2× 4) near 520 °C (after ref. 21).

conditions at a As_4 beam equivalent pressure of 2.4 × 10^{-5} Torr [21]. These data were obtained using the double-modulation version of RDS [22], and hence are independent of experimental artifacts and can be uniquely associated with the indicated surfaces. The long-range order as determined by RHEED is indicated. The RD lineshapes at the lowest and highest sample temperatures are substantially different, but the evolutionary trend of the intermediate spectra indicates that the surface composition is changing continuously within a given reconstruction class. In contrast, the RHEED-defined boundary is relatively abrupt.

The spectral feature at 2.6 eV, which inverts from one temperature limit to the other, has been related by theoretical calculations to transitions involving surface As dimers [23]. The results of these calculations are shown in Fig. 3. The upper data are the same as those in the upper part of Fig. 2 but are given in surface dielectric anisotropy (SDA) form. The lower data are for the Ga-terminated (4× 2) surface that is obtained when Ga is deposited in the absence of As_4. The theoretical spectra were calculated in a tight- binding model assuming complete coverage by surface dimers ((2× 1) and (1× 2) reconstructions.) The lower pair are in acceptable agreement below about 4 eV, the energy range where the calculation is expected to be valid. The transitions that give rise to the various structures can be identified from the details of the calculation. The features at 1.8 and 2.6 eV involve surface Ga and As dimers, respectively. The 1.8 eV feature originates from electronic transitions between filled Ga dimer bonding states and empty Ga lone-pair orbitals, while the 2.6 eV structure is due to electronic transitions between filled As lone-pair orbitals and empty As dimer antibonding states. These results have been used to follow the concentration of the originating species on the surface [22].

The 2.6 eV assignment gains further support from the temperature dependences shown in Fig. 2 [21]. The (2× 4) surface reconstruction is known to arise from an outer layer of As dimers oriented along [$\bar{1}$10], with every fourth dimer missing [23]. The c(4× 4) reconstruction is known to

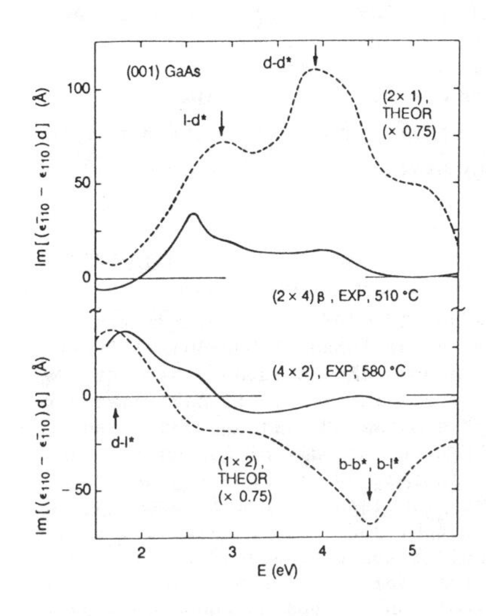

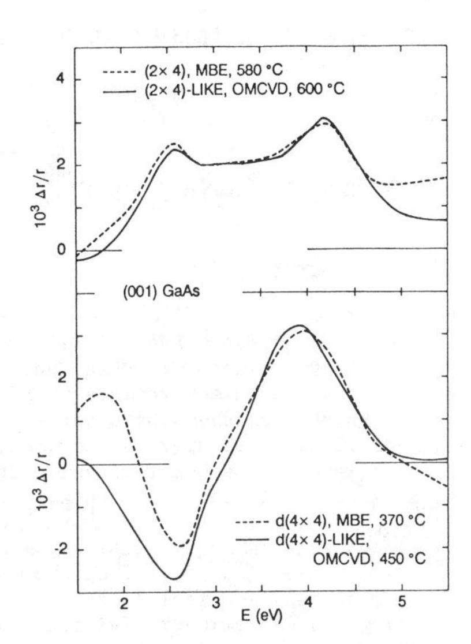

Fig. 3. Comparison between measured (solid line) and calculated (dashed line) SDA spectra. Top: As-terminated SDA spectra: measured for (2× 4) reconstruction at 510 °C and calculated for (2× 1) reconstruction at 580 °C. Bottom: Ga-terminated SDA spectra: measured for (4× 2) reconstruction at 580 °C and calculated for (1× 2) reconstruction at 580 °C (after ref. 3.)

Fig. 4. Comparison of RD spectra of As-terminated (001) GaAs surfaces in uhv (dashed lines) and in atmospheric-pressure H_2. Top: (2× 4) and (2× 4)-like terminations at 580 °C. Bottom: c(4× 4) and c(4× 4)-like terminations at 370 °C.

arise from two outer layers of As with the outer layer of As dimers oriented along [110] [23,24]. The sign reversal of the RD signal is consistent with this 90° orientational difference.

Building on these results, we have obtained the first information about the properties of (001) GaAs in atmospheric-pressure (AP) H_2, of interest to OMCVD, and have been able to interpret the results in terms of probable terminations of these surfaces. The top part of Fig. 4 shows a comparison between RD spectra of a (2× 4)-reconstructed (001) GaAs surface prepared by MBE and a "(2× 4)-like" surface at the same temperature in AP H_2. The latter surface was prepared by controlled exposure to AsH_3. We use the suffix "-like" to describe AP OMCVD surfaces because the notations for MBE surfaces refer to types of long-range order to which RDS does not respond directly. Despite the extreme differences in ambient, the spectra are essentially identical, indicating that (001) GaAs surface in AP H_2 is also predominantly terminated with As dimers oriented along [$\bar{1}$10]. The lower part shows a comparison between a c(4× 4) surface in uhv and a c(4× 4) like surface at the same temperature in AP H_2. Again, a parallel can be drawn between the known configuration in uhv and the previously unknown surface in AP H_2.

Since the c(4× 4) surface is known to involve two outer layers of As [23,24], the above results have substantial implications for ALE [25]. Coverage by *two* layers of As suggest ALE paths that have not previously been considered, but which could resolve some currently puzzling aspects of ALE growth on (001) GaAs. For example, Ga could simply intercalate between the two outer layers of As, with deposition ceasing when the lower As layer is consumed. If this were the case exactly 1 ML of Ga should be deposited per cycle with the outer-layer coverage alternating between 2 and 1 layers of As. An intercalation mechanism would naturally explain why ALE can lead to precisely 1 ML per cycle despite charge-neutrality restrictions [23] that argue against the establishment of complete monolayers on any type of semiconductor surface. Alternatively, Ga deposition could cease only after *all* surface As had been consumed. This would result in the deposition of up to 2 ML of metal per cycle, as recently reported for ALE of AlAs by Ozeki et al. [26]. In this case coverage would alternate between approximately 2 ML of As and 1 of Ga. These possibilities do not exclude still other mechanisms, as for example those involving the (2× 4) or (4× 6) reconstructions that may be appropriate for uhv [27].

Controlling growth: SE

We recently reported a closed-loop system to control the composition x of thick $Al_xGa_{1-x}As$ films during growth by organometallic molecular beam epitaxy (OMMBE) [28]. The system uses a spectroellipsometer to monitor the near-surface composition, reduces the data in real time, then uses this information to regulate the flow of triethylaluminum (TEA) to the growth surface thereby controlling x. The principle of operation can be understood from the data of Fig. 5 [28], which shows the dielectric responses $\epsilon = \epsilon_1 + i\epsilon_2$ of bulk GaAs and of optically thick $Al_xGa_{1-x}As$ layers with x = 0.095, 0.14, and 0.36 at the growth temperature of 600 °C. Specifically, the energy of the E_1, $E_1 + \Delta_1$ critical point threshold depends strongly on x. At 2.6 eV the dependence is nearly linear from x = 0 to 0.4, and can be used to translate differences in the ellipsometrically determinable quantity ϵ_2 into differences in x, thereby linking the optical measurement to the parameter to be controlled.

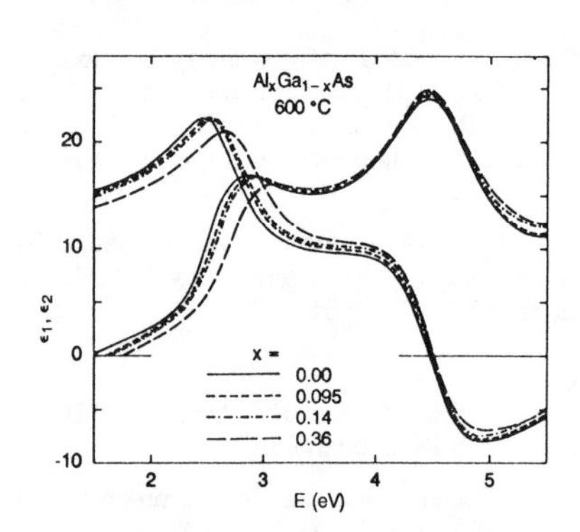

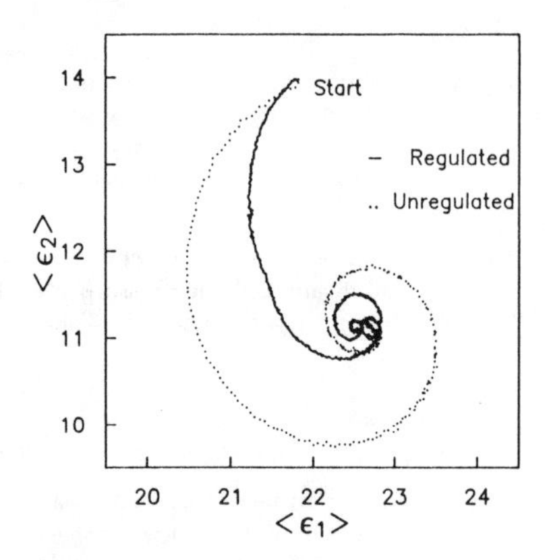

Fig. 5. Dielectric functions of GaAs and $Al_xGa_{1-x}As$ for x = 0.095, 0.14, and 0.36 at 600 °C (after ref. 28).

Fig. 6. Solid line: closed-loop control data for $<\epsilon>$ starting with a TEA flow about half that required to reach the target composition. Dotted line: unregulated equivalent with the TEA flow nearly correct throughout. Each trajectory corresponds to about 20 min of data (after ref. 28).

In practice, x is converted to a target value ϵ_{2t} of ϵ_2 by the expression [29] $\epsilon_2(x) \approx \epsilon_{2s} - 16.9x - 0.33x^2$, where $\epsilon_{2s} = \epsilon_2(0)$ is the dielectric function of GaAs, which is measured at the beginning of a run. The instantaneous value of ϵ_2 is then compared to ϵ_{2t} to correct if necessary the voltage regulating the flow of TEA through the proportional controller. The essential difficulty is to relate the quantity actually determined by the ellipsometer, the pseudodielectric function $<\epsilon> = <\epsilon_1> + i<\epsilon_2>$, to ϵ_2. In general $<\epsilon> = \epsilon$ only if the sample is uniform over the entire penetration depth of light. Otherwise, and particularly for multilayer systems of interest, $<\epsilon>$ is strongly dependent on sample history as could be expected since SE returns information integrated over the penetration depth of light.

Two examples of this dependence are given in Fig. 6 [28]. The solid curve shows the trajectory obtained when the TEA flow was initially set at half that required to reach the target value ϵ_{2t} with the TEA flow increased by the control system itself. The dotted curve shows the trajectory obtained with the control voltage set at the nominal proper value and not adjusted further. While both trajectories begin at the dielectric function of the substrate and converge to the limiting dielectric function of the film as the film becomes uniform and optically thick, the latter trajectory is the simpler one, approximating the asymptotically exponential spiral obtained when a uniform absorbing film is deposited on an absorbing substrate. However, the former trajectory is the more relevant one, because it illustrates the analytic problem that must be faced in an actual control environment where the film composition is not necessarily independent of thickness.

This problem can be solved with a small-term expansion of the relevant Fresnel expressions that essentially takes a time derivative to extract the most recent compositional information from $<\epsilon>$. If r_{so} and r_{oa} are the complex reflectances of the substrate-overlayer (so) and overlayer-ambient (oa) interfaces, respectively, of a three-phase (substrate-overlayer-ambient) system, then the complex reflectance r_{soa} of the system is given by [11]

$$r_{soa} = \frac{r_{oa} + Zr_{so}}{1 + Zr_{so}r_{oa}}, \tag{1}$$

where $Z = \exp(2ikd)$, d is the thickness of the overlayer, and $ck/\omega = (\epsilon - \sin^2\phi)^{1/2}$, where ω is the optical frequency, ϵ is the overlayer dielectric response, and ϕ is the angle of incidence. Equation (1) is valid for both s- and p-polarizations when the appropriate expressions [11] are used for r_{so} and r_{oa}, which are themselves functions of the dielectric functions ϵ_s, ϵ, and $\epsilon_a = 1$ of the substrate, overlayer, and ambient, respectively.

In microscopic terms, r_{so} is simply the ratio of back-reflected to inward-propagating complex field amplitudes within o at the s-o boundary, and as such summarizes the entire previous growth history of the sample. Because this amplitude ratio exists everywhere in o whether or not a physical boundary is actually present, Eq. (1) is also valid for a virtual boundary, and specifically for a running virtual boundary located at a constant depth d beneath the growth surface. Now if $|r_{so}| << 1$ and $|2kd| << 1$, as is the case here, then Eq. (1) can be expanded to first order in r_{so} to yield

$$r_{soa}(d) \approx r_{soa}(0) + 2ikd(r_{soa}(0) - r_{oa}). \tag{2}$$

This expression contains a "compositional" term r_{oa} but not the "history" term r_{so}. Consequently, Eq. (2) can be inverted to obtain the near-surface composition independent of previous growth. For SE applications the equivalent expression is

$$<\epsilon(d)> \approx <\epsilon(0)> + 2ikd(<\epsilon(0)> - \epsilon), \tag{3}$$

where the near-surface value ϵ of the dielectric function is assumed to be essentially constant. Since $<\epsilon(0)>$ and $<\epsilon(d)>$ are the measured values of $<\epsilon>$ before and after deposition of an additional thickness d of overlayer material, ϵ can be calculated directly.

Performance of our system is illustrated in Figs. 7, which show approximately 40 min continuations of both regulated and unregulated data of Fig. 6 on greatly expanded scales [28]. For these data ϵ_2 was calculated from $<\epsilon>$ by evaluating the mean values of $<\epsilon(0)>$ and $(<\epsilon(d)> - <\epsilon(0)>)$ over $N = 31$ data points obtained at 1 s intervals at a deposition rate of 1.7 Å/s (d = 53 Å), with the control voltage corrected at each point by $\eta = 0.7\%$ of the calculated increment for stability. For the regulated case, Fig. 7a shows the convergence of the final remnant of the exponential spiral to the target value $\epsilon_{2t} = 11.27$ to within the noise limit of the ellipsometer, an equivalent precision of x of ± 0.1%. In contrast, the unregulated data of Fig. 7b show a general drift of $<\epsilon_2>$ to decreasing values, indicating an approximately 0.3% drift toward increasing Al composition over the duration of the measurement. Consistent with this is a decrease of the regulated control voltage, not shown, indicating the need to reduce TEA flow to maintain $<\epsilon_2>$ at the target value. This tendency toward increasing Al composition with time also explains the slight offset between the mean measured and target values of $<\epsilon_2>$ in Fig. 7a, a result of the finite time constant of the loop. More to the point, the regulated trajectory of Fig. 6 shows that the control system is able to correct an initial improper setting of the regulation voltage and to establish the desired operating point within the response-time constraints of the system.

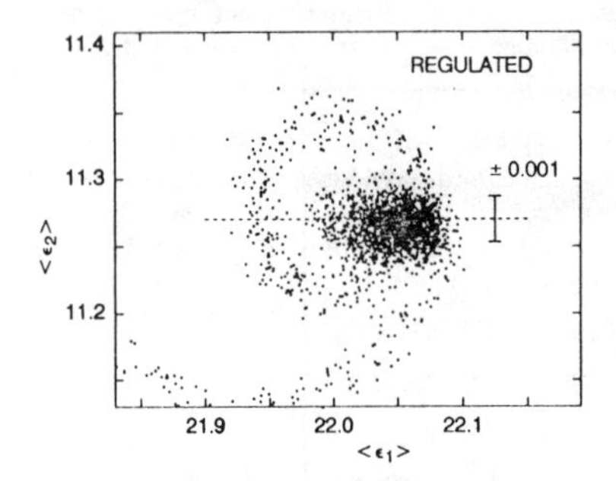

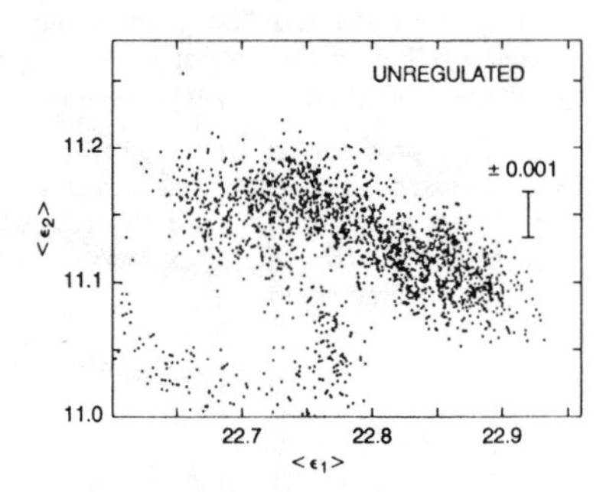

Figs. 7. Left, right: 40 min continuations of the regulated and unregulated, respectively, trajectories of Fig. 6. A compositional uncertainty of ± 0.001 is indicated by the vertical bars (after ref. 28).

We have recently extended the capabilities of this system to deal with structures as well as compositions. In this case the target value ϵ_{2t} takes the form of the desired thickness profile, and the feedback loop adjusts the composition accordingly. Results for two test profiles in the form of half-wave-rectified sine waves, one concave up and one concave down, are shown in Figs. 8. The top parts show the target profile superimposed on the real-time compositions determined by Eq. (3). The middle parts show the fluctuations about the target values. The bottom parts show the regulation voltages needed to achieve these results. Here, compositional fluctuations are larger than those of Figs. 7 because these results were obtained with a 5-point (7 Å) average. This was done to achieve a faster loop response to more accurately track the cusps.

CONCLUSION

The development of optical techniques for monitoring and control of crystal growth is proceeding rapidly. Enough results are now available to permit some assessments of the current situation, speculations about future developments, and suggestions concerning fruitful areas of further research.

The least demanding application is ALE. As reflectance can easily follow dynamic responses to a precision of better than 0.01 ML and has the distinct advantage of experimental simplicity, this approach is best suited for monitoring the completion of alternating monolayers of species. By observing the scattered light, changes in the steady-state morphology can be followed as well. Reflectance systems should be adaptable to closed-loop control of dynamic growth processes, although no configuration has yet been reported.

Conventional growth represents a different challenge since it is ordinarily performed under steady state conditions. Optical probes proven capable of assessing surface information during conventional growth include RDS as well as LLS. As RDS is additionally sensitive to bulk anisotropy, it should also be useful for monitoring and eventually controlling growth of intrinsically anisotropic structures such as vertical superlattices, which should exhibit enormously large (~ 5%) anisotropies in the vicinity of the E_1 transitions even though our attempts to observe these

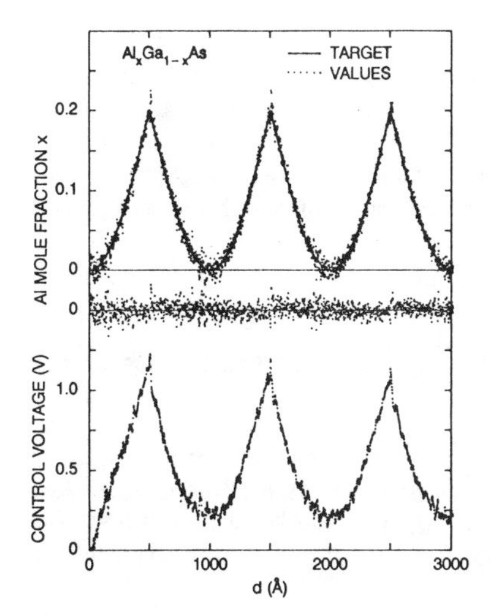

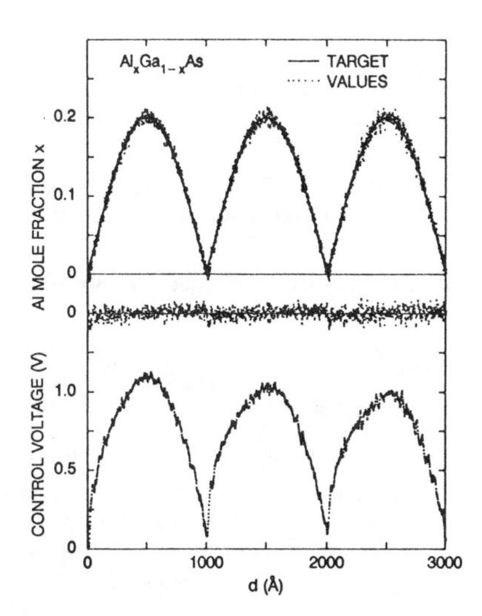

Figs. 8. Left, right: Half-wave-rectified sinusoidal doping profiles grown in $Al_xGa_{1-x}As$ by closed-loop control. Top: values of ϵ_2 calculated from Eq. (3) superimposed on the target profile. Middle: fluctuations about target values. Bottom: control voltages.

anisotropies have not succeeded to date. Other possibilities involving anisotropy concern self-ordering structures such as certain InGaP alloys, which spontaneously order along (111) planes under certain growth conditions, and relaxation of strained pseudomorphic layers, a process that is also known to be anisotropic. Briones and co-workers have recently used RDS to follow the disappearance of antiphase domains during growth of GaAs on Si [18]. We have also recently used RDS to determine carrier concentrations of (001) GaAs layers during OMCVD growth [30].

Inevitably, technology will require full closed-loop control of the growth process to achieve adequate yields as samples become increasingly complex, and to allow identical samples to be grown regardless of specific characteristics of individual machines. Closed-loop control, particularly under steady-state conditions where the surface itself is not changing with time, requires a near-surface (as opposed to a surface) diagnostic capability. Consequently, such systems will involve SE. Our initial work represents a first step toward the complete solution, which will require in addition investigation of loop stability, the achievement of accuracy as well as precision, the use of several wavelengths [31] to monitor several compositions at once as required for growth of quaternary and multinary compounds, the use of the exact Fresnel expressions to obtain thicknesses as well as compositions, and measurement during sample rotation as needed to achieve highest uniformity. Such systems will be computation-intensive, but probably not beyond current capabilities. Future developments should be interesting.

ACKNOWLEDGMENTS

Two of us (I.K. and M.W.) gratefully acknowledge financial support from the Office of Naval Research under Contract no. N-00014-90-J-1267, and from QUEST, respectively. QUEST is a National Science Foundation Science and Technology Center supported by Contract no. DMR 88-10430.

REFERENCES

1. D. E. Aspnes, SPIE Proc. **1285**, 2 (1990); **1361** (in press); Proc. Mater. Res. Soc. **198**, 341 (1990).
2. E. Pelletier, SPIE Proc. **401**, 74 (1983); R. Herrmann and A. Zöller, SPIE Proc. **401**, 83 (1983).
3. N. Kobayashi and Y. Horikoshi, J. Appl. Phys. Jpn. **28**, L1880 (1989).
4. M. Makimoto, Y. Yamauchi, N. Kobayashi, and Y. Horikoshi, J. Appl. Phys. Jpn. **29**, L207 (1990); N. Kobayashi, T. Makimoto, Y. Yamauchi, and Y. Horikoshi, J. Cryst. Growth **107**, 62 (1991).
5. Y. Horikoshi, H. Yamaguchi, F. Briones, and M. Kawashima, J. Cryst. Growth **105**, 326 (1990).
6. A. J. Pidduck, D. J. Robbins, A. G. Cullis, D. B. Gasson, and J. L. Glasper, J. Electrochem. Soc. **136**, 3083 (1989); A. J. Pidduck, D. J. Robbins, D. B. Gasson, C. Pickering, and J. L. Glasper, J. Electrochem. Soc. **136**, 3088 (1989); A. J. Pidduck, D. J. Robbins, I. M. Young, and G. Patel, Thin Solid Films **183**, 255 (1989).
7. T. Stehlin, M. Feller, P. Guyot-Sionnest, and Y. R. Shen, Optics Lett. **13**, 389 (1988).
8. T. F. Heinz, M. M. T. Loy, and S. S. Iyer, Mat. Res. Symp. Proc. **75**, 697 (1987); M. E. Pemble, D. S. Buhaenko, S. M. Francis, P. A. Goulding, and J. T. Allen, J. Cryst. Growth **107**, 37 (1991).
9. Y. R. Shen, Nature **337**, 519 (1989).
10. W. Richter, P. Kurpas, R. Lückerath, M. Motzkus, and M. Waschbüsch, J. Cryst. Growth **107**, 13 (1991).
11. D. E. Aspnes, in **Optical Properties of Solids, New Developments**, ed. B. O. Seraphin (North-Holland, Amsterdam, 1976), p. 799; Proc. SPIE **946**, 112 (1988); R. M. A. Azzam and N. M. Bashara, **Ellipsometry and Polarized Light** (North-Holland, Amsterdam, 1977).
12. B. Drevillon, Proc. SPIE **1186**, 110 (1990).
13. D. E. Aspnes, R. Bhat, E. Colas, L. T. Florez, J. P. Harbison, M. K. Kelly, V. G. Keramidas, M. A. Koza, and A. A. Studna, Proc. SPIE **1037**, 2 (1989); E. Colas, D. E. Aspnes, R. Bhat, A. A. Studna, M. A. Koza, and V. G. Keramidas, Proc. SPIE **1186**, 96 (1989); E. Colas, D. E. Aspnes, R. Bhat, A. A. Studna, J. P. Harbison, L. T. Florez, M. A. Koza, and V. G. Keramidas, J. Cryst. Growth **107**, 47 (1991).
14. D. E. Aspnes, J. P. Harbison, A. A. Studna, and L. T. Florez, J. Vac. Sci. Technol. **A6**, 1327 (1988).
15. A. A. Studna, D. E. Aspnes, L. T. Florez, B. J. Wilkens, and R. E. Ryan, J. Vac. Sci. Technol. **A7**, 3291 (1989).
16. D. E. Aspnes, J. P. Harbison, A. A. Studna, and L. T. Florez, Phys. Rev. Lett. **59**, 1687 (1987).
17. See, e.g., F. Briones and Y. Horikoshi, J. Appl. Phys. Jpn. **29**, 1014 (1990); O. Acher, S. M. Koch, F. Omnes, M. Delour, M. Razeghi, and B. Drévillon, J. Appl. Phys. **68**, 3564 (1990); L. Samuelson, K. Deppert, S. Jeppesen, J. Jönsson, G. Paulsson, and P. Schmidt, J. Cryst. Growth **107**, 68 (1991); F. Briones and A. Ruiz, J. Cryst. Growth (in press).
18. Y. Gonzâlez, L. Gonzâlez, and F. Briones, J. Cryst. Growth (in press).
19. F. Manghi, R. Del Sole, A. Selloni, and E. Molinari, Phys. Rev. **B41**, 9935 (1990).
20. Y. C. Chang and D. E. Aspnes, Phys. Rev. **B41**, 12002 (1990).
21. M. Wassermeier, I. Kamiya, D. E. Aspnes, L. T. Florez, J. P. Harbison, and P. M. Petroff, J. Vac. Sci. Technol. (submitted).

22. D. E. Aspnes, Y. C. Chang, A. A. Studna, L. T. Florez, H. H. Farrell, and J. P. Harbison, Phys. Rev. Lett. **64**, 192 (1990).

23. H. H. Farrell, J. P. Harbison, and L. D. Peterson, J. Vac. Sci. Technol. **B5**, 1482 (1987); D. J. Chadi, J. Vac. Sci. Technol. **A5**, 834 (1987).

24. M. Sauvage-Simkin, R. Pinchaux, J. Massies, P. Calverie, N. Jedrecy, J. Bonnet, and I. K. Robinson, Phys. Rev. Lett. **62**, 563 (1989).

25. The formation by AsH_3 of (001) GaAs surfaces terminated by more than 1 ML of As has also recently been observed by Creighton (remarks given in talk D1.2, this Symposium).

26. M. Ozeki, K. Mochizuki, N. Ohtsuka, and K. Kodama, Thin Solid Films **174**, 63 (1989).

27. J. Nishizawa, T. Kurabayashi, H. Abe, and A. Nozoe, Surface Sci. **185**, 249 (1987); T. H. Chiu, Appl. Phys. Lett. **55**, 1244 (1989); J. R. Creighton, Surface Sci. **234**, 287 (1990); B. Maa and P. D. Dapkus (to be published).

28. D. E. Aspnes, W. E. Quinn, and S. Gregory, Appl. Phys. Lett. **57**, 2707 (1990).

29. D. E. Aspnes, W. E. Quinn, and S. Gregory, Appl. Phys. Lett. **56**, 2569 (1990).

30. H. Tanaka, E. Colas, I. Kamiya, D. E. Aspnes, and R. Bhat (to be published).

31. Y. T. Kim, R. W. Collins, and K. Vedam, Surface Sci. **233**, 341 (1990); A. R. Heyd, I. An, R. W. Collins, Y. Cong, K. Vedam, S. S. Bose, and D. L. Miller, J. Vac. Sci. Technol. (in press).

REAL-TIME ANALYSIS OF IN-SITU SPECTROSCOPIC ELLIPSOMETRIC DATA DURING MBE GROWTH OF III-V SEMICONDUCTORS

B. JOHS*, J.L. EDWARDS**, K.T. SHIRALAGI**, R. DROOPAD**, K.Y. CHOI**, G.N. MARACAS**, D. MEYER*, G.T. COONEY*, and JOHN A. WOOLLAM*
*J.A. Woollam Co., 650 'J' St. Suite 39, Lincoln, NE, 68508.
**Center for Solid State Electronics Research, Arizona State University, Tempe, AZ 85287.
+Research supported by DARPA contract DAAH01-89-C-0357

ABSTRACT

A modular spectroscopic ellipsometer, capable of both in-situ and ex-situ operation, has been used to measure important growth parameters of GaAs/AlGaAs structures. The ex-situ measurements provided layer thicknesses and compositions of the grown structures. In-situ ellipsometric measurements allowed the determination of growth rates, layer thicknesses, and high temperature optical constants. By performing a regression analysis of the in-situ data in real-time, the thickness and composition of an AlGaAs layer were extracted during the MBE growth of the structure.

INTRODUCTION

Ellipsometry is a non-invasive optical technique which is readily adapted for in-situ measurements. It is based on measuring the polarization state of reflected light when linearly polarized light is made incident on the sample [1]. Traditionally, the polarization state of the reflected light is characterized by two parameters, Ψ and Δ, which are defined in terms of the complex Fresnel reflection coefficients of the sample:

$$\tan\Psi\, e^{i\Delta} = \frac{r_p}{r_s} \tag{1}$$

Ellipsometric data can be taken at multiple wavelengths (spectroscopic ellipsometry, denoted SE) and also at different angles of incidence (variable angle of incidence spectroscopic ellipsometry, known as VASE). These additional ellipsometric measurements provide much more information about the sample than can be obtained from a single wavelength and angle measurement. SE is more suited for in-situ applications [2-4], while VASE allows for a more comprehensive ex-situ characterization [5].

Whereas the measured ellipsometric parameters Ψ and Δ are influenced by the structure (layer thicknesses, compositions, microstructure, etc.) and optical constants of the sample, analytic expressions for the structure and optical constants in terms of Ψ and Δ can be derived only in a few simple cases. To analyze more complex samples, it is necessary to construct an optical model for the sample which can generate predicted values for Ψ and Δ. The model requires a knowledge of the optical constants of the materials in the structure and is parameterized by the layer thicknesses, alloy fractions, etc. of the structure. A regression analysis algorithm [6] is used to adjust the model parameters to minimize the difference between the experimentally measured Ψ's and Δ's and the model generated Ψ's and Δ's. Due to the inherent complexity of the model computation and regression analysis procedure, the analysis of ellipsometric data can be a time intensive task.

In this paper, we report the use of in-situ SE and ex-situ VASE to characterize GaAs/AlGaAs structures grown by MBE. In-situ SE data was acquired in real-time during the growth of the structures, under *real* growth conditions, i.e., normal growth temperatures with substrate rotation enabled. Post-deposition analysis of the in-situ data allowed extraction of layer thicknesses, growth rates, and optical constants of the structure at the growth temperature. The sample was removed from the growth chamber and ex-situ VASE provided an independent determination of the AlGaAs layer thicknesses and compositions. Using these analysis results, a model for the growth of an AlGaAs layer on GaAs, parameterized by the layer thickness and composition, was developed. In a subsequent growth run, this model was used to analyze the in-situ ellipsometric data in real-time. The real-time analysis of in-situ ellipsometric data can provide the crystal grower with instant feedback on the thickness and composition of a AlGaAs layer during the MBE growth.

ACQUISITION OF ELLIPSOMETRIC DATA

Figure 1 shows a schematic of the ellipsometer hardware that was used in this work. A light beam chopped variation of the standard rotating analyzer ellipsometer (RAE) is employed [7]. This configuration is insensitive to polarization in the light source, and allows data to be taken in ambient lighting. The tilt stages and stainless steel bellows provide the degrees of freedom necessary to effect system alignment. The alignment procedure ensures that all of the optical elements are normal to the light beam by adjusting the tilt stages to achieve a beam bounce-back. Special strain-free windows are used to minimize birefringence effects. The hardware is also modular; the input and output arms of the instrument can easily be removed from the MBE chamber and mounted on an ex-situ VASE base. The system is fully automated and controlled by an 80386-based personal computer, which is also used for the data analysis.

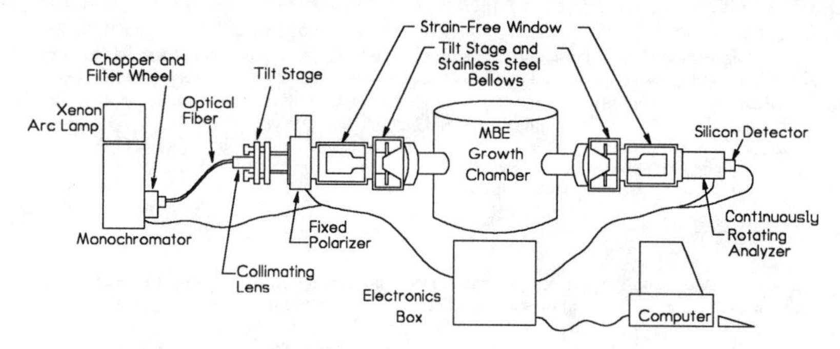

Figure 1. In-situ chopped beam rotating analyzer ellipsometer configuration.

Acquiring ellipsometric data in the MBE growth environment is a non-trivial task. The above instrument design, along with the beam bounce-back alignment procedure, provides accurate data under static, low substrate temperature conditions. However during growth, the substrate and K-Cells are at elevated temperatures, and the substrate is rotated to promote growth uniformity. Signal noise due to the intense black body radiation emitted from the chamber is minimized by inserting a cold filter in front of the analyzer. Effects due to the substrate rotation, which causes the beam to precess about the detector iris, are minimized by overfilling the iris with a large beam diameter. In addition, the system alignment and calibration procedures are also performed under the growth conditions.

During the MBE growth of the structures, the monochromator was scanned to collect ellipsometric data at 3 photon energies: 2.07, 2.61, and 3.54 eV. Acquisition of the data at the 3 energies required just under 10 seconds. Faster acquisition was possible, but a longer averaging time was used to improve the signal to noise ratio.

EXPERIMENTAL RESULTS

The ellipsometric data from two separate growth runs is reported in this paper. The nominal growth structures are shown in figure 2.

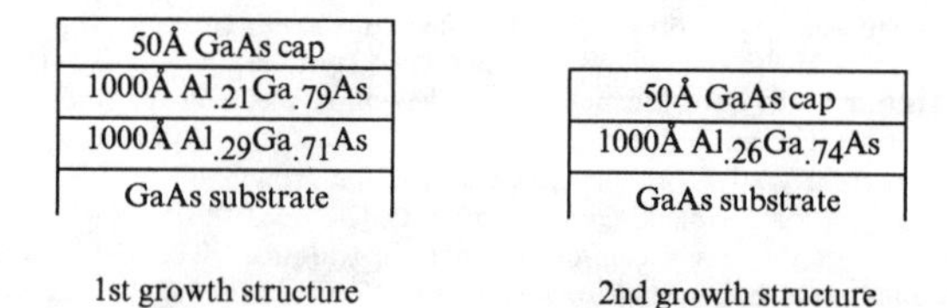

Figure 2. Nominally MBE grown structures for this experiment.

Both structures were grown at a substrate temperature of 626°C, as determined by an optical pyrometer. The 50Å GaAs cap layers were grown to prevent oxidation of the AlGaAs layers; however, the growth of the cap layers was not monitored in this experiment. The first structure was grown to determine the high temperature optical constants of AlGaAs at two compositions. These results were integrated into a simple interpolation model that could generate ellipsometric data for an AlGaAs layer of any composition. In the second growth, a layer of intermediate composition was grown, and real-time regression analysis of the in-situ ellipsometric data, using the model derived from the first growth, was used to determine the thickness and composition of the AlGaAs layer during the growth.

Ellipsometric Data from the 1st Growth Run

The plot in figure 3 displays experimental and fit data from the growth of the first AlGaAs structure. The growth of the 1st AlGaAs layer occurred between time = 59 - 64 minutes and the 2nd AlGaAs layer was grown during time = 66 - 72 minutes. The fit parameters for this growth, extracted from a post-deposition regression analysis of the in-situ experimental data, are given in table 1 (the angle of incidence was assumed to be 75°).

Figure 4 displays a Ψ plot of the VASE experimental and fit data. Table 2 summarizes the fit model that was obtained from the analysis of the ex-situ VASE data. These results are in good agreement with the in-situ fit and the nominal growth parameters. The regression fit algorithm did indicate a relatively high correlation between the two AlGaAs layer thicknesses of table 2, but the rest of the fit parameters were uncorrelated.

In-situ Fit Model

	In-situ Fit Model
Growth rate = 212Å/minute	967Å $Al_xGa_{1-x}As$
Growth rate = 194Å/minute	960Å $Al_xGa_{1-x}As$
	GaAs substrate

Optical Constants (T=626° C)

Layer	n, 2.06eV	k, 2.06eV	n,2.61eV	k,2.61eV	n,3.54eV	k,3.54eV
substrate	4.2827	-0.4425	4.6058	-1.5763	3.5974	-2.0172
1st layer	4.0569	-0.2787	4.6224	-0.9302	3.7783	-1.9722
2nd layer	4.1355	-0.3195	4.6708	-1.0746	3.7451	-1.9713

Table 1. Fit parameters extracted from the in-situ data, 1st growth structure.

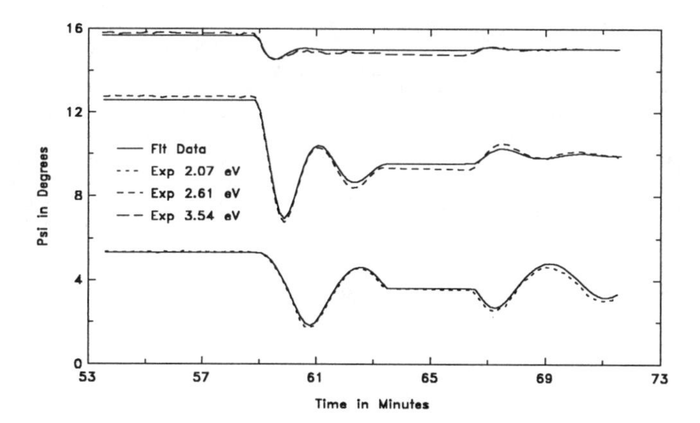

Figure 3. Experimental and Fit ellipsometric Ψ data from the 1st growth run.

VASE fit Model
7.5Å GaAs Oxide
27.6Å GaAs
914Å $Al_{.212}Ga_{.788}As$
979Å $Al_{.292}Ga_{.708}As$
GaAs substrate

Table 2. Fit parameters obtained from VASE analysis of the 1st growth structure.

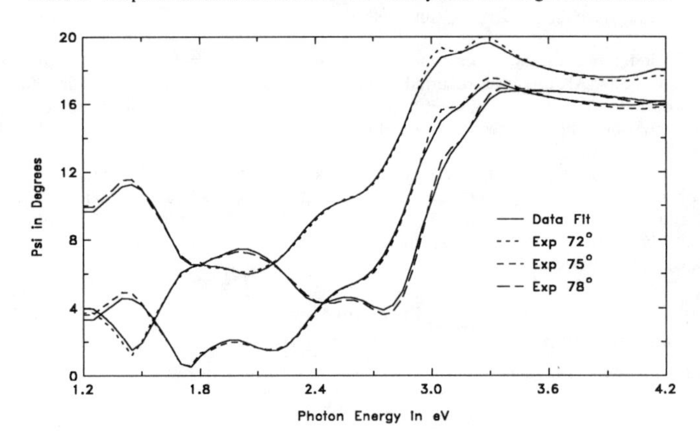

Figure 4. Ex-situ VASE data for the 1st growth structure.

Ellipsometric Data from the 2nd Growth Run

The high temperature, composition dependent optical constant data extracted from the 1st growth was used create a composition dependent model for $Al_xGa_{1-x}As$ optical constants at the growth temperature of 626° C. The optical constants for a given composition 'x' were calculated by interpolating between the optical constants of the layers of known composition.

During the growth of the 2nd structure, the simple model shown in table 3 was used to fit the experimental data in real-time. The three fit parameters were: AlGaAs layer thickness, composition, and angle of incidence. The angle of incidence was allowed to vary, accounting for changes due to sample loading and substrate rotation. Throughout the growth, the angle was fit to be $75.05° \pm 0.03°$. As multiple parameters were used to effect the real-time fit, it was absolutely necessary to overdetermine the model by acquiring data at multiple wavelengths.

The graphs in figure 5 display the fit parameters that were extracted, in real-time, during the growth of the AlGaAs layer. The values are all quite reasonable, except for the glitch that appears in the composition near time=61 minutes. This might be due to bad ellipsometric data points. However, upon removable from the vacuum chamber, the surface of the grown structure was observed to be slightly spotty. It is therefore possible that the spike in composition may have been caused by an instability in the aluminum K-cell.

Figure 6 shows the in-situ Δ data acquired during the 2nd growth run. Notice the discrepancy in the data near time=61 minutes. This corresponds to the spike observed in the real-time alloy fraction fit.

Real-time Fit Model
x Å $Al_yGa_{1-y}As$
GaAs substrate

Table 3. Model used for the real-time analysis of data in the 2nd growth run.

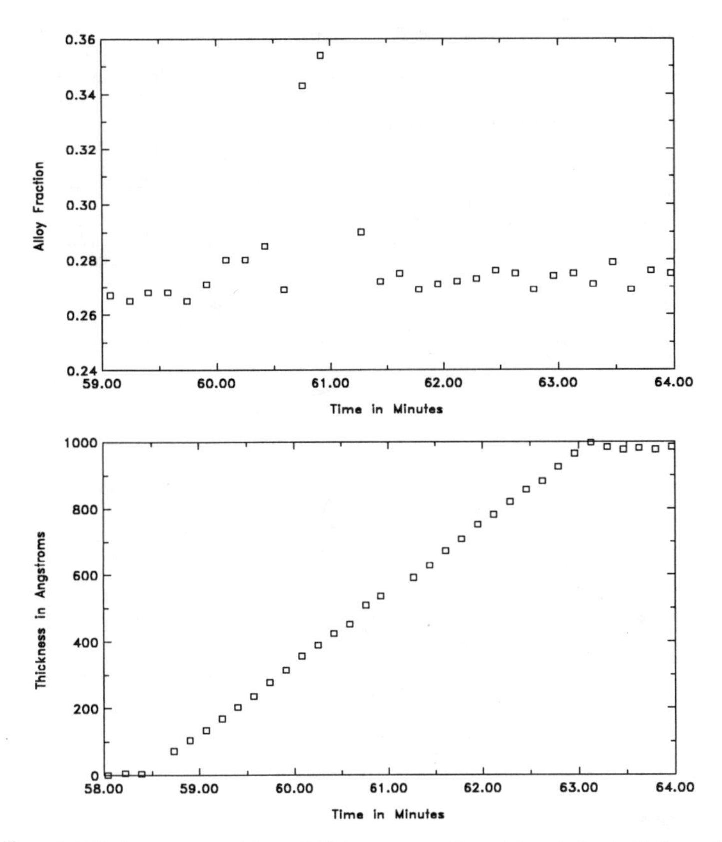

Figure 5. AlGaAs layer composition and thickness extracted in real-time during the 2nd growth.

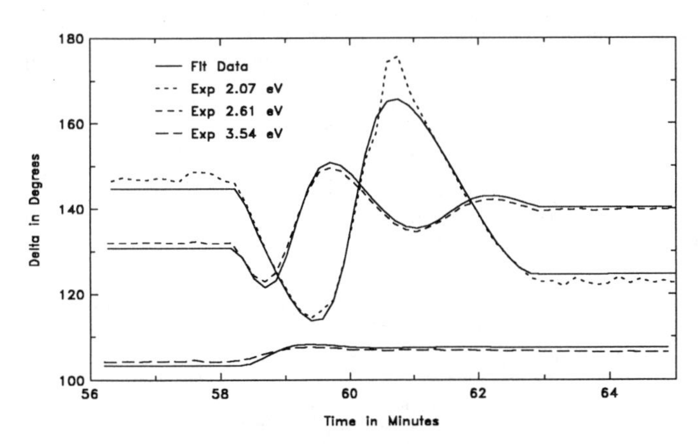

Figure 6. Post-deposition analysis of in-situ data from the 2nd growth.

Table 4 compares the analysis results obtained from ellipsometric data on the 2nd growth structure. The resulting fit models are in good agreement, and also confirm the values extracted in real-time during the MBE growth of the structure. Although the AlGaAs layer thickness is right on the nominal value of 1000Å, the layer composition is slightly higher than the nominally grown value of 0.26. Experimental and fit ex-situ VASE data for the 2nd structure is also displayed in figure 7.

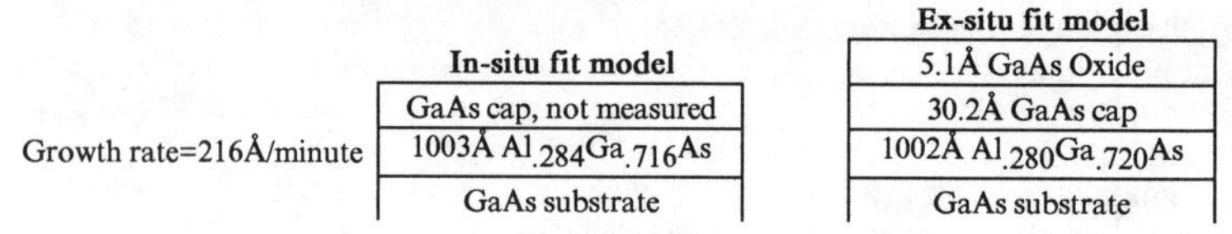

	In-situ fit model	Ex-situ fit model
		5.1Å GaAs Oxide
	GaAs cap, not measured	30.2Å GaAs cap
Growth rate=216Å/minute	1003Å $Al_{.284}Ga_{.716}As$	1002Å $Al_{.280}Ga_{.720}As$
	GaAs substrate	GaAs substrate

Table 4. Comparison of the in-situ and ex-situ fit models for the 2nd growth structure.

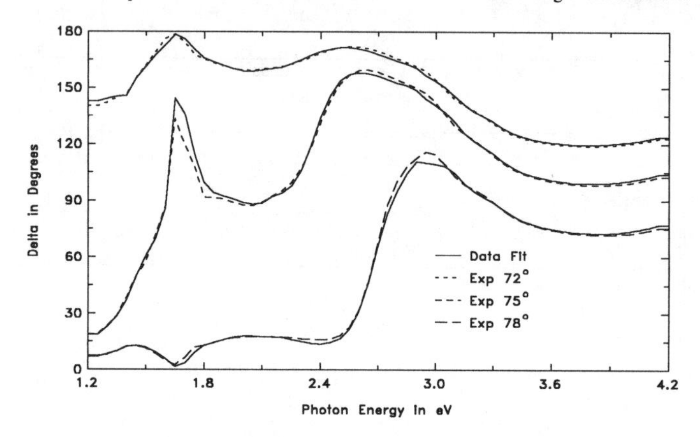

Figure 7. VASE data from the 2nd growth structure.

CONCLUSIONS

Acquisition of accurate in-situ ellipsometric data has been obtained during the MBE growth of GaAs/AlGaAs structures. This data, along with ex-situ VASE data, was used to extract the layer thicknesses, compositions, growth rates, and high temperature optical constants of the grown structures. Real-time analysis of the in-situ ellipsometric data was also performed, providing the layer thickness and composition of an AlGaAs layer during the growth.

The model used to analyze the in-situ data in real-time was kept simple. While the real-time analysis software uses the exact Fresnel expressions for the optical model and can generate and analyze structures of arbitrary complexity, the acquisition rate of the ellipsometer hardware limits the number of fit parameters that can be defined. The next generation of hardware, which will acquire ellipsometric data at multiple wavelengths simultaneously, will allow for the real-time analysis of more complex growth structures. The high-end personal computers that have recently become available (and affordable) have adequate computing power to perform the real-time analysis of ellipsometric data.

References

1. R.M.A. Azzam and N.M. Bashara, *Ellipsometry and Polarized Light* (North-Holland, Amsterdam, 1977).
2. R.W. Collins, Y. Cong, Y.T. Kim, R. Messier, and K. Vedam, in *Abstracts of the 40th Meeting of the International Society of Electrochemistry*, 17-22 Sept. 1989, Kyoto, Japan, p. 464.
3. D.E. Aspnes, W.E. Quinn, and S. Gregory, Appl. Phys. Lett. **56**, 2569 (1990).
4. Huade Yao, Paul G. Snyder, and John A. Woollam, in *2nd Int'l Conf. on Elec. Mats.* (Mater. Res. Soc. Proc., Pittsburgh, PA) p. 501.
5. John A. Woollam and Paul G. Snyder, Mat. Sci. and Eng. **B5**, 279 (1990).
6. W.H. Press, B.P. Flannery, S.A. Teukolsky, and W.T. Vetterling, *Numerical Recipes in C* (Cambridge University Press, New York, 1988), p. 542.
7. R.W. Collins, Rev. Sci. Instrum. **61**, 2029 (1990).

FTIR STUDIES OF ORGANOMETALLIC SURFACE CHEMISTRY RELEVANT TO ATOMIC LAYER EPITAXY.

Ananth V. Annapragada, Sateria Salim and Klavs F. Jensen.
Department of Chemical Engineering, Massachusetts Institute of Technology,
Cambridge, MA 02139.

ABSTRACT

The adsorption and surface reactions of trimethylgallium and tertiarybutylarsine on GaAs(100) surfaces have been investigated by Fourier transform infrared spectroscopy. Adsorbed methyl groups resulting from the dissociative chemisorption of trimethylgallium on GaAs(100) are shown to form As-H and CH_2 species on the surface. The CH_2 groups are stable on the surface at temperatures as high as 550 °C. The surface coverage is low (~0.2% of a monolayer) and is reduced by the presence of hydrogen on the surface. This dehydrogenation of surface methyl groups could be a possible route to carbon incorporation in GaAs grown by atomic layer epitaxy. Tertiarybutylarsine is shown to decompose primarily by homolysis to form a tertiary butyl group and AsH_2. At temperatures below 400 °C on trimethylgallium dosed surfaces, the decomposition products appear to cause the hydrogenation of methylene groups remaining from prior surface dosing with trimethylgallium. At high temperatures, the tertiarybutyl radical appears to undergo dehydrogenation reactions to an unsaturated species which is stable on the surface. In contrast, the dehydrogenation does not appear to occur on surfaces treated with tertiarybutylarsine. The data for trimethylgallium and tertiarybutylarsine support the general assertion that surface As-H species play a critical role in the removal of hydrocarbon species from the growth surface.

INTRODUCTION

Understanding the surface chemistry of organometallic compounds of gallium and arsenic is of critical importance to the further development of atomic layer epitaxy (ALE) of GaAs. The reactions of trimethylgallium (TMG) on GaAs have been investigated by a range of surface science techniques, including molecular beam scattering, temperature programmed desorption, and x-ray photoelectron spectroscopy (XPS) [1,2,3]. It is generally recognized that TMG chemisorbs dissociatively and that methyl radicals and TMG fragments desorb upon heating of the GaAs substrate. However, reactions of adsorbed methyl groups on surfaces encountered in ALE are still unclear.

In contract to TMG, few studies have been conducted on the surface chemistry of arsenic precursors. Tertiarybutylarsine (t-BAs)is of particular interest since it has emerged as one of the main organometallic As-sources for the replacement of AsH_3. Conventional organometallic vapor phase epitaxy (OMVPE) with this compound has produced GaAs with electrical properties comparable to those obtained with arsine [4] and its use in ALE has also been demonstrated [5].

In this work, we present Fourier transform infrared (FTIR) spectroscopy investigations of the surface reactions of TMG on t-BAs treated GaAs(100) surfaces and of t-BAs on TMG treated GaAs(100) surfaces. This switching between gallium and arsenic rich surfaces emulates the ALE process. Infrared (IR) spectroscopy is a natural choice for the study of surface chemistry of organometallic compounds since adsorbed alkyl groups and surface hydrides have very distinct signatures in the IR spectrum.

EXPERIMENTAL

The experimental set-up used in this study has been described elsewhere [6]. Briefly, it consists of a load-locked UHV chamber with the capability to perform infrared multiple internal reflection (IR-MIR) experiments while dosing the sample. A mass spectrometer allows detection of desorbing species. The sample is mounted on a resistively heated molybdenum plate which can be heated to ~800 °C.

Sample preparation

A 40×14×2 mm, 45° trapezoid GaAs MIR element was degreased in boiling trichloroethane and mounted on the holder. After outgassing in the load lock, the sample was transferred into the dosing/analysis chamber. In the dosing/analysis chamber, the sample was heated to 660 °C in 2 × 10^{-5} torr t-BAs for 15 minutes. Maa *et al.* [5] have reported that a similar treatment of GaAs results in the formation of an arsenic rich surface with a (2×4) RHEED pattern. Following this exposure, one of the following procedures was followed:

(A) Cooling the sample below a temperature of 150 °C while maintaining the t-BAs dose resulted in a *hydrogen-terminated arsenic-rich* surface.

(B) Shutting the t-BAs flow off above 300 °C and then cooling the sample to room temperature resulted in the formation of a *hydrogen-free arsenic-rich* surface.

(C) Dosing the hydrogen free arsenic rich surface with TMG resulted in the formation of a gallium rich surface.

The presence of As-H species on the surface (A) and their absence on (B) is discussed below.

RESULTS AND DISCUSSION

TMG on GaAs(100)

Figure 1(a) shows the C-H stretching region of the IR spectrum after dosing of 30,000 Langmuirs (L) of TMG on a hydrogen-free arsenic-rich surface. A broad smeared peak is visible in the 2800-3000 cm^{-1} region. This peak cannot be assigned solely to the C-H stretching frequencies of adsorbed methyl groups, but it is consistent with the presence of both CH_3 and non-terminal CH_2 species on the surface. At the signal levels of these spectra, it is difficult to reliably deconvolute the broad peak into its components. The CH_2 species in these spectra are unlikely to be linear CH_2 species since that would imply the presence of C_2 hydrocarbons on the surface. We are not aware of reports of such species being present on GaAs (100) after dosing with TMG or being impurities in TMG. The CH_2 species in these spectra are therefore assigned to bridging CH_2 species, possibly bridging two gallium atoms on the surface.

Holding the surface at 5×10^{-8} torr for 30 minutes results in the spectrum shown in Figure 1(b), the only difference between Figures 1(a) and 1(b) being very small negative peaks close to 3000, 2970, 2916, and 2870 cm^{-1}, which correspond to the gas-phase C-H stretching frequencies for TMG [7]. This behavior is consistent with the loss of small amounts of physisorbed TMG from the surface [6]. The stability of the broad peak in vacuum after physisorbed TMG has left the surface suggests that the primary methyl and methylene C-H stretching signals stem from products of the chemisorption of TMG.

Subsequent heating under vacuum to 200 °C and cooling to 57 °C produces the spectrum shown in Figure 1(c). The broad peak present before heating has almost completely disappeared and a new peak, assigned to terminal CH_2 species [8], has appeared at 3016 cm^{-1}. Additional experiments have demonstrated that the terminal CH_2 species does not desorb at temperatures as high as 550°C. The presence of this species during the heterogeneous decomposition of TMG [9] and TMAs [10] has been suggested in the literature to explain particular experimental observations, but no direct observation of such a species has been reported to date for OMVPE and ALE related systems. The estimated surface concentration of terminal CH_2 based on the area under the 3016 cm^{-1} peak in Figure 2(d), suggests a surface coverage of 0.1-0.2% of a monolayer. This is significantly lower than typical detection limits for electron spectroscopies (*e.g.*, XPS, Auger).

Figure 2 (b) shows IR spectra in the As-H stretching region after high doses of TMG on a hydrogen-free arsenic-rich surface. The two peaks at 2170 and 2110 cm^{-1} are assigned to the AsH_2 asymmetric and symmetric stretches, respectively [7, 11]. The presence of these peaks in the absence of a hydrogen source (including hot filaments) combined with the existence of CH_2 species on the surface is consistent with hydrogen removal from adsorbed methyl species. Heating the sample after the TMG dose to 200 °C in vacuum and cooling to 57 °C resulted in the spectrum, Figure 3(c), where the majority of the surface As-H has disappeared. This is consistent with the known desorption temperature of hydrogen from GaAs(110) of ~170 °C reported by Mokwa *et al* [12].

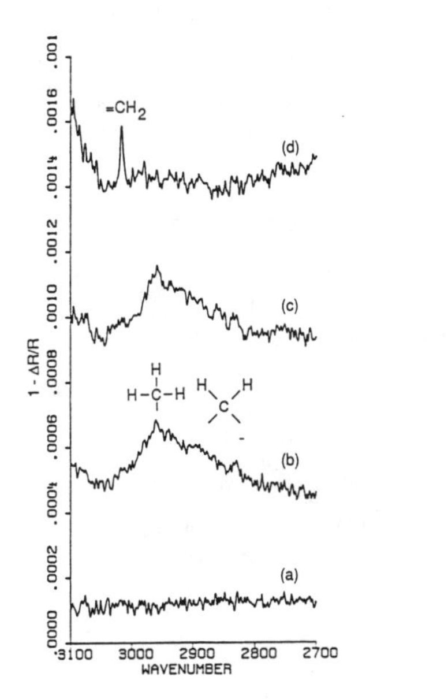

Figure 1. IR spectra of TMG on GaAs (100)
(a) baseline 52°C
(b) after dosing 30,000 L, 51°C
(c) 30 min. at 5×10^{-8} Torr after dosing, 43°C
(d) heating to 200 °C and cool to 57°C.

Figure 2. As-H region IR spectra of TMG on GaAs (100)
(a) baseline 52°C
(b) after dosing 54000L 51°C
(c) heating to 200 °C and cool to 57°C

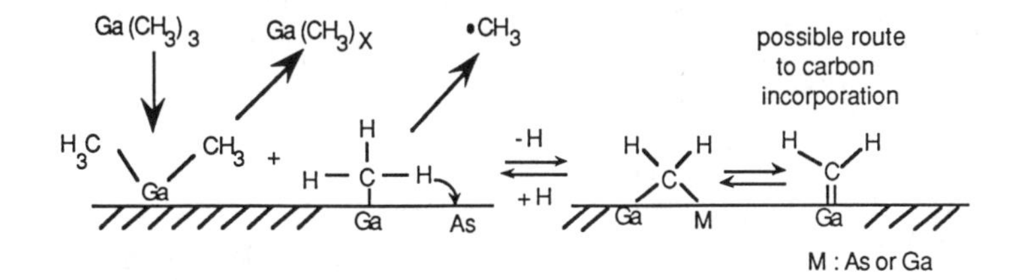

Figure 3. Proposed mechanism for TMG decomposition consistent with IR data.

The formation of terminal CH_2 and As-H species during TMG adsorption on a hydrogen-free arsenic rich GaAs(100) surface and subsequent sample heating suggests the mechanism shown schematically in Figure 3. TMG primarily chemisorbs dissociatively and the majority of the resultant reaction products, methyl radicals and either mono- or dimethyl-gallium, desorb. This behavior is consistent with previous observations of the surface reactions of TMG [1,2,3]. A small fraction of methyl groups is dehydrogenated to bridging and terminal CH_2 species. The dehydrogenation of the methyl radicals results in the formation of AsH and AsH_2 species on the surface. The terminal CH_2 species are stable on the surface to temperatures as high as 550 -600 °C. Because the direct desorption of CH2 would involve the simultaneous breaking of two surface bonds, the removal of this species from the surface most likely has to proceed through hydrogenation to methyl groups, which then desorb.

Further dehydrogenation of CH_2 species represents a possible route to carbon incorporation which is consistent with the selective incorporation of carbon onto the arsenic sublattice during the OMVPE and ALE of GaAs. This carbon incorporation mechanism could also explain the relatively large amount of carbon typically incorporated during ALE with TMG.

t-BAs on GaAs(100)

Figure 4 displays the C-H stretch region of the IR spectra for a TMG dosed GaAs(100) surface during exposure to 2×10^{-5} torr t-BAs at various temperatures. These spectra are all referenced to the spectrum of 300 L TMG on the surface. As the temperature is increased to 410 °C, two negative peaks at 2916 and 2840 cm^{-1} are apparent in the spectrum, corresponding to the loss of bridging methylene species from the surface presumably through the hydrogenation to methyls. The temperature is increased further to 660 °C and spectra are once again obtained on the cooling cycle of the t-BAs dose, as shown in the upper portion of Figure 4. Starting at 380 °C, C-H stretches corresponding to CH_3 and CH_2 species are clearly visible. The area under these peaks decreases continuously as the temperature is lowered. The structure of these alkyls can be inferred from the spectra in the C-H bending region shown in Figure 5. Here, a doublet around 1600 cm^{-1} is assigned to a C=C stretch conjugated to an electron-rich bond (such as another double bond or a carbon-metal bond), the strong peak at 1509 cm^{-1} is assigned to a CH_2 deformation and the peak at 1465 cm^{-1} to a CH_3 deformation. In addition, weak signals characteristic of the skeletal vibrations of the t-butyl group are evident at 1370 and 1352 cm^{-1}. Again, the areas under these peaks decreases continuously with temperature.

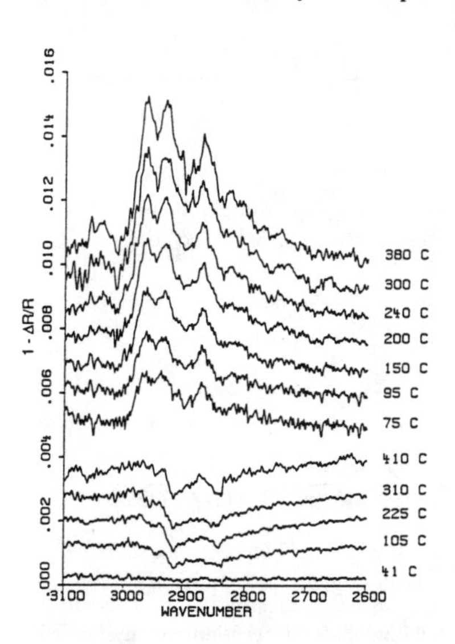

Figure 4. C-H stretch region of IR spectra for a TMG dosed GaAs (100) surface during exposure to t-BAs at different temperatures.

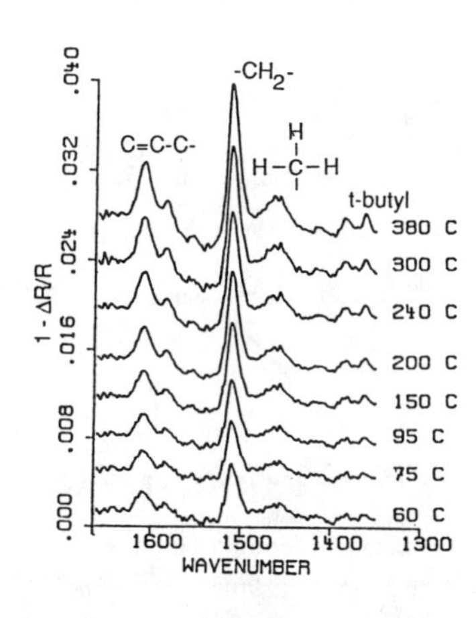

Figure 5. C-H bending region of IR spectra for a TMG dosed GaAs (100) surface during exposure to t-BAs at different temperatures.

Figure 6 shows the corresponding spectra in the As-H region as a function of temperature. As the temperature is increased from 41 °C, no As-H species are evident on the surface. On the cooling cycle from the high temperature however, asymmetric and symmetric As-H stretches are clearly visible at 2170 and 2110 cm^{-1} respectively. These spectra demonstrate that cooling the sample down to room temperature in a t-BAs atmosphere results in the formation of a *hydrogen terminated* arsenic-rich surface. The uppermost trace in Figure 6 is the spectrum of the surface at 57 °C when the t-BAs dose is cut at 300 °C. The absence of As-H stretching peaks in this spectrum shows that cooling in a vacuum after cutting the t-BAs above the desorption temperature of hydrogen results in the formation of a *hydrogen free* surface.

The concentration of surface hydrogen is estimated from the areas under the As-H peaks in Figure 6, and is plotted as a function of surface temperature in Figure 7. The data agrees well with the desorption temperature of hydrogen from GaAs(110) (~170 °C) reported by Mokwa [12]. Above this temperature, the surface concentration is controlled by the decomposition rate of t-BAs, and therefore increases with increasing temperature. Below ~170 °C, the desorption rate is very low, and results in the accumulation of hydrogen on the surface. This data clearly shows that the decomposition of t-BAs is sufficient to maintain surface As-H species even at temperatures significantly above the hydrogen desorption temperature.

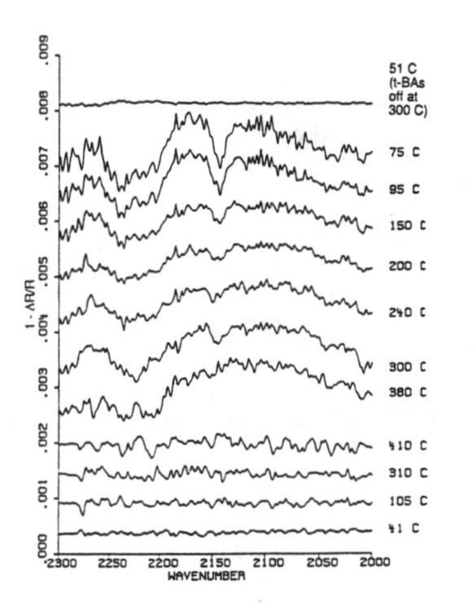

Figure 6. As-H region of IR spectra for a TMG dosed GaAs (100) surface during exposure to t-BAs at different temperatures.

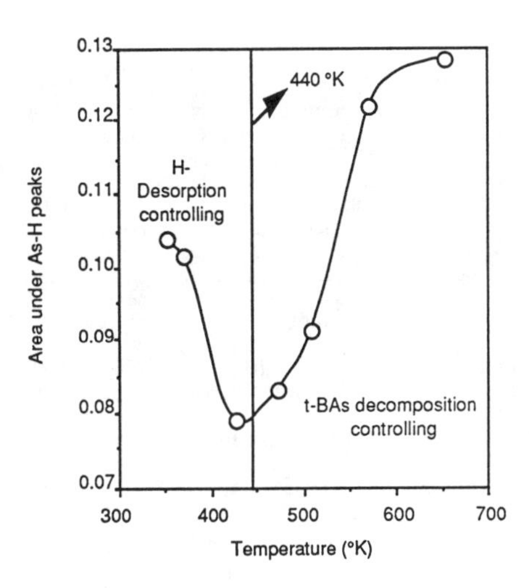

Figure 7. Relative change in concentration of surface hydrogen as a function of temperature during surface decomposition of t-BAs.

It has been demonstrated that the heterogeneous decomposition of t-BAs on GaAs(100) appears to proceed predominantly by homolysis of the C-As bond [6]. The main products of the surface decomposition observed by mass spectrometry were the t-butyl radical and AsH_2. This information and the IR data shown in Figures 6 and 7 suggest the mechanism displayed schematically in Figure 8 for the decomposition of t-BAs on TMG dosed GaAs(100) surfaces. t-BAs undergoes homolysis to form a t-butyl group and AsH_2 on the surface. The majority of these species desorb, but a small fraction of the t-butyl groups undergo successive dehydrogenation steps on the surface to form an unsaturated intermediate. This species contains a double bond conjugated with a metal carbon bond and a linear CH_2 consistent with the C-H bending modes shown in Figure 5. The mechanism for the dehydrogenation of the t-butyl radical on the surface is similar to the mechanism reported by Brainard and Madix[13] for the reaction of t-butyl alcohol with Ag surfaces. The observation that significant amounts of surface alkyl species from t-BAs decomposition appear only on gallium rich surfaces [6] is in agreement with the proposed mechanism of dehydrogenation. On arsenic rich hydrogen terminated surfaces, the surface is presumably saturated with As-H bonds, thus preventing the dehydrogenation of the surface t-butyl groups before desorption.

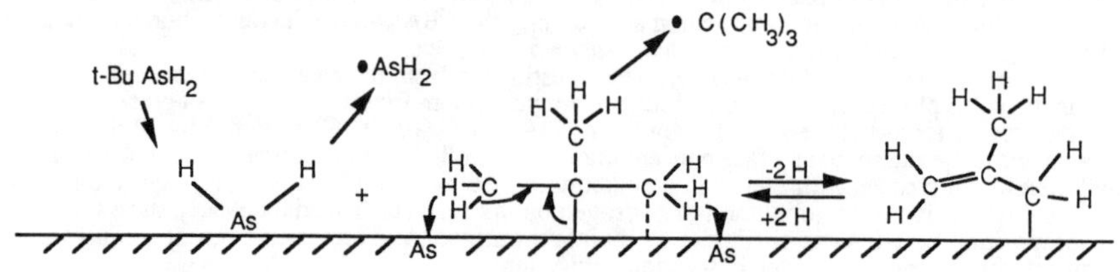

Figure 8. Proposed mechanism for t-BAs decomposition consistent with the IR data.

CONCLUSIONS

The IR data provide experimental evidence for the general assertion that As-H species are required for the removal of carbon from the growth interface. Moreover, the data show direct evidence that t-BAs brings hydrogen to the GaAs surface as previously conjectured from growth studies [14]. The IR spectra of TMG on GaAs indicate that methyl radicals lose hydrogen to form As-H and bridging/terminal CH_2 species on the surface. Subsequent dehydrogenation of these species is a potential route to carbon incorporation which is consistent with the experimentally observed selective incorporation of carbon onto the As-sublattice during OMVPE and ALE growth of GaAs. The observation that CH_2 formation is more predominant on hydrogen free arsenic rich surface than on arsenic rich hydrogen terminated surfaces is a possible explanation for the relatively larger amounts of carbon observed in ALE as compared to conventional OMVPE. The presence of large quantities of surface As-H species during OMVPE because of the large V/III ratio would prevent the formation of CH_2 species and block this route to carbon incorporation. In contrast the arsenic rich surface in ALE is exposed to purge cycles at temperatures significantly above the surface hydrogen desorption temperature (~170°C). Thus, the surface exposed to the TMG cycle would be expected to be hydrogen deficient. The importance of bringing hydrogen to the surface is further evidenced by the observed reduction in the concentration of CH_2 species by the dosing with t-BAs. This removal of CH_2 species from the surface is also in agreement with the low carbon contents of GaAs films grown with t-BAs by conventional OMVPE [4].

References

1. U.Memmert and M.Yu, *Appl. Phys. Lett.* **56** 1883 (1990)
2. V.M.Donnelly and J.A.MacCaulley, *Surf Sci.* **238** 34 (1990)
3. J.R.Creighton,*Surf Sci.* **234** 287 (1990)
4. G. Haacke, S.P. Watkins, and H. Burkhard, *Appl. Phys. Lett.* **54**, 2029 (1989).
5. B.Y.Maa and P.D.Dapkus, *Mat.Res.Soc. Symp.Proc.*, this symposium
6. A.V.Annapragada and K.F.Jensen, *Mat.Res.Soc. Symp.Proc.* **204** 53 (1991).
7. E. Maslowsky Jr., Vibrational spectra of organometallic compounds, Wiley Interscience (1977)
8. L.J. Bellamy, The Infrared Spectraof Complex Molecules, Chapman and Hall (1980)
9. F.Lee, A.L.Backman, R.Lin, T.R.Gow, and R.I.Masel, *Surf. Sci.* **216** 173 (1989).
10. J.R.Creighton, *Mat.Res.Soc. Symp. Proc.* **131** 129 (1989).
11. L.P Sadwicke, K.L. Wang, D.L. Joseph, and R.F. Hicks, *J.Vac. Sci. Tech.* **B7** 273 (1989).
12. W. Mokwa, D. Kohl, and G. Heiland, *Phys. Rev.* **B29**(12) 6709 (1984)
13. R.L.Brainard and R.Madix, *J. Am.Chem.Soc.* **111** 3826 (1989)
14. R.M. Lum, J.K. Klingert, and M.G. Lamont, *J. Crystal Growth* **89**, 137 (1988).

IN-SITU MEASUREMENT OF GaAs OPTICAL CONSTANTS AND SURFACE QUALITY, AS FUNCTIONS OF TEMPERATURE

HUADE YAO AND PAUL G. SNYDER
University of Nebraska, Center for Microelectronic and Optical Materials Research, and Department of Electrical Engineering, Lincoln, NE 68588-0511

ABSTRACT

In-situ spectroscopic ellipsometry (SE) was applied to monitor GaAs (100) surface changes induced at elevated temperatures inside an ultrahigh vacuum (UHV) chamber ($<1\times10^{-9}$ torr base pressure, without As overpressure). The real time data showed clearly the evolution of the native-oxide desorption at ~577 °C, on a molecular-beam-epitaxy (MBE)-grown GaAs (100) surface. In addition, surface degradation was found *before* and *after* the oxide desorption. A clean and smooth surface was obtained from an arsenic-capped, MBE-grown GaAs sample, after the arsenic coating was evaporated at ~350 °C inside the UHV. Pseudodielectric functions $\langle\varepsilon\rangle$ of GaAs, from 1.6 eV to 4.5 eV, were obtained through the SE measurements, from this oxide-free surface, at temperatures ranging from room temperature (RT) to ~610 °C. These $\langle\varepsilon\rangle$ data were used as reference data to develop an algorithm for determining surface temperatures from *in-situ* SE measurements, thus turning the SE instrument into a sensitive optical thermometer.

INTRODUCTION

SE is a sensitive, nondestructive optical technique to determine thin-film thicknesses, multilayer structures, optical constants of bulk materials and surface changes [1,2]. This convenient surface-sensitive technique has been recently applied for *in-situ* characterization [3-5]. We report here the results of our *in-situ* SE measurements of GaAs optical constants and surface changes, at elevated temperatures, inside a UHV chamber.

THEORY

Ellipsometry measures the change in the state of polarization of light reflected from a sample surface. A linearly polarized, collimated light beam is incident on the sample, at a known angle of incidence. After reflection from the sample, the light is in general elliptically polarized. Optical properties, thin-film thicknesses and sample surface changes can be studied by analyzing the changes of the reflected wave polarization, as a function of wavelength and angle of incidence.

Ellipsometric measurement determines the ratio of complex reflectance R_p to R_s, where R_p and R_s are the reflection coefficients of light polarized parallel to (p) or perpendicular to (s) the plane of incidence. The ratio is defined as:

$$\rho = R_p / R_s = \tan(\psi)e^{i\Delta}, \qquad (1)$$

where the values of $\tan(\psi)$ and Δ are the amplitude and phase of the complex ratio. Results of the SE measurements are expressed as $\psi(h\nu_i, \Phi_j)$ and $\Delta(h\nu_i, \Phi_j)$ where $h\nu$ is the photon energy and Φ is the external angle of incidence. The ability of measuring the phase changes $\Delta(h\nu_i, \Phi_j)$ in particular gives the ellipsometer great sensitivity to the surface changes and presence of thin films on the reflecting surface [6].

The pseudodielectric function $<\varepsilon>$ is obtained from the ellipsometrically measured values of ρ, assuming a two-phase model (ambient/substrate) [1]:

$$<\varepsilon> = <\varepsilon_1> + i <\varepsilon_2> = \varepsilon_a \left[\left(\frac{1-\rho}{1+\rho} \right)^2 \sin^2\Phi \tan^2\Phi + \sin^2\Phi \right], \quad (2)$$

regardless of the possible presence of surface overlayers. The ε_a in Eq.(2) represents the ambient dielectric function (i.e., ε_a=1 in vacuum). It has been shown [2,7] that the peak value of the imaginary part $<\varepsilon_2>$, which corresponds to the critical point E_2 position (i.e., ~4.8 eV for GaAs at RT), is readily interpretable: the biggest value relates to the cleanest and smoothest unperturbed surface. Decreasing values of $<\varepsilon_2>$ indicate surface degradation: i.e., the presence of surface oxide or other overlayers, microscopic roughness or mixture of constituents in the near surface region, etc. In addition, the penetration depth of light at the E_2 position (e.g., ~55 Å for GaAs at RT) is nearly minimized so that the surface effects are maximized. Therefore, the ellipsometric probe at or near the E_2 photon energy is a sensitive test of the surface condition of semiconductors.

EXPERIMENTAL

A GaAs (100) sample was clamped on a resistor-heater plate that could be rotated and tilted by a rotary drive, inside the UHV chamber. The optical system, which consisted of a rotating-polarizer ellipsometer attached to the UHV chamber, fitted with a pair of low-strain fused-quartz windows, was described in detail in references 4, 5 and 8. The exact angle of incidence was determined by measuring a known GaAs sample at RT [8]. Temperatures were measured and controlled by two k-type thermocouples, which were calibrated by an infrared (IR) optical pyrometer. The typical base pressure of the UHV was $<1\text{x}10^{-9}$ torr and all the measurements were made without arsenic overpressure.

MEASUREMENTS AND ANALYSIS

Surface Studies

A MBE-grown GaAs (100) virgin surface (unheated and untreated), with native oxide layer typically ~22 Å thick, was installed into the UHV chamber. Surface changes at elevated temperatures were studied, in real time, by taking ellipsometric data (Ψ and Δ) periodically in time (about once every 25-seconds) at a photon energy near the E_2 critical point. These data were converted, at the same time, to a pseudodielectric function $<\varepsilon>=<\varepsilon_1>+i<\varepsilon_2>$. Due to the nature of $<\varepsilon_2>$ at E_2 energy, surface changes induced at elevated temperatures can be monitored in real time, by measuring the changes of $<\varepsilon_2>$.

Fig. 1 shows $<\varepsilon_2>$ as a function of time at two different temperatures. The small change of photon energy from ~500 °C to ~577 °C was made because the E_2 energy decreases with increasing temperature [9]. At ~577 °C, an obvious increase of $<\varepsilon_2>$ (region B) after heating the sample for ~30 minutes indicates the desorption of the oxide layer at that temperature. A flat plateau followed the oxide desorption (region C) indicates an oxide-free GaAs surface. However, a steady declining of $<\varepsilon_2>$ in region A suggests that the GaAs substrate has been degraded before the oxide desorption. One possible explanation is that an interfacial chemical reaction, between the oxide layer and the substrate, roughens and degrades the substrate

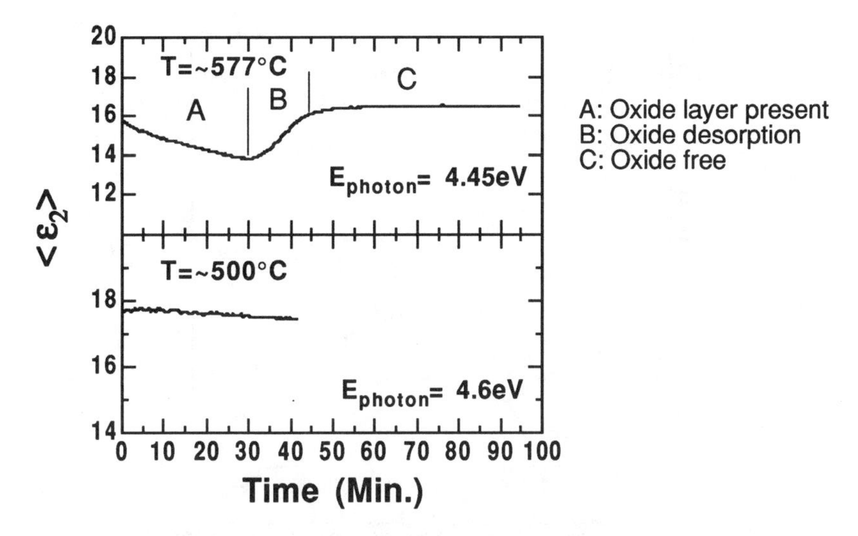

FIG. 1 Real time $<\varepsilon_2>$ data at two different temperatures, near the E_2 critical-point energy.

surface [10]. A recent work by Aspnes *et al* also came to a similar conclusion, through a different approach [3]. Similar evidence (declining $<\varepsilon_2>$) of substrate-surface degradation can be seen at lower temperatures, ~500 °C, as shown in Fig. 1.

SE measurements taken at RT, from a sample which had been heated to desorb the oxide (~577 °C) are shown in Fig. 2. Although the surface is oxide free, these data differ significantly, especially in the phase changes $\Delta(h\nu_i)$, from those for the known GaAs bulk data [11] (shown in squares). This provides further evidence of surface degradation after the oxide desorption.

A very clean and smooth surface was obtained from another MBE-grown GaAs (100) surface, which was capped with an arsenic layer immediately after the material was grown in the MBE chamber. The arsenic cap protected the sample from being oxidized in air during transport to the measurement chamber. Before the SE measurement, the arsenic cap was evaporated in the UHV chamber, by heating the sample surface to ~350 °C. SE data at RT from this surface were comparable with the data from ref. 11, which were obtained from a wet-chemically etched GaAs surface [5,8]. This oxide-free, smooth surface was the necessary component for an accurate measurement of the dielectric functions of bulk GaAs at elevated temperatures. When no overlayers are present on a smooth surface, the bulk dielectric function is given by the pseudodielectric function measured by SE.

Pseudodielectric Functions

Pseudodielectric functions measured from 1.6 eV to 4.5 eV, at the temperatures ranging from RT to 610 °C, are shown selectively in Fig. 3. Actual SE measurements were made at increments of ~50 °C [8]. Starting at 500 °C, RT SE data were also taken each time, after measurements made at the elevated temperature, to check the surface quality. It was found that the sample surface remained smooth and clean until the temperature reached ~577 °C. At this and higher temperatures the surface became slightly roughened. The surface roughness could be modeled as a top non-

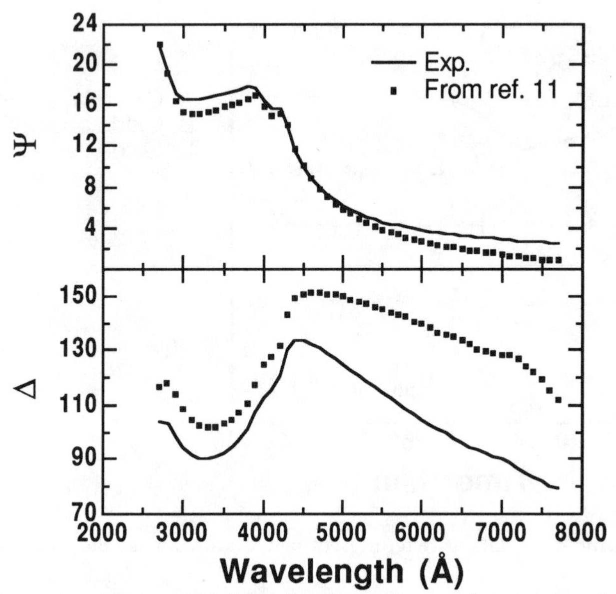

FIG. 2 Room temperature SE data after the native oxide was desorbed at ~577 °C. Squares are the data expected for a very smooth, oxide-free bulk sample [11].

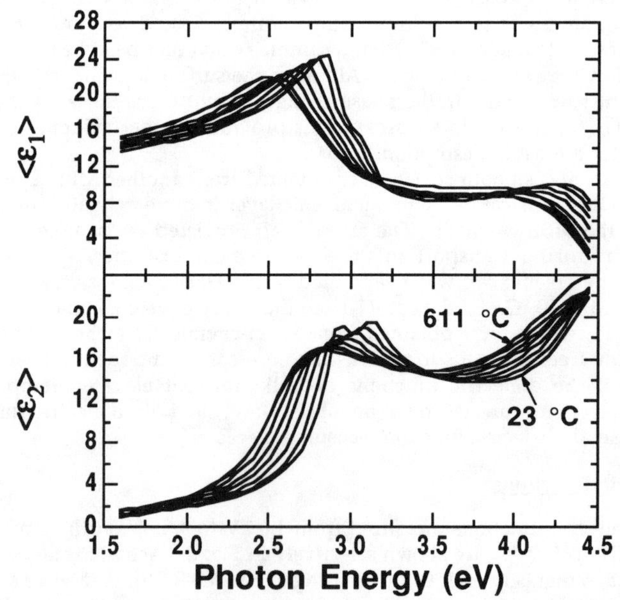

FIG. 3 Representative pseudodielectric functions $<\varepsilon>$ of GaAs at 23, 100, 200, 300, 400, 500, 577, and 611 °C. The 577 and 611 °C curves have been corrected for slight surface roughening effects.

dense GaAs layer, that is, containing voids [8]. The thickness of the non-dense layer increased with temperature. Corrections were made mathematically to remove the nondense layer effects from the pseudodielectric functions at 577 °C and 611 °C shown in Fig. 3.

We believe this surface roughening is caused by the congruent evaporation of Ga and As at high temperatures between ~577 °C and ~657 °C [12]. Independent Auger measurements on the heated sample afterwards, indicated that the stoichiometry of the GaAs sample did not change after being heated to 611 °C [4].

Optical Thermometry

The temperature dependent pseudo-optical constants of GaAs, measured at a series of fixed temperatures ranging from RT to ~610 °C, serve as a set of reference functions that can be used to determine unknown sample surface temperatures. In this application, the SE acts as an optical thermometer.

An algorithm was developed to compute the dielectric function spectrum at an arbitrary temperature [13], by linear interpolation of the pseudorefractive indices at two reference temperatures separated by ~50 °C. Thus, an unknown surface temperature can be determined by fitting the measured data, with temperature as an adjustable parameter. We have tested this algorithm in our UHV chamber, and it has also been tested in a MBE-growth chamber [14]. Shown in Fig. 4, as an example, is a temperature-fit of the SE data from a GaAs (100) surface measured at ~364 °C in our UHV system. The fit temperature (~363 °C) and the calibrated TC reading are quite consistent. Generally, the temperatures obtained by SE as an optical thermometer were within ±10 °C of the conventional (thermocouple) temperature measurements. Sensitivity of the SE temperature measurement is high throughout the RT to ~600 °C range, making it complementary to optical pyrometry which is not effective below ~440 °C [15].

SUMMARY

We have presented the *in-situ* measurements of GaAs optical constants and surface quality, as functions of temperature. The real time data of $<\varepsilon_2>$ showed clearly the evolution of the native-oxide desorption at ~577 °C. Surface degradation was found *before* and *after* the oxide desorption, on a MBE-grown GaAs (100) surface. Pseudodielectric functions $<\varepsilon>$ of GaAs, from 1.6 eV to 4.5 eV, were obtained through the SE measurements, ranging from RT to ~610 °C, from an oxide-free, smooth surface obtained from a pre-arsenic-capped, MBE-grown GaAs. These $<\varepsilon>$ data were used as references to determine the unknown sample surface temperatures from *in-situ* SE measurements. It has been shown that the SE instrument can be utilized as a sensitive optical thermometer, covering a wide temperature range from RT to growth temperatures.

ACKNOWLEDGMENTS

This work was supported by DARPA under the contract No. DAAH 01-89-C, and by NASA-Lewis Grant NAG-3-154. For the MBE-grown GaAs sample we thank T. Bird and K. Stair of the Amoco research center.

REFERENCES

1. R.M.A. Azzam and N.M. Bashara, *Ellipsometry and Polarized Light*, (North-Holland, Amsterdam, 1977).

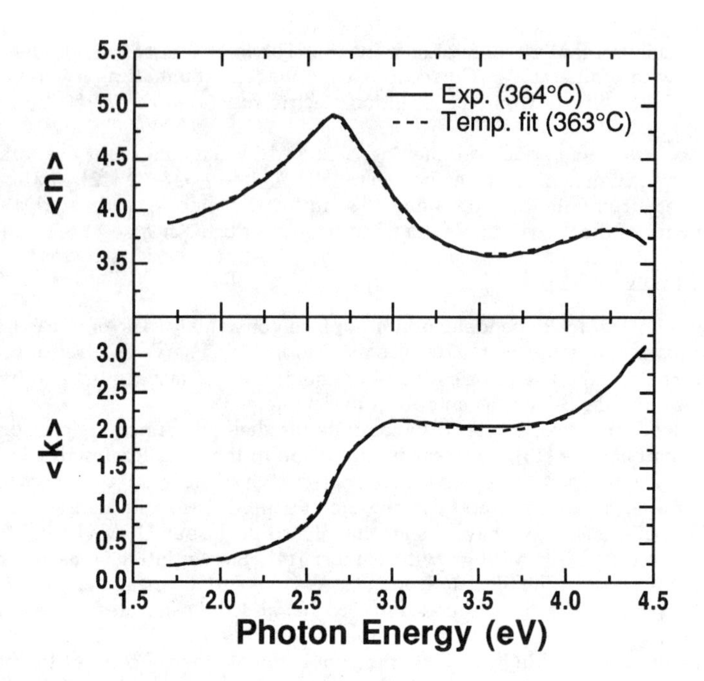

FIG. 4 Temperature-fit to *in-situ* SE data.

2. D.E. Aspnes, in *Handbook of Optical Constants of Solids*, edited by E.D. Palik, (Academic Press, New York, 1985) p89.
3. D.E. Aspnes, W.E. Quinn, and S. Gregory, Appl. Phys. Lett. **56**, 2569 (1990).
4. H.D. Yao, P.G. Snyder and J.A. Woollam, Proceedings of ICEM-**2**, 501 (1991).
5. H.D. Yao, P.G. Snyder and J.A. Woollam, Mat. Res. Soc. Symp. Proc. **202**, 339 (in press, 1991).
6. D.E. Aspnes, in *Properties of Solids-New Developments*, edited by B.O. Saraphin (North-Holland, Amsterdam, 1976) p799.
7. D.E. Aspnes, J. Vac. Sci. Technol. **17**, 1057 (1980).
8. H.D. Yao, P.G. Snyder, and J.A. Woollam, J. Appl. Phys. (in press, 1991).
9. P. Lautenschlager, M. Garriga, S. Logothetidis, and M. Cardona, Phys. Rev. B **35**, 9174 (1987).
10. D.E. Aspens, G.P. Schwartz, G.J. Gualtieri, A.A. Studna, and B. Schwartz, J. Electrochem. Soc. **128**, 590 (1981).
11. D.E. Aspnes and A.A. Studna, Phys. Rev. B **27**, 985 (1983).
12. C.T. Foxon, J.A. Harvey and B.A. Joyce, J. Phys. Chem. Solids **34**,1693 (1973).
13. The algorithm was programmed by B. Johs, of the J.A. Woollam Co.
14. B. Johs, D. Meyer, G. Cooney, H.D. Yao, P.G. Snyder, J.A. Woollam, J. Edwards and G. Maracus, Mat. Res. Soc. Symp. Proc. **216**, 459 (in press, 1991).
15. S.L. Wright , R.F. Marks and A.E. Goldberg, J. Vac. Sci. Technol. B **6**, 842 (1988).

STUDIES OF OXIDE DESORPTION FROM GaAs BY DIFFUSE ELECTRON SCATTERING AND OPTICAL REFLECTIVITY

T. VAN BUUREN, T. TIEDJE*, M.K. WEILMEIER, K.M. COLBOW, AND J.A. MACKENZIE
University of British Columbia, Department of Physics, Vancouver Canada, V6T2A6
*Also: Electrical Engineering Department

ABSTRACT

We have determined that the temperature for desorption of gallium oxide from GaAs increases linearly with oxide thickness, for oxide layers between about 6Å and 26Å thick. Different thicknesses of oxide layers were created by varying the exposure time of the GaAs wafers to a low pressure oxygen plasma. In addition, we show by diffuse light scattering that highly polished GaAs substrates roughen during the oxide desorption. These results are interpreted in terms of a model in which the oxide evaporates inhomogeneously. The oxide desorption was also studied by monitoring the secondary electrons produced by the high energy electrons from the RHEED gun. After the gallium oxide desorption there is a reversible, order of magnitude, increase in the number of secondary electrons produced. We interpret this result as evidence for the formation of microscopic gallium droplets on the GaAs surface.

INTRODUCTION

A sacrificial oxide layer is normally grown on GaAs to remove surface contaminants [1] and to passivate the surface prior to growth of epitaxial layers. Before growth can commence, the wafer must be heat treated to remove the surface oxide. Since the oxide layer is more volatile than the GaAs itself, it can be removed by heating with minor damage to the wafer. The surface oxide comes off in two steps: the arsenic oxide evaporates first (in the 400-500°C range) followed at higher temperatures by the gallium oxide. The gallium oxide comes off over a narrow temperature interval, typically above 580°C [2,3]. Because the gallium oxide peak is relatively sharp, it is sometimes used as a reference point for substrate temperature calibration. This calibration is complicated by the fact that the desorption temperature depends on how the oxide is prepared [3].

In general, one would like to find conditions which produce an oxide which desorbs at the lowest possible temperature in order to minimize evaporation of the GaAs substrate. To understand the origin of variations in the oxide desorption temperature we have measured the gallium oxide desorption temperature for a series of oxides with different thickness.

During the oxide desorption process one would expect the work function and the surface morphology to change. These changes should be detectable in the secondary electron emission caused by the high energy electrons from the RHEED gun. In this paper we report on the first measurements of this process.

EXPERIMENTAL

The substrates used in this experiment are two inch, semi-insulating gallium arsenide wafers cut 2° off (100) toward (110). The oxides for these experiments were grown in a low pressure (200 mTorr) remote plasma discharge in flowing O_2 gas. The plasma is confined away from the substrate to prevent surface damage from the oxygen ions, therefore the substrate is oxidized by oxygen atoms and ozone, which are, dexcitation products of the plasma. Substrates had the initial manufacturers surface oxide removed and then were exposed between 30 seconds and two hours in the low pressure plasma reactor. The resulting oxide was then desorbed in a VG V80H MBE system under an As_4 beam equivalent pressure of 1.5×10^{-5} Torr, during a 10°C/min programmed temperature ramp. The wafers are polished on the front surface and textured on the back to increase radiation coupling between the wafer and the heater foils. The ramp rate was controlled by a feedback thermocouple positioned between the heater foils and the substrate.

In the experiments described here a new optical temperature monitoring device was employed[4]. The wafer temperature was inferred from the optical band gap which was determined from the diffuse reflectivity of a front-side polished, back-side textured substrate. A tungsten lamp is the light source for the temperature measurement apparatus, and provides high intensity illumination of the substrate.

The optical temperature measurement has a sensitivity and reproducibility of about ±1°C between room temperature and 700°C. The absolute accuracy is limited to about ±10°C, the accuracy with which the bandgap of GaAs is known as a function of temperature[5].

A quadrupole mass spectrometer with a direct line of sight to the GaAs wafer was used to detect the mass of the desorbed species during the oxide removal. During the experiment the mass spectrometer scanned repetitively through the 150-160 AMU mass range which includes Ga_2O, the main mass peak which appears during the oxide evaporation.

The oxide desorption was also studied by monitoring the secondary electrons produced by the high energy electrons from the RHEED gun. The secondary electrons are detected by a simple plate detector. The detector was capable of being biased to ±500 volts with respect to the MBE chamber walls. Experiments were conducted with a positive 100 volt bias to enhance secondary electron collection.

RESULTS

The gallium oxide desorption temperature will be defined as the temperature at which the mass spectrometer peak corresponding to Ga_2O at mass 154 reaches maximum amplitude. In Fig. 1, the measured oxide desorption temperature is plotted as a function of exposure time in the oxidation reactor. In the inset to Fig. 1 we see that the thin oxides are completely evaporated by the time the thicker oxides begin to desorb. The width of the gallium oxide desorption peak is smaller than can be accounted for by an exponential increase in the equilibrium vapour pressure associated with a layer by layer desorption process. These observations suggest that the oxide desorption is associated with a phase transition or thin film instability.

We used x-ray photoelectron-emission spectroscopy to determine the oxide thickness on the GaAs substrates. The method used to determine the oxide thickness has been reported elsewhere[6]. In Fig. 2, we show the oxide desorption temperature as a function of the thickness of the oxide. The relationship between the desorption temperature and the oxide thickness is approximately linear. In principle this linear increase in the oxide desorption temperature could be due to a progressive change in the oxide composition associated with increased oxidation time in the plasma reactor. However this model does not readily explain the observed surface roughening effect discussed below.

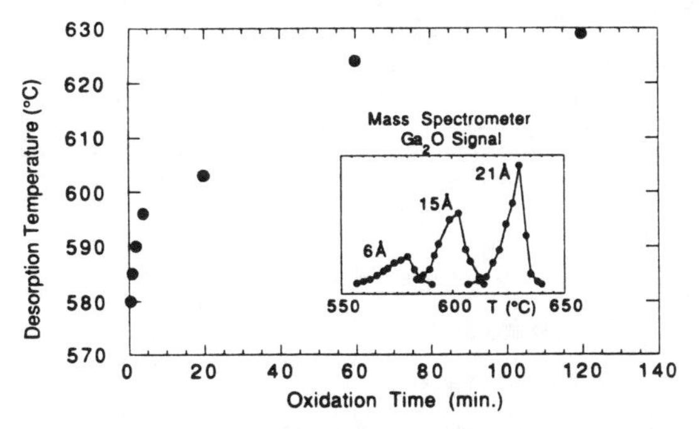

Fig. 1. Gallium oxide desorption temperature as a function of oxidation time. .The inset shows the Ga_2O signal from the mass spectrometer during the desorption for three different oxide thickness during identical 10°C/min temperature ramps.

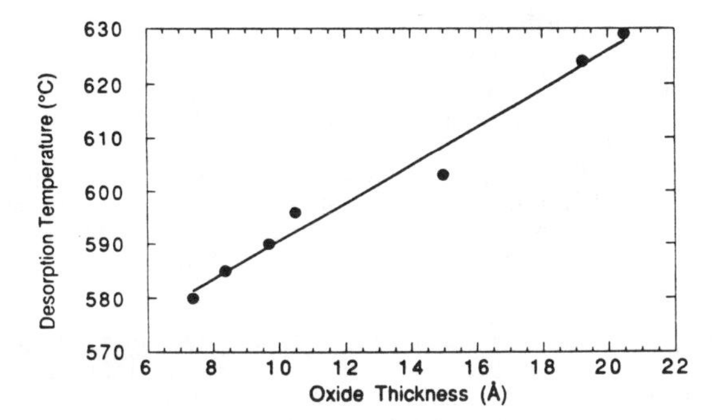

Fig. 2. Gallium oxide desorption as a function of the thickness of the surface oxide.

Under the bright lamp of the optical temperature monitor, it was possible to observe the formation of a faint haze on the polished front surface of the substrate during the gallium oxide desorption. The haze was thought to be caused by surface roughening during the oxide desorption process. The formation of the haze is shown more quantitatively in Fig. 3, where we plot the intensity of the diffuse reflection of a HeNe laser beam from the front surface of the wafer during a temperature ramp. The sharp rise in the diffuse reflectivity coincides exactly with the Ga_2O signal seen in the mass spectrometer. Once the oxide desorption is complete diffuse reflectivity continues to increase with temperature but at a much slower rate. Also we find this process is not reversible and the intensity of the diffuse reflection remains constant as the substrate is cooled.

We interpret the diffuse reflectivity measurements as evidence that the surface roughens on a lateral length scale comparable with the wavelength of light during the oxide desorption. A model that explains all the experimental observations is that the oxide desorption is associated with a thin film instability. In this model we assume the oxide first develops cracks or holes where bare substrate is exposed, and then evaporates preferentially around the perimeter of the holes. During the oxide desorption some areas of the substrate surface will be clean while other areas still have an oxide cap. Roughening will occur because of differential evaporation of GaAs from the clean surface relative to the capped surface, shown schematically in Fig. 3. One would expect the surface to continue to roughen to a lesser degree after the oxide desorption due to conversion of the rough surface to a lower energy faceted structure.

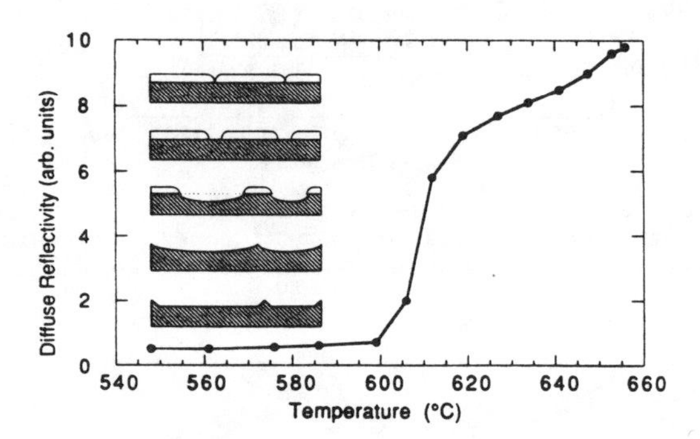

Fig. 3. Diffuse reflectivity of a polished GaAs wafer during a linear temperature ramp in which a 17Å oxide is desorbed. The inset shows schematically our model of the inhomogeneous oxide desorption which explains the origin of the surface texture which produces the observed increase in the diffuse reflectivity.

Since the evaporation rate of Ga and As from GaAs increases exponentially with temperature one would expect increased roughening from thicker oxides because they desorb at higher temperatures. Examination of a series of samples under a phase contrast microscope proves that surface roughness increases with oxide thickness[7].

Preliminary results show the diffuse reflectivity drops during epitaxial growth of GaAs on these substrates[8]. We expect that the growth time needed to smooth the surface will be proportional to the thickness of the initial oxide.

We found the secondary electron current produced by the the RHEED electrons incident on the sample remained constant, until the wafer temperature reached the point where gallium oxide began to evaporate. During and after the gallium oxide desorption, the current increased in a exponential manner as the substrate temperature is ramped even after all trace of the surface oxide was gone. As the sample was cooled, the secondary electron current drops back to its original value.

In Fig. 4 we show the increase in the secondary electron current observed during a 10°C/min temperature ramp of a substrate with a 7Å oxide. At 580°C there is a kink in the secondary electron current, this kink occurs at the same time as the Ga_2O peak is observed in the mass spectrometer. We observe a kink in the secondary electron current during gallium oxide desorption in all oxidized samples, but a kink in the current is not observed during the temperature ramp of oxide free samples. In Fig. 5 we show the reversible nature of the exponential increase in the secondary

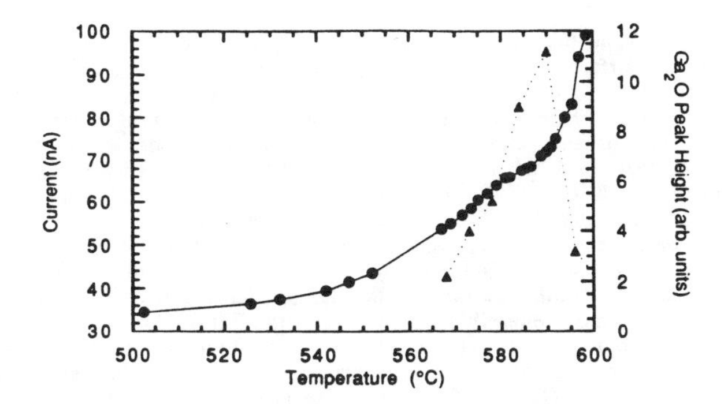

Fig. 4. Mass spectrometer data and secondary electron data measured during desorption of a 7Å oxide

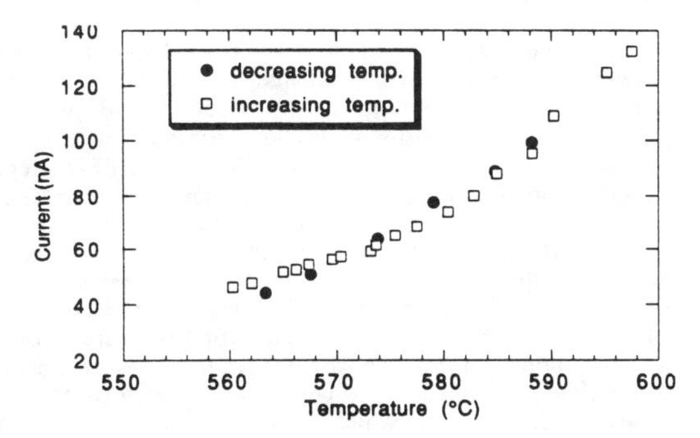

Fig. 5. Secondary electron current measured during a slow stepped ramp of an oxide free sample. Substrate was allowed to equilibrate before the current was measured.

electron current for a sample with no oxide. In this experiment the wafer temperature is stepped at 10°C intervals every 5 min, therefore the current was allowed to reach an equilibrium value before measurement. Note that the current in Fig. 4 is lower after the oxide desorption because with the 10°C/min ramp rate the secondary electron current does not reach its full equilbrium value.

The exponential increase in the secondary electron current is thought to be caused by microscopic roughness on the GaAs surface, that is, roughness on the 10-100Å scale. We conclude that the is roughness is microscopic in size because the reversible effect which caused the electron scattering was not detected in the light scattering experiment.

DISCUSSION

In the above interpretation, the oxide desorption process is triggered by the nucleation of defects in the surface oxide, which causes holes to form in the surface oxide. Additional surface area is created in the formation of holes or cracks in the oxide, hence there must also be an increase in the surface energy of the system. This increase in the surface energy will act as activation barrier to hole formation. If the cross-sectional shape of the minimum energy hole in the oxide is independent of the oxide thickness, other than an increase in size, then the surface energy of the hole will scale linearly with the oxide thickness. Therefore the activation energy for the initiation of the oxide desorption will increase linearly with the oxide thickness. In this case one would expect the desorption temperature to scale linearly with the oxide thickness, because the rate of surmounting the nucleation barrier will be a function of the ratio of the nucleation barrier to the temperature.

We interpret the reversible nature of the secondary electron effect as evidence for microscopic roughening due to the formation of gallium droplets on the substrate surface. The gallium oxide desorption temperature, is very close to the congruent sublimation for GaAs (~640°C but depends on the arsenic flux). Above the congruent sublimation temperature the vapour pressure of As over GaAs exceeds that of Ga over GaAs. The Ga that is now freed from the GaAs lattice due to the loss of arsenic can desorb, recombine with arsenic to form GaAs, or remain on the surface in metallic form. It is known that Ga free on the surface will collect in droplet form. Once the temperature is decreased , the rate of As loss from GaAs drops below the rate of Ga loss, hence the supply of Ga to the droplets is reduced. The remaining Ga droplets are quickly consumed by evaporation or by reaction with the incoming As_4 flux. The possibility of microscopic gallium droplets present on the surface, under an As_4 flux conditions has been proposed previously[9]. A RHEED electron incident on a Ga droplet will produce more secondaries than the same electron incident on the GaAs substrate surface for the following reasons. The Ga droplets increase the effective surface area available for electron emission and have a lower work function (3.9eV) compared with bulk GaAs (4.4eV). Other material dependent differences in the secondary electron emission might also be expected.

In conclusion, a linear increase in oxide desorption temperature with oxide layer thickness and an associated surface roughening effect are interpreted in terms of an inhomogeneous oxide desorption process. These results highlight the desirability of a thin surface oxide layer which will evaporate at lower temperatures. A reversible order of magnitude increase in the secondary electron emission from GaAs at temperatures above 570°C is explained in terms of the formation of gallium droplets on the substrate surface.

REFERENCES

1. J.A. McClintock, R.A. Wilson and N.E. Byer; J. Vac. Sci. Tech. **20**, 241 (1982).
2. S.J. Ingrey, W.M. Lau and N.S. McIntyre; J. Vac. Sci. Tech. **A4**, 984 (1986).
3. A.J. SpringThorpe, S.J. Ingrey, B. Emmerstorfer, P. Mandeville and W.T. Moore; Appl. Phys. Lett. **50**, 77 (1987).
4. M.K. Weilmeier, J.M. Colbow, T. Tiedje, T. Van Buuren and Li Xu; Can. J. Phys. (to be Published 1991).
5. C.D. Thurmond; J. Electrochem. Soc. **122**, 1133 (1975)
6. T. Van Buuren, T. Tiedje, et. al., Appl. Phys. Lett. (to be Published 1991)
7. L. Isernia; Johnson-Matthey Electronics (Private Communication)
8. C. Lavoie, (Private Communication)
9. E.M. Gibson, C.T. Foxon, J. Zhang and B.A. Joyce; Appl. Phys. Lett. **57**, 1203 (1990).

PART III

III-V Semiconductor Studies

RECENT PROGRESS IN ATOMIC LAYER EPITAXY OF III-V COMPOUNDS

S. M. Bedair
North Carolina State University, Dept. of Electrical and Computer Engineering, Raleigh, North Carolina 27695-7911

ABSTRACT

The potential applications of Atomic Layer Epitaxy of III-V compounds will be outlined. These include the growth of special structures and devices such as ordered alloys, ultra-thin quantum wells, non-alloyed contacts, planar doped FET's and HBT's. Also, the main challenges facing ALE will be outlined along with possible solutions. These include reactor design, control of carbon doping and the growth of ternary alloys. A general assessment of the ALE technology will be provided.

CURRENT CHALLENGES FACING THE ALE TECHNIQUE

The ALE technique has suffered from several shortcomings that we believe have slowed down its potential applications and the interest of many researchers. The first problem is the very low growth rates where in some cases growth rates as slow as 0.02 μm/hour were reported. Recently improvement in the growth rate has been achieved and a growth rate of about 0.2 μm/h was reported, which we still believe to be discouragingly slow[1]. The main reason for such a low growth rate is the commonly used approach based on exposure/purging each of the reactants with a vent/run manifold configuration. The finite gas residence time in the reactor and valve switching times will always lead to growth of only a small fraction of a micron per hour.

The second problem facing ALE is gas phase reactions which limit ALE growth to low temperatures. The premature decomposition of trimethylgallium (TMG), for example, in the gas phase results in the growth of more than one monolayer per ALE cycle. The process in this case is not controlled by surface reactions.

Another problem facing ALE is the synthesis of ternary alloys such as AlGaAs and InGaAs. This problem is due to the narrow temperature range of ALE in most systems and thus the lack of compatible group III precursors that will provide a self-limiting process at the same temperature.

Some of the above problems have been eliminated by optimization of ALE reactor design. The approach adopted in our laboratory relies on rotating the substrate between the different source gas streams that are continuously flowing through a specially designed vertical reactor. The growth rate will depend on the substrate rotation speed. Growth rates in the range of 0.4 to 0.7 μm/h can be achieved with this approach. Such growth rates are comparable with those reported by MBE. For high temperature growth and to achieve compatible growth of ternary alloys the thickness of the thermal boundary layer must be minimized. The approach taken in our laboratory is to mechanically shear off the gaseous boundary layer when the substrate rotates between the reactive gas streams. One ALE reactor used is a modified Emcore 3200 system operating at 30 torr. The reactor chamber is partitioned into six compartments to further separate the reactive gases and to assist in shearing off the boundary layer. The chamber is divided by 0.01" molybdenum sheets (baffles). The detailed design is discussed elsewhere[2].

ALE growth of GaAs using TMG and AsH_3, with and without the baffles was studied. We have observed that monolayer per cycle growth is achieved only when the baffles are

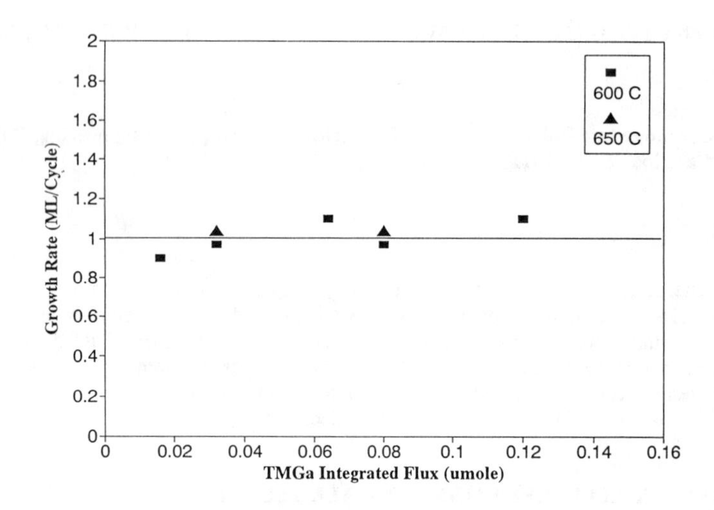

Figure 1 Growth of GaAs as a Function of TMG Integrated Flux.

used. Also, thickness uniformity across the wafer is only observed under the ALE growth conditions. The range of self-limiting growth is limited, due to incomplete removal of the boundary layer. Another system used in our laboratory operates at atmospheric pressure and uses a graphite susceptor. This system allows better removal of the boundary layer and the self-limiting process was observed over a broader range of growth conditions. Figure 1 shows that the self-limiting conditions are met for growth at 600 and 650°C over a wide range of TMG integrated flux.

The two systems were used to achieve device quality GaAs, AlGaAs, InGaP and other films. Also several devices were built based on these ALE materials such as δ-doped FETs, heterojunction bipolar transistors, resonant tunneling diodes and others. In the following discussions we will outline some of these results.

ALE GROWTH OF GaInP

Details of the growth process of $Ga_{0.5}In_{0.5}P$ grown on GaAs substrates were previously reported[3]. The growth relies on the sequential exposure to TMG, PH_3, TEI and PH_3 to grow Ga-P-In-P, ... layers. Double crystal x-ray diffraction was performed and confirms that the ALE grown $In_xGa_{1-x}P$ films are lattice matched to GaAs with x = 0.5 for films grown on either (100) or (100) 2° off oriented substrates. Cross-sectional TEM was also used to study the crystal structure of the ALE GaInP films. Ordering was not observed for films grown on the (100) nominal substrates. However, ordering was observed for all samples grown on the 2° misoriented substrates. The ordering is found to be CuPt, where the Ga and In atoms alternate on (111) planes on the column III sublattice. This can be explained based on the atomic arrangement which favours that phosphorus atoms be attached to three In atoms (or Ga and one Ga (or In) rather than two In and two Ga atoms. Thus, a highly strained InP-GaP monolayer superlattice is formed as a result of the ordered arrangement of this structure. Photoluminescence (PL) at liquid helium temperature was also performed on the ordered and

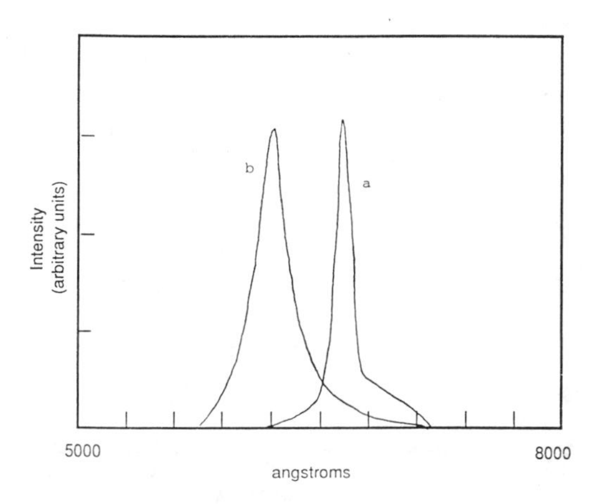

Figure 2 **4K PL Spectrum of GaInP for; a) Ordered Structure, b) Disordered Structure.**

disordered structures and results are shown in Figure 2. The PL shows emission peaks at 1.85 and 2.01 eV for the ordered and disordered structures respectively as shown in the figure. This difference in the value of E_g is confirmed by photoreflectance measurements[4], where ordered structures showed a record low bandgap of 1.76 eV at room temperature. This value of the bandgap is the lowest reported for this ternary and may be a result of achieving such a high degree of ordering by the ALE technique.

To assess the quality of the GaInP grown by ALE, a series of GaAs quantum wells (QWs) were grown using approximately 350 Å of GaInP on either side for the high bandgap barrier. Since the ordered material forms domains and has a suppressed bandgap, the disordered alloy was used for the barriers. Growth temperature was 480°C to minimize intermixing at the GaAs/GaInP heterointerfaces. Phosphorous based compounds tend to have a surface exchange reaction when exposed solely to arsine at typical growth temperatures. Therefore, to prevent this intermixing the sample was hidden from the column V flux for 15 s while the two gases were switched. This was accomplished in our system by rotating the sample the sample underneath the fixed part of the susceptor, effectively hiding it from exposure. No detrimental effect to the surface morphology was detected upon examination of QWs grown in this manner.

PL at 25 K was performed on the samples to detect QW emission. Three QWs with 5.65, 20 and 31 Å were grown, based on thicknesses produced by ALE of 2, 7 and 11 GaAs monolayers, respectively. Figure 3 shows the 4K PL of the 5.65 Å quantum well. The full width at half maximum (FWHM) is around 30 meV. The FWHMs measured here are comparable to those of gas source molecular beam epitaxial material[5].

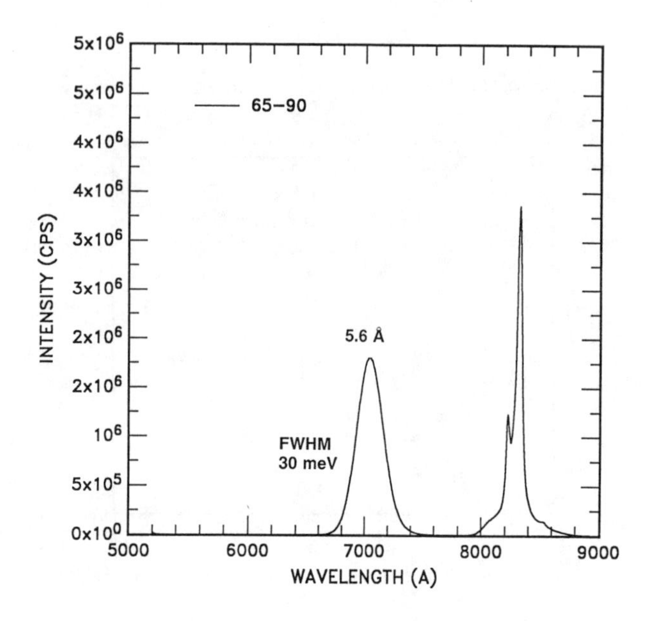

Figure 3 **4K PL Spectrum of 5.65 Å GaAs Quantum Well with 350 A GaInP Barriers.**

ALE GROWTH OF PLANAR DOPED FET

Atomic Layer Epitaxy (ALE) was used for growth of the δ-doped device structures. ALE offers an attractive approach for the synthesis of δ-doped structures, since it is a low-temperature growth process which minimizes dopant diffusion. ALE also allows an accurate control of epilayer thickness between the dopant plane and the gate over a large area wafer, resulting in uniform device performance. TMG and AsH_3 have been used for the growth of GaAs. Hydrogen selenide (H_2Se) was used as the n-type dopant source. A cross-section of the device is shown in Figure 4. The growth starts with an undoped GaAs buffer layer grown at 600°C. The GaAs buffer layer is necessary for good device pinch-off and low output conductance. The ALE grown GaAs buffer is n-type with a background carrier concentration in the low $10^{15}/cm^3$ to the high $10^{14}/cm^3$, and a room temperature mobility of 4200 $cm^2/V.sec$. Following this layer, the substrate temperature was lowered to 480°C to optimize the incorporation of selenium into GaAs. At temperatures much above or below 480°C the fraction of selenium atoms which are electrically active tends to drop resulting in low free carrier concentrations. The H_2Se (30ppm) was introduced during the arsenic exposure cycle of the growth with minimum arsine flow. The wafer was kept under H_2Se for one minute with 20 sccm of H_2Se flowing. This procedure resulted in a doping profile with a peak carrier concentration of about 6 x $10^{18}/cm^3$ and a full width at half maximum (FWHM) of about 40 Å as obtained by C-V measurement. The undoped top GaAs layer was grown at 600°C to achieve a better quality undoped film and thus improve the breakdown voltage characteristics. The final layer was the contacting layer which resulted in uniform and reproducible nonalloyed low resistance ohmic contacts. This contacting layer consists of 10 sheets of Se planar doping separated from each other by 50 A of undoped GaAs. The flow of H_2Se was set to its

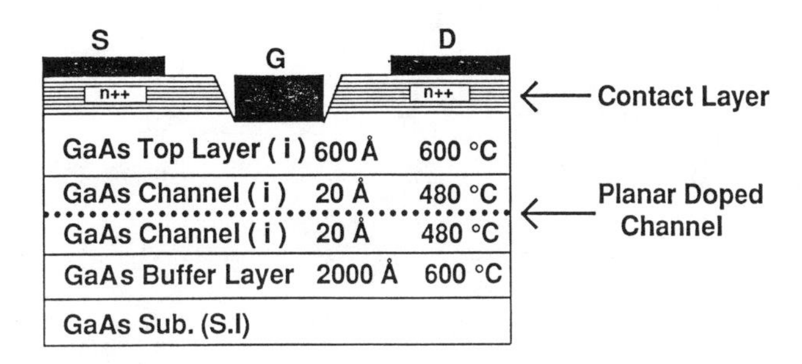

Figure 4 **Cross Section of GaAs δ-doped MESFET.**

maximum of 170 sccm and the substrate was held under H_2Se for 30 sec at 480°C after being exposed to TMG. The resulting peak carrier concentration in each plane was 2 x 10 ^{19}/cm^3 as obtained by capacitance-voltage profiling. The lowest specific contact resistance achieved by this structure was 7 x 10^{-7} $\Omega \cdot cm^2$ obtained from a transmission line measurement (TLM). The device fabrication used a standard FET process starting with a mesa etch, followed by source and drain metallization consisting of AuGe (800 Å)/Ni(200 Å)/Au (2000 Å), and finally a gate recess and gate metallization of Ti(500 Å)/Au(1500 Å). The gate length and gate width of the device were 1.5 μm and 75 μm, respectively.

The I-V characteristics of this device are shown in Figure 5. A maximum transconductance of 144 mS/mm was obtained at a current density of 460 mA/mm. The extremely high gate-drain breakdown voltage exceeded 25V. This is one of the highest values reported for a gate to drain spacing of 1.6 μm in spite of the heavily doped n^+ region between gate and drain[6]. The I-V characteristics show excellent pinch-off with a very small output conductance which is the result of good quality GaAs buffer layers. The transconductance showed a broad maximum range as a function of gate-to-source voltage, typical of delta-doped FET structures.

pnp HETEROJUNCTION BIPOLAR TRANSISTOR WITH CARBON-DOPED COLLECTOR AND EMITTER

Carbon acts as a p-type dopant that exhibits very low diffusivity and a very high solubility limit in GaAs. These qualities make carbon doping attractive for the base of npn HBT's as well as emitter and collector of pnp HBT's. ALE has been used for the growth of AlGaAs/GaAs pnp HBT's, with carbon as the p-type dopant. Carbon doping was controlled by adjusting the growth conditions such as growth temperature, group III and group V fluxes

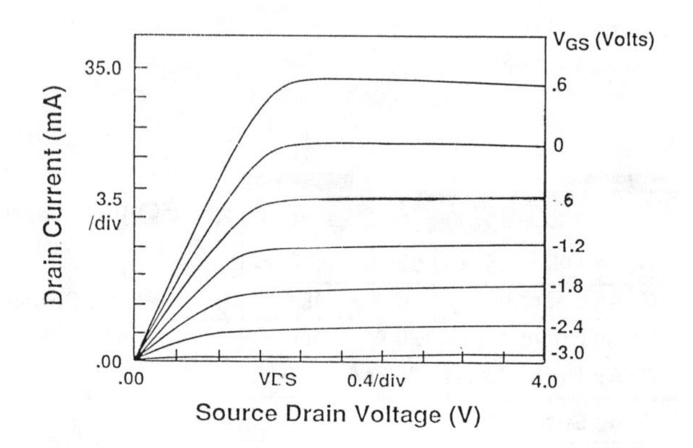

Figure 5 Drain Current Characteristics of a 1.5 x 75 μm^2 GaAs δ-doped MESFET.

and exposure times. P-type carrier concentrations in the range 10^{15}-$10^{20}/cm^3$ were achieved for the ALE grown GaAs. However for AlGaAs, due the strong Al-C bond, the lowest background p-type doping was in the high $10^{16}/cm^3$ range.

Table I shows the HBT structure with the corresponding carrier concentrations and thicknesses. SiH_4 was used as the n-type dopant for GaAs. Wet chemical etching was used to form the base and the emitter mesas, as well as to provide for device isolation. Ti/Pt/Au was used to contact the emitter and collector. AuGe/Ni/Au was used for the base contact. A common emitter current gain over 100 was obtained with excellent I-V characteristics. Early voltage is a relatively high 30V. Details of the HBT results will be published elsewhere[7].

LAYER	Doping (cm^{-3})	Thickness (μm)
p^+-GaAs	2 x 10^{18}	0.15
p-$Al_{.3}Ga_{.7}As$	1 x 10^{18}	0.075
n^+-GaAs	8 x 10^{17}	0.12
p-GaAs	5 x 10^{16}	0.4
p^+-GaAs	1 x 10^{18}	0.5
GaAs Buffer	Undoped	0.1
S.I. SUBSTRATE	Undoped	500

Table 1 pnp AlGaAs/GaAs HBT Structure Grown by ALE.

CONCLUSION

The ALE growth of III-V compounds has suffered from several shortcomings such as a limited range of growth temperatures, high carbon background and difficulties in the synthesis of ternary alloys. We have demonstrated that these problems can be minimized by special reactor and susceptor designs that allow the substrate to rotate between streams of the precursor gases. Such an approach was used to grow ternary alloys such as AlGaAs, InGaP and InGaAs, and to deposit ordered ternary alloys. ALE was also used for the deposition of highly doped structures for nonalloyed contacts, planar-doped FET's, and pnp HBT's. We believe that ALE is now capable of providing material quality comparable with other techniques, while offering control of the deposition process on the monolayer level.

ACKNOWLEDGMENTS

I would like to thank, J.R. Hauser, N. El-Masry, B. McDermott, Kim Reid, A. Dip, P. Colter, S. Huessin, M. Hashemi, T. Henderson and B. Bayraktaroglu for achieving these new ALE results. This work is supported by ONR/SDIO, SERI and NSF.

REFERENCES

1. K. Mochizuki, M. Ozeki, K. Kodama and N. Ohtsuka, J. Crystal Growth, 93, 557-561 (1988).

2. P. C. Colter, S. A. Hussien, A. Dip, W. M. Duncan and S. M. Bedair, Appl. Phys. Lett. (accepted).

3. B. T. McDermott, N. A. El-Masry, B. L. Jiang, F. Hyuga, W. M. Duncan and S. M. Bedair, J. Crystal Growth, 107, 96-101 (1991).

4. B. T. McDermott, K. G. Reid, N. A. El-Masry, S. M. Bedair, W. M. Duncan, X. Yin and Fred Pollak, Appl. Phys. Lett. 56, 1172-1174 (1990).

5. M. J. Hafich, J. H. Quingley, R. E. Owens, G. Y. Robinson, D. Li and N. Otsuka, Appl. Phys. Letter 54, 2686-2688 (1989).

6. M. Hashemi, B. McDermott, U. R. Mishra, J. Ramdani, A. Morris, J. R. Hauser and S. M. Bedair, to be published in Elect. Dev. Lett.

7. T. Henderson, B. Bayraktavoglu, S. Hussien, A. Dip, P. Colter and S. M. Bedair, Electronic Lett. 27, 692 (1991).

LASER ASSISTED ATOMIC LAYER EPITAXY -A VEHICLE TO OPTOELECTRONIC INTEGRATION

Q.CHEN, J.S.OSINSKI, C.A.BEYLER, M.CAO*,AND P.D.DAPKUS, J.J.ALWAN** AND J.J.COLEMAN**
Departments of Materials Sciences and Electrical Engineering and Center for Photonic Technology, University of Southern California, Los Angeles, CA 90089.
* Laser Institute, Huazhong University of Science and Technology, Wuhan 430074, Hubei, P.R.China.
**Department of Electrical Engineering, University of Illinois, 1406 W. Green Street, Urbana, IL 61801.

ABSTRACT

Two implementations of laser assisted atomic layer epitaxy(LALE) for selective area growth of GaAs using trimethylgallium and AsH_3 as precursors are described. A wide range of growth parameters lead to self-limiting monolayer/cycle growth which is suited for precise layer thickness control. By combining LALE with conventional metalorganic chemical vapor deposition, $Al_{0.3}Ga_{0.7}As$/GaAs double heterostructures including LALE GaAs have been grown, permitting electrical and optical characterization to be performed on the thin and small areas of the LALE deposits. The information is used in a growth parameter optimization process resulting in device quality GaAs. Quantum well lasers with active region grown by LALE are demonstrated for the first time. The application of LALE to optoelectronic integration is demonstrated by depositing small area quantum wells as the gain medium in an otherwise transparent waveguide.

INTRODUCTION

The ability to integrate lasers, modulators, waveguides and electronic devices on a single semiconductor wafer is a valuable asset for optical communication and optical computing devices. However, these optical, optoelectronic, and electronic devices have vastly different vertical layer structures, which complicates both crystal growth and device processing. To date, limited success in monolithic integration of AlGaAs/GaAs lasers, modulators and waveguides etc. has been achieved through repeated etch and regrowth[1]. A localized area deposition technique with good selectivity, accurate layer thickness control, and device material quality is highly desirable. Laser assisted atomic layer epitaxy (LALE) has been shown to offer selective area deposition of GaAs [2]-[6] with layer thickness control down to the atomic scale due to the self-limiting nature of the process. Because it is a low temperature technique, the selectivity in localized deposition exceeds that of laser assisted metalorganic chemical vapor deposition (MOCVD) and shows none of the abnormal edge growth seen in the selective area growth in the openings of a dielectric mask. Other advantages of LALE are the smooth edges of deposit for low scattering waveguides and, when combined with conventional MOCVD, it is virtually an *in situ* process compatible with high throughput.

GaAs material grown by LALE has shown reasonable photoluminescence (PL) response at low temperatures[4][7]. LALE GaAs has also been characterized by Raman spectroscopy[8], indicating uniform and bulk-like crystalline material. More recently, we have formed double heterostructures(DHs) and quantum wells(QWs) by sandwiching LALE grown GaAs between conventionally grown $Al_xGa_{1-x}As$. This hybridization approach permits the capacitance-voltage (CV) and PL measurements performed on the thin and small area LALE deposits. A growth parameter optimization leads to the LALE of GaAs of device quality as evidenced by QW laser diodes with active region grown by LALE[9]. It is clear however that a substantial effort is required to improve the purity of the material and the design of the system to accommodate selective area growth of arbitrary patterns at fine resolution.

In this paper, our previous work using the rotating susceptor reactor will be reviewed first. In addition, LALE with a H_2-flushed window reactor, whereby the input gas stream is switched, is described. Direct patterning is made possible using this reactor. We have also employed a novel optics setup that allows the utilization of almost all the optical power for deposition. Finally, the application of LALE to optoelectronic integration is further exploited by forming QW laser structures with their active regions grown by LALE, leaving portions of the cavity as a transparent waveguide. The result is promising.

ATOMIC LAYER EPITAXY(ALE) AND LASER ASSISTED ALE

ALE of GaAs using trimethylgallium(TMGa) and AsH_3 was first reported by Nishizawa *et al*[10] and followed by others with number of variations[5][6][11][12]. The basic idea in ALE is to expose the substrate alternately to the group III and group V precursors to avoid the direct reactions between them in the gas phase or at the surface. This is shown schematically in Fig.1 for ALE at atmospheric pressure. After exposure to TMGa, the substrate is covered by an 'atomic' layer of Ga containing species (suitable for 1/2 monolayer of GaAs). A H_2 purge step follows to ensure no intermixing between the TMGa and AsH_3. Upon exposure to AsH_3 a complete monolayer of GaAs is formed (2.827 Å on {100} substrate). Another step of H_2 purge prepares the substrate for the next ALE cycle. A complete cycle typically takes 4-10 seconds. It is found that the growth rate is independent of AsH_3 exposure above a minimal required quantity and is insensitive to TMGa exposure within a range of molecular flux (about an order of magnitude) which is the now-well-known self-limiting growth behavior. At a given set of flow rates and exposure times, monolayer self-limiting growth occurs over a small range of temperatures for thermally driven ALE[2][6][13]. On the other hand, monolayer self-limiting growth was reported at various temperatures from 400 °C to 500 °C by adopting suitable flow rates and exposure times[14]-[16]. With a special design of the reactor that largely eliminates the gas phase reactions, monolayer self-limiting growth is clearly observed at temperature as high as 560 °C[12]. At lower temperatures where thermally driven ALE can not proceed appreciably within

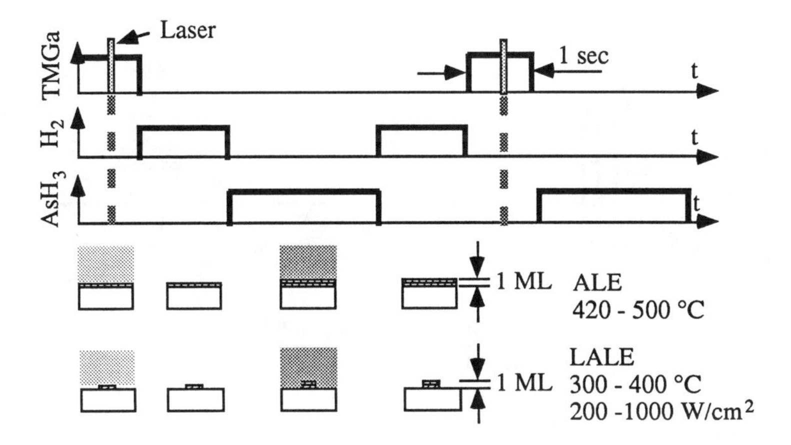

Fig.1 Schematics of ALE and LALE

a reasonable amount of time, 300 - 400°C, ALE is locally promoted by the irradiation with a laser beam. The 514.5 nm line of an argon ion laser is used most frequently and successfully for monolayer self-limiting LALE using both TMGa and triethylgallium (TEGa). The use of the 1.06 μm radiation from a YAG laser resulted in no self-limiting growth for both TMGa[7] and TEGa[14] cases. The 248 nm line of a KrF excimer laser was also used successfully for the TMGa case[17]. Since the gas phase TMGa molecules have an absorption edge extending to ~210 nm-220 nm[18][19], shorter wavelength are not recommended for LALE carried out at atmospheric pressure.

EXPERIMENTAL APPROACHES

In our experiments, maintaining a clean optical path during the alternative exposures to TMGa and AsH_3 is achieved through two different reactor designs: a rotating susceptor design and a switched flow design as shown in Fig.2. In the rotating susceptor design, the substrate is rotated on a pyramidal graphite susceptor. TMGa and AsH_3 are passed continuously into their respective chambers separated by two H_2 purge chambers. A laser beam with λ=514.5nm from an argon ion laser is directed through the quartz window on the TMGa chamber onto the substrate while it is exposed to TMGa. A complete LALE cycle occurs every revolution, writing single line where the path of the beam traverses the substrate. Notice here the average residence time of the laser spot is D/v, where the D is the beam diameter and v the linear scan speed, which is about 0.1 second in most of our experiments.

As an alternative approach, the LALE in the switching-flow reactor proceeds as the TMGa and AsH_3 is switched into the manifold with a continuous

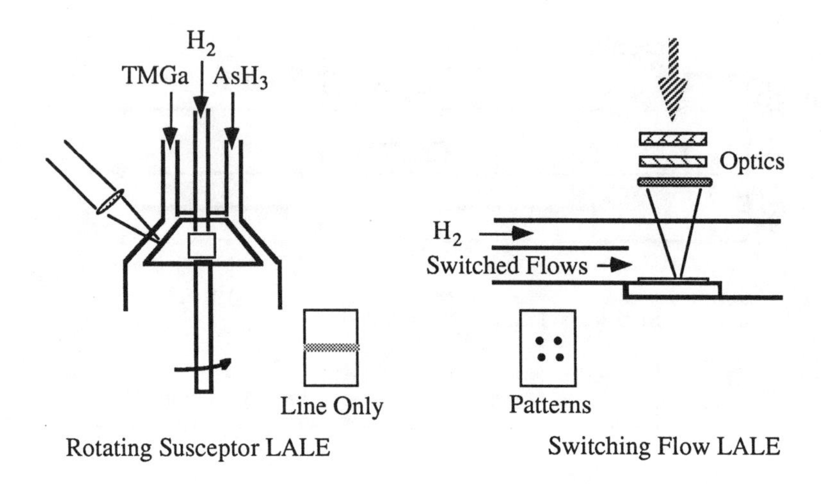

Fig.2 LALE Experiment Setup

H_2 carrier gas flow and a laser beam is directed through a quartz window flushed with another H_2 flow to keep the window from fogging. With this approach, patterns of arbitrary shapes can be deposited. Optics have been employed to form a desirable pattern, such as a shadow mask-imaging lens combination or a planar beam splitter-focusing lens combination. In the latter case, about 94% of the laser power is utilized. Besides, the ability to change the delay time between the time when the laser is on with respect to the time when TMGa is switched into the reactor allows us to obtain information related to the mechanisms of LALE. An important feature that comes with both reactor designs is their compatibility with conventional MOCVD. This allows us to grow DHs and QWs with the central layer grown by LALE, and yet perform CV and PL measurements on the thin and small area deposits.

For growth rate measurements, stripes about 1130 Å thick are deposited and thickness is measured with a stylus profilometer. For material characterization, DHs and QWs are grown by a three step process: first the lower barrier of $Al_{0.3}Ga_{0.7}As$ is grown at 750 °C by conventional MOCVD and then the temperature is lowered to 380 °C at which LALE is carried out. The top barrier is again grown at 750 °C by conventional MOCVD to complete the DH. For CV measurements, Ag dots are evaporated to form a Schottky barrier.

RESULTS AND DISCUSSIONS

1. LALE growth behavior

Shown in Fig.3 is the dependence of LALE growth rate upon laser power for growth on freshly grown $Al_{0.3}Ga_{0.7}As$ at a linear scan speed of 3.18 mm/sec and substrate temperature of 380 °C. An appreciable LALE power window is observed. This power dependence translates into a relatively flat-top deposition profile irrespective of the Gaussian distribution of the laser beam intensity distribution.

In conventional MOCVD, the growth rate of GaAs is in direct proportion to the input TMGa concentration. In LALE, however, the growth rates are insensitive to increasing TMGa input over more than an order of magnitude in flow rates as is shown in Fig.4. In this data, the lower end is practically limited by the minimum flow of the mass flow controller, so it is expected that the flow rate to achieve monolayer growth can still be reduced. At low flow rates the width of the flat portion of the stripe profiles decreases with decreasing TMGa flow.

In the early high temperature LALE experiment of Doi *et al*[20] (500 °C),

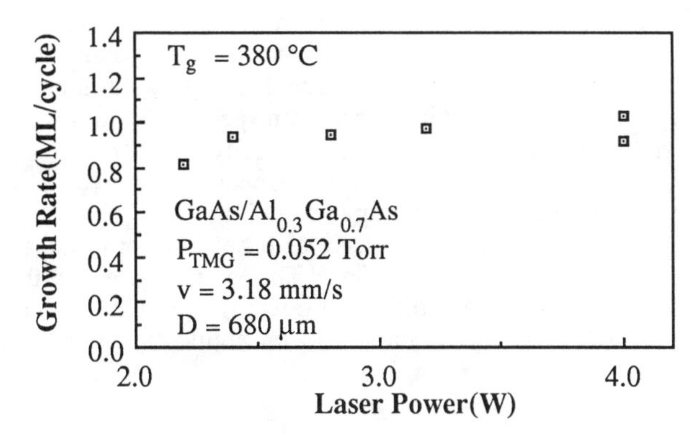

Fig.3 Dependence of growth rate on laser power.

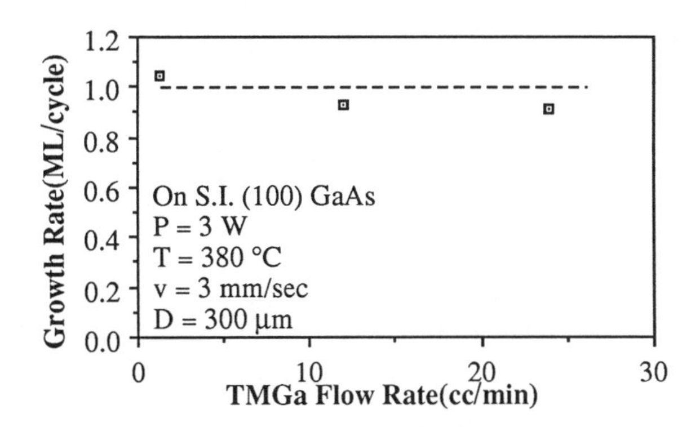

Fig.4 LALE growth rate as a function of TMGa flow rate

it was found that no enhancement of the growth was observable when the laser beam was synchronized with the AsH_3 injection. We have performed an experiment whereby the timing of a the 0.1sec pulse of the laser illumination is delayed with respect to the time when TMGa is switched into the switched flow reactor. The result is shown in Fig.5 where the growth rate is plotted as a function of this delay time. From 0~0.4 seconds, there is a delay related to the time constant of the gas handling system and the reactor for the TMGa to reach the substrate. The growth rate is one monolayer/cycle whenever TMGa is present on the substrate. There is then a fast fall-off after TMGa is turned off. The 1/e fall-off time is about 0.6 seconds. Although the details of the surface chemistry are not accurately known, we can infer from this data that any Ga containing species will desorb within a short time after TMGa is cut off from the main stream, perhaps less that 1 sec., unless they are chemisorbed with the assistance of the laser energy input.

LALE stripes of various widths have been written in the rotating susceptor reactor on bare semi-insulating (100) GaAs and freshly grown $Al_{0.3}Ga_{0.7}As$ starting layers at scan speeds up to 10 mm/sec. Fig.6 shows a microphotograph of a GaAs stripe written on $Al_{0.3}Ga_{0.7}As$ at scan speed of 6.3 mm/sec. The flat-top characteristic of the self limiting growth is readily seen. The top of the stripe is smooth although stripes written at higher speed developed cross-hatched patterns possibly due to the rapid heating and cooling during the process. Using our switching flow LALE reactor, we have also been able to image a shadow mask onto the substrate as is shown in Fig.7. The finer features are the result of diffraction and interference since the small openings (~1mm) of the mask are illuminated with a coherent light source. In an application that requires the

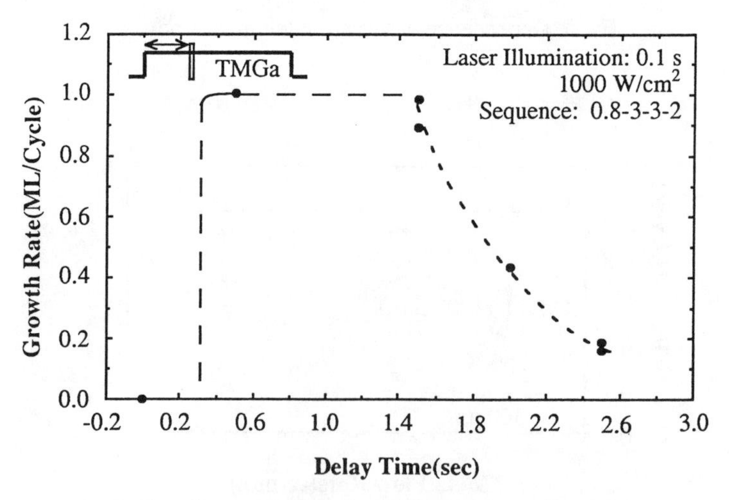

Fig.5 The effect of delay in laser illumination on LALE growth

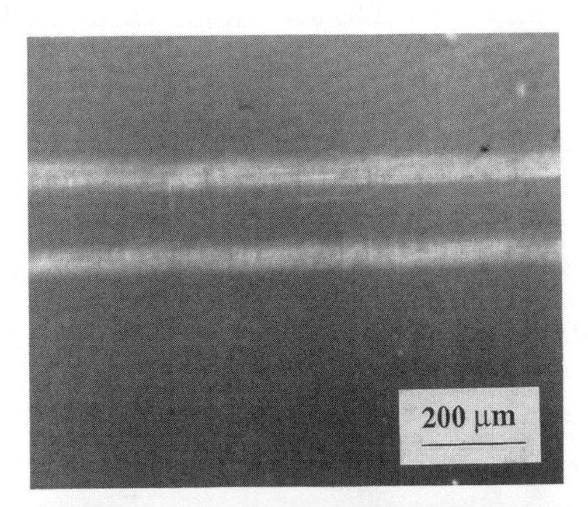

Fig.6
LALE GaAs Stripe
written at 6.3 mm/s.

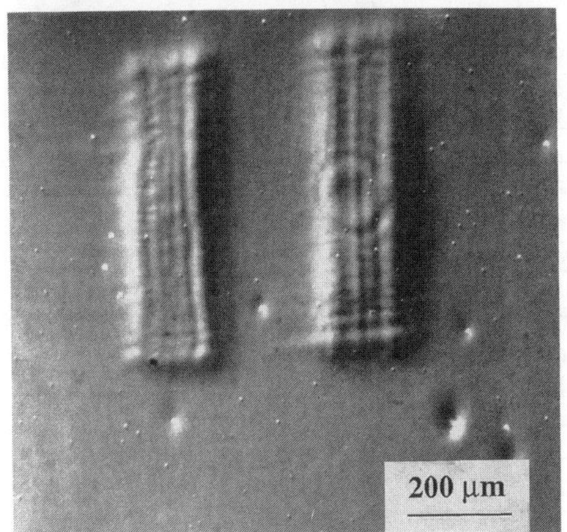

Fig.7
Direct patterning through
a shadow mask.

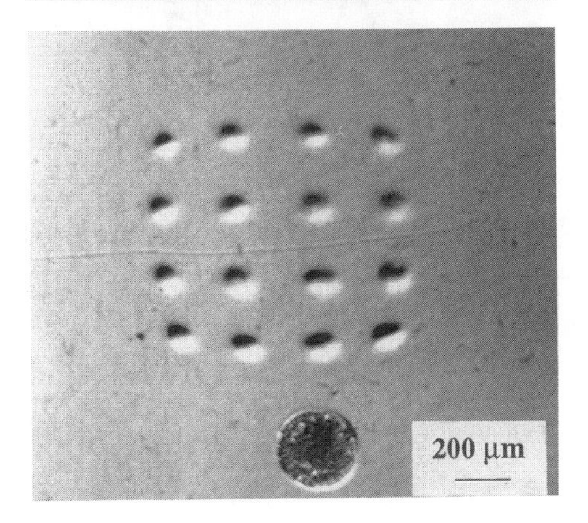

Fig.8
2D array of GaAs dots
on $Al_{0.3}Ga_{0.7}As$.

imaging of a complex pattern, this effect can be alleviated by expanding the original beam, using masks of wider openings, and then reducing onto the substrate or alternatively by using an incoherent light source. Nevertheless, this result demonstrates the feasibility of direct patterning by LALE.

We have also employed a novel optics setup wherein each one of the specially designed planar beam splitters divides the input beam into multiple equally spaced and equally intense parallel beams (2 or 4 beams) with ~94% efficiency. The use of two of such devices in series generates a 2D array of parallel beams. These beams are then focused onto the substrate forming 2D array of dots. Shown in Fig.8 is the microphotograph of such a GaAs dot array deposited on freshly grown $Al_{0.3}Ga_{0.7}As$. The variation in the spacing and the dot sizes are due to the distortion of a simple lens.

2. Material Characterization and Optimization

Current laser power limitations limit the area of deposition by LALE. In addition, current reactor designs limit the thickness of layers that can be grown during practical experiment. These factors make the characterization of LALE films difficult. As a result, little is known of the purity or transport properties of GaAs grown by LALE. By taking advantages of our unique reactor designs, we have formed DHs with the GaAs grown by LALE and the $Al_{0.3}Ga_{0.7}As$ by MOCVD. The thin GaAs layer (850~1100 Å) is probed in CV measurements by advancing the depletion layer under a Schottky barrier through the cladding layer into the LALE grown material. Similar DHs are also used for PL measurements but the top barrier is considerably thinner (~1000 Å) for better excitation efficiency. Shown in Fig.9 is a room temperature PL spectrum from such a DH. This is comparable to the DHs grown by thermal ALE. Our CV measurement gives hole concentrations ranging from $2x10^{17}$ to $3x10^{18}/cm^3$. The p-type impurity is caused by carbon acceptors as confirmed by low

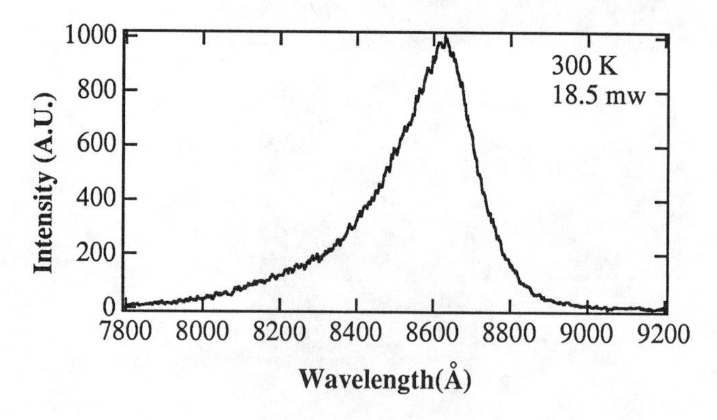

Fig.9 Room temperature from a LALE DH.

temperature PL and SIMS direct identification. CV, PL, and to a lesser extent SIMS measurements are used for a growth optimization process to obtain pure and optically efficient GaAs. The observed trends are summarized in Table I. With the laser power increasing from the low end to the high end of the LALE power window, room temperature PL efficiency increases and the carrier concentration decreases. These behaviors can be consistently attributed to a more complete decomposition of TMGa and a slightly higher temperature that promotes perfect lattice site registry as a result of local photon fluence. While LALE shows strong self-limiting with respect to TMGa exposure, our optimization process reveals that excess TMGa is detrimental to the quality of the GaAs film as is seen by a decrease in PL efficiency and an increase in the carrier concentration and the carbon concentration measured by SIMS when the TMGa delivery is increased from the low end of the TMGa exposure window. The effect of AsH_3 exposure time on PL efficiency is studied with 26 monolayer QW's. There is a noticeable improvement in the PL efficiency when AsH_3 exposure time is lengthened from 3 sec to 5 sec. Further increase in the AsH_3 exposure time to 10 sec does not seem to improve the PL efficiency, but the SIMS C count shows still a slight decrease. This may possibly be explained by the following: at longer AsH_3 exposure time, contamination other than carbon will dominate and deteriorate the interface and growth front. This counteracts the improvement expected from the carbon reduction offered by the partially decomposed AsH_3. It is also expected that higher substrate temperature will result in better material. The upper limit of the substrate temperature is imposed by the point at which the laser assistance provides little enhancement of the growth rate. We have obtained LALE in a substrate temperature range from 300 °C to 400 °C. Within this rage, there is only a slight reduction in C count as substrate temperature increases. Consequently, most of our DHs and laser structures are grown at 380 °C. Our optimization shows that it is desirable to grow at the low end of the LALE TMGa exposure range and the high end of the LALE power window in order to obtain GaAs film of device quality. Although

Table I. Summary of the growth optimization

Effect On / Parameters	PL Efficiency	Carrier Concentration	C Counts From SIMS
↑ Laser Power	↑	↓	
↑ TMGa Mole Fraction	↓	↑	↑
↑ AsH_3 Exposure Time	↗—		↓
↑ Bias Temperature			—

all the growth parameters (laser power, beam diameter, scan speed, TMGa flow rate and duration, AsH_3 exposure time, and purge time) are inter-related to an extent, this study has helped to delineate the roles of the parameters and to understand the LALE process.

3. LALE device applications

Our optimization process resulted in routine growth of GaAs with good PL response by LALE. Thin films (~24-26 monolayers) are confined by conventionally grown $Al_xGa_{1-x}As$ layers to form a graded-index separate-confinement heterojunction laser structure. Broad area laser bars of 35 μm or 50 μm contact width are formed parallel to the LALE stripe. Shown in Fig.10 is the lasing spectrum along with the electroluminescence spectrum below threshold. The testing conditions used are 450 ns pulses at 10 kHz repetition rate. Threshold current density as low as 650 A/cm^2 is obtained from 580 μm long devices indicating device quality of the LALE GaAs. This result is the first demonstration that GaAs by LALE can be used as the active medium of a semiconductor laser.

By aligning the laser cavity perpendicular to a 300 μm wide LALE stripe, an important device concept is demonstrated: lasers with passive waveguides as part of the cavity. Such a light source is readily integrable with waveguides, modulators, and optical amplifiers with slight modification. As is drawn in the inset of Fig.11, a 5 μm-wide ridge waveguide is etched perpendicular to the LALE stripe. The 293 μm long contact selectively pumps the 300 μm LALE stripe, leaving sections of passive waveguides 35 μm long at one end and 650 μm long at the other end. Fig.11 shows the lasing spectrum under pulsed testing condition along with the electroluminescence spectra of the device before and after cleaving off the 650 μm waveguide length to 11 μm. The threshold current for devices with the 650 μm long waveguide present is 225

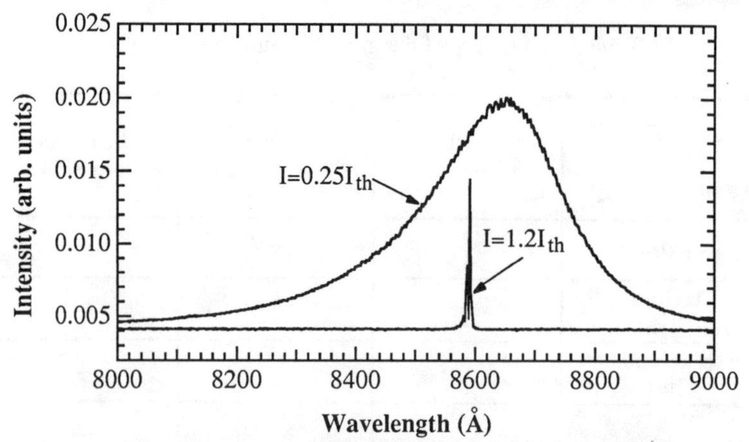

Fig 10 Lasing and electroluminescence spectra of QW laser with active region grown by LALE.

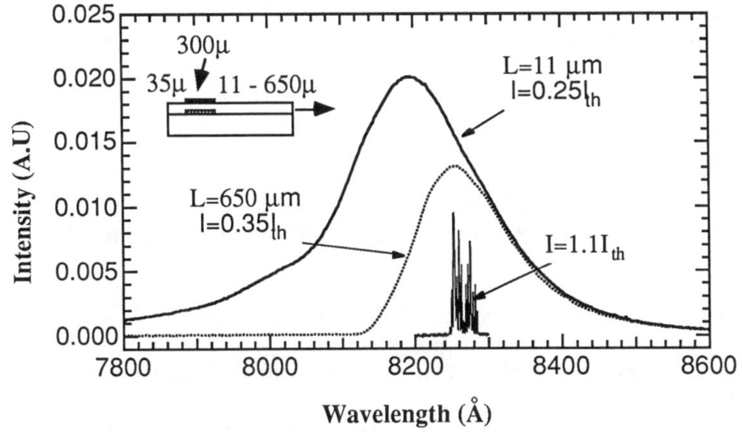

Fig.11 Lasing and electroluminescence spectra of LALE QW lasers with transparent waveguide as part of the cavity.

mA and reduces to 155 mA after cleaving. These high values are caused by losses in the waveguide as can be inferred from the electroluminescence spectra. Excessive losses are caused in this case by the deposition of a thinner QW active region than that of the device in the Fig.10, resulting in the overlap of the QW emission with the band edge of the waveguide region. Since the waveguide, acting as an integrated edge filter, strongly absorbs wavelength shorter than ~8200 Å, the laser is forced to oscillate at a longer wavelength.With the material quality attainable, threshold current can reduced by adopting an optimal structural design. This result shows for the first time that LALE is an attractive technique in the realization of optoelectronic integration.

SUMMARY AND CONCLUSION

In summary, we have successfully implemented LALE of GaAs by both direct writing and pattern projection. A novel optics setup is employed allowing the use of nearly all the laser power for localized deposition of 2D array of GaAs dots. LALE of GaAs using TMGa and AsH_3 is shown to have wide ranges of laser power and input TMGa flux for monolayer self-limiting growth and is to be used to the advantage of precise layer thickness control down to atomic dimension.

GaAs grown by LALE is characterized by CV, PL, and SIMS, indicating good electrical and optical quality. In particular, the LALE GaAs films measured are p type ranging from $2x10^{17}$ to $3x10^{18}$ /cm^3. The responsible impurity identified by low temperature PL and SIMS measurement is carbon resulting from incomplete decomposition of TMGa at low temperature. The knowledge acquired from these characterizations is used in a growth optimization process. Device quality GaAs can be grown by LALE through optimizing the growth parameters.

Finally, for the first time, QW lasers with active region grown by LALE are demonstrated. The usefulness of LALE in optoelectronic integration is shown by including a waveguide as a portion of the laser cavity.

ACKNOWLEDGEMENT

This work is supported by the Office of Naval Research, the Solar Energy Research Institute, and the SDIO/IST program through the Army Research Office.

1. K.Aiki, M.Nakamura, J.Umeda, IEEE J.QE-13, 597 (1977).
2. A.Doi, Y.Aoyagi, S.Namba, Appl. Phys. Lett. 49, 785 (1986).
3. S.Iwai, T.Meguro, A.Doi, Y.Aoyagi, and S.Namba, Thin Solid Films 163, 405 (1988).
4. N.Karam, R.Liu, I.Yoshida, and S.M.Bedair, Appl. Phys. Lett. 52, 1144 (1988).
5. S.P.DenBaars, P.D.Dapkus, J.S.Osinski, M.Zandian, C.A.Beyler, and K.M.Dzuko, in Proceedings of the 15th International Symposium on GaAs and Related Compounds, Atlanta, GA, 1988, Inst. Phys. Conf. Ser. 96, edited by J.S.Harris (Inst. Phys., London, Bristol, 1989), chap.3, pp89.
6. S.P.DenBaars and P.D.Dapkus, J. Cryst. Growth 98, 195 (1989).
7. S.P.DenBaars, Ph.D. thesis, University of Southern California, (1988).
8. T.Miyoshi, S.Iwai, Y.Iimura, Y.Aoyagi, and S.Namba, Jpn. J. Appl. Phys. 29(8), 1435 (1990).
9. Q.Chen, J.S.Osinski, and P.D.Dapkus, Appl. Phys. Lett., 57(14), 1437 (1990).
10. J.Nishizawa, H.Abe, and T.Kurabayashi, J. Electrochem. Soc. 132(5), 1197 (1985).
11. S.M.Bedair, M.A.Tischler, T.Katsuyama, and N.A.El-Masry, Appl. Phys. Lett. 47(1), 51 (1985).
12. M.Ozeki, K.Mochizuki, N.Ohtsuka, and K.Kodama, Appl. Phys. Lett. 53(16), 1509 (1988).
13. Y.Kawakyu, H.Ishikawa, M.Sasaki, and M.Mashita, Jpn. J. Appl. Phys. 28(8), L1439 (1989).
14. W.G.Jeong, Ph.D Thesis, Uninversity of Southern California, (1990).
15. H.Ohno, S.Ohtska, A.Ohuchi, T.Matsubara, and H.Hasegawa, J. Cryst. Growth 93, 342 (1988).
16. K.Mochizuki, M.Ozeki, K.Kodama, and N.Ohtsuka, J. Cryst. Growth 93, 557 (1988).
17. S.Iwai, T.Meguro, A.Doi, Y.Aoyagi, and S.Namba, Thin Solid Films 163, 405 (1988).
18. M.Sasaki, Y.Kawakyu, and M.Mashita, Jpn. J. Appl. Phys. 28(1), L131 (1989).
19. Y.Eytz-Froidevaux, R.P.Salathe, and H.H.Gilgen, Mat. Res. Soc. Symp. Proc. 17, 29 (1983).
20. A.Doi, Y.Aoyagi, and S.Namba, Appl. Phys. Lett. 48(26), 1787 (1986).

BEAM ASSISTED ATOMIC LAYER CONTROLLED EPITAXY AND ETCHING OF GaAs

T. Meguro and Y. Aoyagi
RIKEN (The Institute of Physical and Chemical Research)
Hirosawa, Wako, Saitama 351-01, Japan

Atomic layer epitaxy (ALE) using laser irradiation and digital etching of GaAs are described herein.

Epitaxy: We have succeeded in the laser-assisted ALE (laser-ALE) of GaAs using visible wavelength Ar^+ laser irradiation and an alkylgallium source. Visible wavelength photon irradiation induces surface decomposition but not volume decomposition of alkylmetal molecule source gases. ALE is realized by the enhancement of decomposition of alkylgallium molecules only on the As-terminated surface but not on the Ga-terminated surface. This site-selectivity of alkylgallium decomposition is induced by the optical absorption band broadening, which is due to the chemisorption of alkylgallium at the As-terminated surface.

Etching: In ditigal etching, etchant gas pulses and an energetic beam sequentially impinge onto the substrate surface. In the Ar^+/Cl_2 system, the etch rate is found to be independent of both Cl_2 flux and Ar^+ beam density, and the etch rate saturates at a level below one monolayer per cycle. By using Cl radicals as etchants instead of Cl_2, the self-limited etching characteristics of digital etching are obtained within both the Ar^+ incidence time and Cl_2 feed time of the etching cycle.

1. Introduction

Various fabrication techniques for low dimensional quantum structures have been proposed in recent years, but the complicated procedures involved and/or resulting damage produced in substrates are still serious problems which must be overcome. Basic requirements for fabrication of these novel structures are i) monoatomic layer controllability, ii) uniformity over large area, and iii) damage-free processing.

Atomic layer epitaxy (ALE)[1] has been realized as one of the most attractive candidates for atomic-scale additive structuring of materials, and many approaches have been reported up to now.[2] In the trimethylgallium (TMGa)/AsH_3 system, a clear saturation of the growth rate at one monolayer (ML) per cycle for various growth parameters has been easily obtained. In the case of the triethylgallium (TEGa)/AsH_3 system, however, clear saturation has not been realized in thermal ALE. ALE under visible wavelength cw Ar^+ laser irradiation (laser-ALE) can overcome this problem and has been extensively investigated.[3-18] In laser-ALE, complete self-limited growth is attained over a wide range of growth conditions, and a clear growth rate saturation at one monolayer (ML)/cycle is observed in both of the TMGa/AsH_3 and TEGa/AsH_3 systems.

Many investigations on atomic-scale additive techniques such as ALE have been carried out as mentioned above, but few reports have been made on atomic-scale subtractive techniques. Maki and Ehrlich have reported bilayer etching of GaAs under continuous gas feed and simultaneous pulsed laser irradiation.[19] We have proposed a digital etching technique in which the etch rate is independent of etching parameters, and proceeds with submonolayer controllability under low energy Ar^+ or electron irradiation.[20-23] A similar approach has been examined for Si etching.[24,25] In digital etching, etchants are introduced in sequence with a necessary purge period between the subsequent etchant pulse and energetic beam pulse, in contrast to conventional etching methods.

The beam induced surface modification processes have the advantage that the reaction is selective to the area irradiated by the beam and can be applied in a direct pattern drawing technique. In this paper, basic characteristics and surface processes in laser-ALE and digital etching of GaAs are reviewed and recent results are described.

2. Specific features of laser-ALE and digital etching

2.1. Laser-ALE

In conventional Ar^+ laser-activated vapor phase epitaxy of GaAs with simultaneous feed of group III and V source gases, the Ar^+ laser irradiation enhances the growth rate by one or two orders of magnitude.[26] This result could not be interpreted simply by elevating the temperature of the irradiated area, and suggests that the Ar^+ laser irradiation initiates photochemical effects at the growing surface. The detailed analysis of carbon content in grown layer shows that the laser irradiation enhances the decomposition of alkylgallium used as a source gas for group III element.[27] The growth process in laser-ALE is interpreted as the preferential decomposition of alkylgallium chemisorbed only on the As-terminated surface (site-selective decomposition).[5,7,9] The enhancement ratio of the decomposition rate on the As-terminated surface to the Ga-terminated surface was estimated to be over 100.[7] The fact that the Ar^+ laser irradiation enhances the decomposition of alkylgallium on the As-terminated surface was experimentally confirmed from the observation of reflection of high energy electron diffraction (RHEED) using a chemical beam epitaxy (CBE) system.[28]

The most noticeable difference in growth characteristics between thermal and laser-ALE is the suitability of source gases. In thermal ALE, TMGa is superior to TEGa, because ALE growth is difficult to obtain using TEGa, while both TMGa and TEGa can be utilized in laser-ALE. The growth mechanism of thermal ALE is considered to be caused by a site-blocking by residual alkyls and/or a site-selectivity in adsorption or desorption of alkylgallium, not in decomposition. [2,23] Therefore source gases with higher vapor pressures are more suitable for thermal ALE. In addition, since the decomposition temperature of TEGa is lower than that of TMGa, TEGa easily decomposes on the surface at a growth temperature suitable for thermal ALE. Thus ,the ideal ALE characteristics are not obtained in thermal ALE using TEGa. In laser-ALE, the ideal ALE characteristics are obtained by the site-selective decomposition of alkylgallium, so that TEGa (which decomposes into Ga at a lower temperature) is more suitable. In laser-ALE, both TMGa and TEGa provide the ideal growth characteristics of ALE, and ALE is achieved over wider experimental conditions in the TEGa/AsH_3 system.

Laser-ALE also has advantages in direct writing of atomically controlled, pattern defined epitaxial growth, thereby allowing maskless processing. Using typical growth conditions for laser-ALE, the surface morphology of the laser irradiated area has a flat-top shape in spite of steep intensity distribution of the incident laser beam. No growth is observed in the surrounding areas either. As expected from these growth characteristics of laser-ALE, the growth rate is independent of incident photon number over a wide range. In the case of direct pattern drawing using laser scanning, laser-ALE provides a uniform thickness with a flat-top surface even in areas of overlapping laser irradiation. [8,12] This is in contrast in conventional laser MOVPE, where the thickness reflects the beam profile of the incident laser and the thickness of overlapping areas becomes propotionally thicker.

2.2. Digital etching

The digital etching sequence is divided into four steps, as schematically shown in Fig.

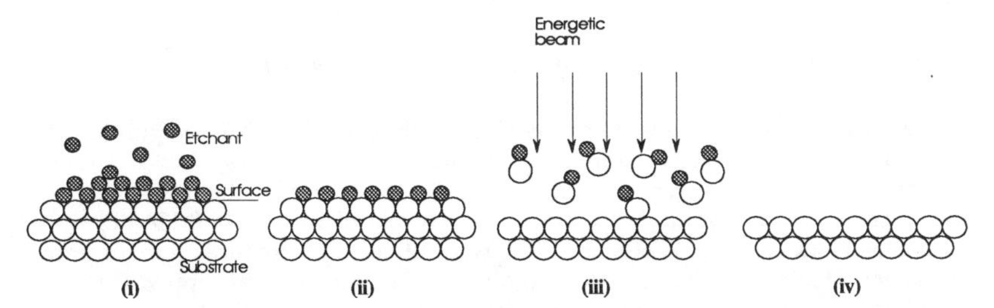

Fig. 1 The expected ideal surface processes during digital etching. In digital etching, an etchant and an energetic beam, which induces a surface chemical reaction, are alternately directed to the substrate surface; (i) Etchant feed, (ii) Etchant purge, (iii) Beam irradiation, and (iv) Products purge.

1. In digital etching, an etchant and an energetic beam, which induces a surface chemical reaction, are alternately directed to the substrate surface as described as follows: (i) the etchant is introduced into the etching chamber and adsorbs at surface sites of the substrate to be etched (the etchant does not spontaneously etch the substrate); (ii) excess etchant is purged; (iii) an energetic beam, such as Ar^+, electron or photon, is directed to the surface, promoting cascaded desorption of surface atoms bonded with etchant, owing to beam-induced chemical sputtering (physical sputtering owing to energetic beam bombardment should be suppressed to maintain a self-limiting condition); and (iv) etching products are purged, completing 1 cycle of the digital etching process. Physical sputtering and spontaneous chemical etching should be negligible in ideal digital etching.

If the ideal layer-by-layer etching such as shown in Fig. 1 is realized, the digital etching technique can be applied to the atomic-scale flattenning of surfaces. The ideal atomic layer etching would proceed by the layer-by-layer removal of surface atoms bonded with etchants, making the initial surface roughness decrease as shown by the model in Figs. 2(a) - (h).

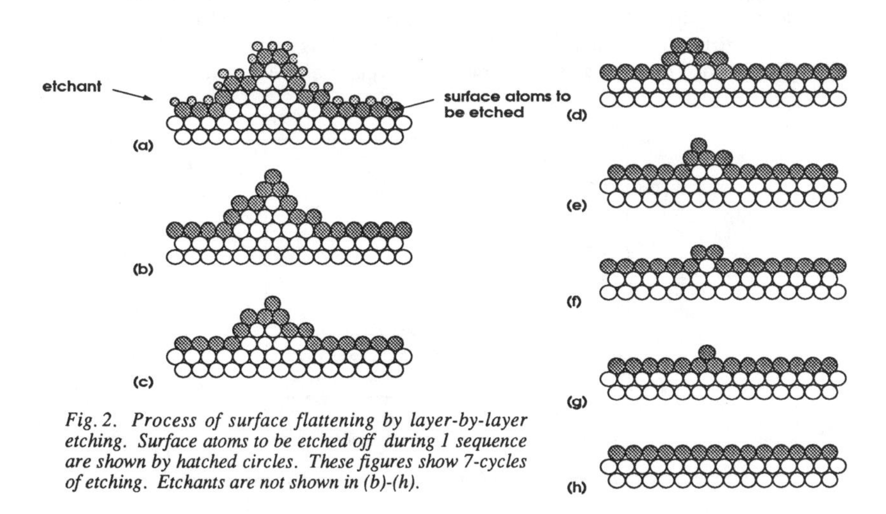

Fig. 2. Process of surface flattening by layer-by-layer etching. Surface atoms to be etched off during 1 sequence are shown by hatched circles. These figures show 7-cycles of etching. Etchants are not shown in (b)-(h).

3. Process characteristics and surface mechanism

3.1. Laser-ALE: Microscopic surface processes in site-selective decomposition

In this section, the microscopic mechanism of the site-selective decomposition in laser-ALE is discussed. The details of experiments have been described in previous papers[5,6] and are briefly reviewed here. The growth system used in the laser-ALE experiment was an ordinary vertical type metalorganic vapor phase epitaxy (MOVPE) system with a load lock chamber. TEGa and AsH_3 were alternately introduced into the growth chamber for 1 second with a 1 second purge interval between each gas supply. During the intervals H_2 is fed into the chamber to purge residual source gases. The total duration of 1 growth cycle was 4 seconds and leads to 1 ML of GaAs (0.283 nm) being grown. The laser irradiation was synchronized with the TEGa introduction because no growth was observed by light synchronization with other periods.[10] The laser used was typically a cw Ar^+ laser with 514.5 nm oscillation. A cw Ar^+ laser with 355.0 nm and cw YAG laser with 1.064 μm were used for modeling laser-ALE. Substrates were (001)±0.5° oriented GaAs wafers. An AlAs buffer layer was grown by thermal CVD using trimethylaluminum (TMAl) and AsH_3 at 600 °C prior to the laser-ALE growth in experiments shown in the section 3.1.2.

3.1.1. Models of the growth process (Dependence on wavelength of the incident irradiation)

Figure 3 shows the homoepitaxial growth rate of GaAs on a GaAs substrate as a function of the incident laser power density under various laser irradiations. In the case of UV or visible laser irradiation, the ideal growth characteristics of laser-ALE was obtained. The threshold intensity needed to achieve ALE under UV laser irradiation is lower than that obtained under visible laser irradiation, suggesting that a high quantum yield is expected in the shorter wavelength region for laser-ALE. With IR laser irradiation, no plateau was observed, because the absorption coefficient of the GaAs substrate in the IR region is low. This causes substrate heating due to the bulk absorption of the IR laser beam, leading to a purely thermal effect for this growth. This result stresses that the origin of laser-ALE is not a thermal effect but a photo-induced surface chemical effect.

Two plausible microscopic surface models can be proposed from this wavelength dependence, the direct surface absorption model and the charge transfer model.

Direct surface absorption model is based on the direct absorption of photons by TEGa chemisorbed on the As atoms. The chemisorption of TEGa molecules on the As-

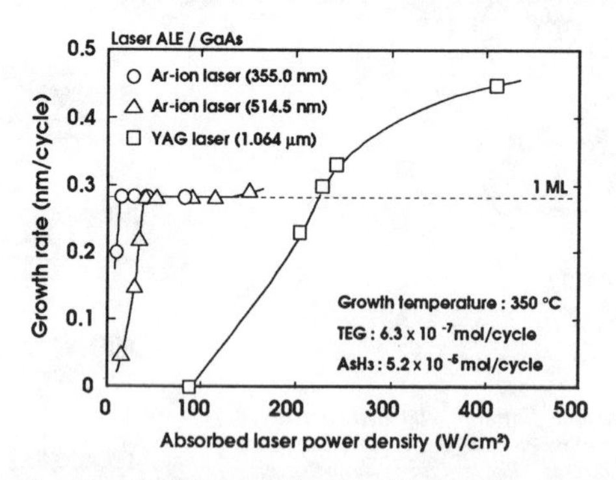

Fig. 3 Growth rate of GaAs as a function of incident laser power density.

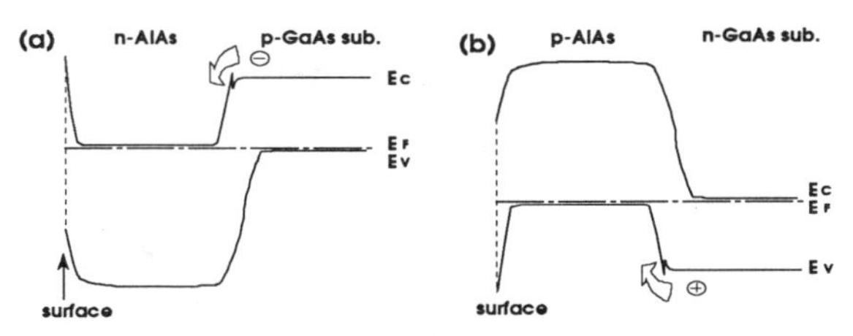

Fig. 4. Schematic band diagram of samples with AlAs buffer layer.

terminated surface causes the optical absorption band of TEGa to extend toward longer wavelengths, even though TEGa has no absorption in the vapor phase. For TEGa molecules adsorbed on Ga atoms, no or little optical absorption is observed[30], thus the site-selective decomposition is expected.

Charge transfer model is based on photo-carrier generation in the bulk. Laser photons absorbed at the surface generate free carriers which are easily transferred to TEGa adsorbed on As atoms because of the polar chemical bonding nature between TEGa and the As atoms. This charged TEGa easily decomposes into Ga or Ga compounds. In the case of TEGa adsorbed on the Ga-terminated surface, charge transfer does not occur because of the weak non-polar bonding involved. This model is supported by theoretical considerations[31] and explains the site-selective decomposition.

3.1.2. Contribution of photo-generated carrier transfer (Laser ALE of GaAs on AlAs buffer layer)

To understand the details of the microscopic kinetics of the site-selective decomposition in laser-ALE, the effect of photo-generated carriers must be discussed. The wavelength of the visible Ar^+ laser is 514.5 nm (2.410 eV), and the photon energy of this line is smaller than the direct band gap of AlAs at 350 °C, $E_{g,dir}$ (Γ_{15v}-Γ_{1c}: 3.00 eV). Though the indirect band gap of AlAs at 350 °C, $E_{g,ind}$ (Γ_{15v}-X_{1c}: 2.01 eV) is smaller than the incident photon energy, an affect by an of indirect transition can be ruled out because of its small transition probability compared with the direct transition. Therefore, the amount of carriers generated in the AlAs buffer layer is estimated to be several orders smaller than that in GaAs.

In the AlAs/GaAs heterostructure, photo carriers generated in GaAs by the Ar^+ laser irradiation through the AlAs buffer layer hardly diffuse into the AlAs layer due to the energy gap schematically shown in Figs. 4a and 4b. In the case of an n-AlAs/p-GaAs structure (type A), only electrons can diffuse toward the surface, while only holes can diffuse in the case of p-AlAs/n-GaAs structure (type B). So, if the thickness of the AlAs buffer layer is enough to suppress the tunneling transition of carriers from the GaAs substrate to the growing surface, photo-generated carriers which can diffuse from the sub-

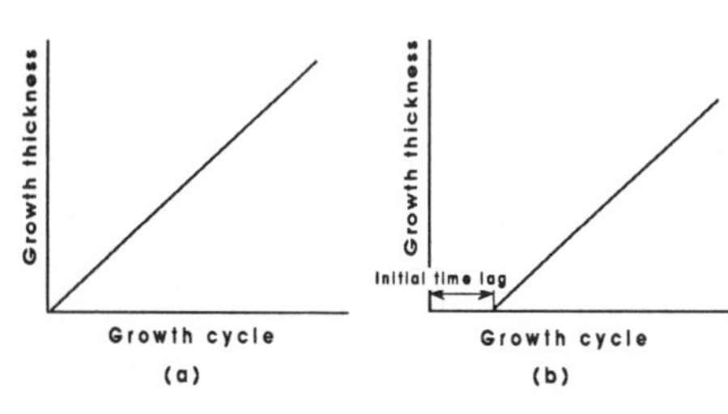

Fig. 5. Schematic illustrations of expected growth thickness as a function of growth cycles. If direct photo-absorption dominates the surface process in laser-ALE growth, the growth characteristics show a linear dependence (a). If the photo-carrier generation is dominant, the growth cycle dependence shows an initial time lag because of the initiation of the ALE growth by a thermally grown GaAs layer on the AlAs buffer.

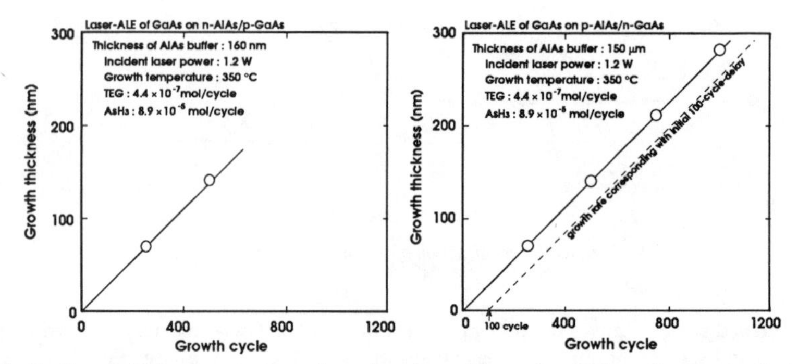

Fig. 6. Growth thickness vs. numbers of growth cycles on (a) type A and (b) type B substrate. Broken line in (b) indicates the growth tendency with the initial 100-cycle delay.

strate towards surface (*e.g.* electron in type A) can not reach due to surface band bending. Thus, the contribution of photo-carriers generated in a GaAs substrate during laser-ALE is suppressed.

Here, if the laser-ALE growth is achieved by the direct photo-absorption, the ALE growth arises from the first monolayer on the AlAs buffer and the growth thickness dependence on growth cycle should show the linear dependence as shown in Fig. 5(a). On the contrary, if the laser-ALE growth is achieved by photo-carrier generation, no growth would be observed on the AlAs buffer. However, there is the one possibility that a thermally grown GaAs layer initiates the laser-ALE growth. Namely, the first monolayer of GaAs is grown on the AlAs buffer due to the thermal decomposition of a TEGa, and after one monolayer grows, laser-ALE growth can arise by photo-carrier generation in the GaAs layer grown on the AlAs buffer. In this case, the growth cycle dependence should show an initial time lag owing to the thermal decomposition rate as shown in Fig. 5(b), since the thermal decomposition rate of alkylgallium is a few orders slower than the photo decomposition rate.[7,26]

Figures 6(a) and (b) shows growth thickness of GaAs as a function of growth cycles and it is clear that no delay is observed both on type A and B substrates. Consequently, the ALE growth is initiated by the laser irradiation on the AlAs buffer layer and it is concluded that laser-ALE is realized by the direct absorption of photons by chemisorbed alkylgallium. As shown by the laser power density dependence of growth rate in Figs. 7(a)

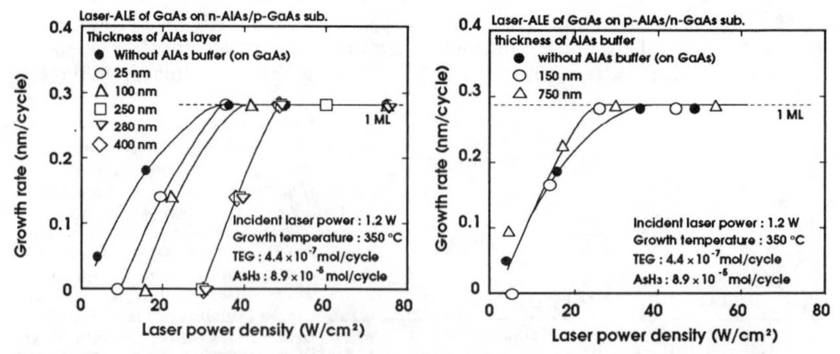

Fig. 7. Growth rate vs. incident laser power density on (a) type-A substrate and (b) type-B substrate.

and (b), the threshold laser power density slightly shifts with AlAs buffer layer thickness below 100 nm but is independent of thickness above 250 nm of AlAs thickness. With type A structures, the threshold power density for ALE increases, and the threshold value becomes lower in type B structures. Though the detailed mechanism is not understood, it is probable that the photo-carriers generated in the ALE grown GaAs layer on the AlAs buffer layer accumulate at the surface and induce the surface absorption specrtum shift in the adsorbate.

3.2. Digital etching using alternative incidence of Ar^+ and Cl etchants

The digital etching study was carried out in an electron-beam-excited plasma (EBEP)[32] system, which is described in previous reports.[20,21,32] The most favorable features of EBEP are a stable high ion or electron current being obtained even at low beam energies (<100 eV) and that charged species type (ion or electron) and energy can be switched simply by the bias voltage applied to the sample stage. Typical etching characteristics of Si and GaAs by Ar ion beam etching (IBE) or Cl_2 reactive ion beam etching (RIBE) have been previously reported.[33,34]

3.2.1. Process scheme

Figures 8(a) and (b) show the etching sequence using the Cl_2 molecules and Cl radical as the etchant, respectively. When Cl_2 molecules were used as etchants, the etching sequence was controlled by simply alternating the Cl_2 inlet valve operation and the Ar plasma generation. The Cl_2 ventilation valve is operated in inverse phase to the introduction valve and the pressure in the ventilation line is balanced with that in the etching chamber to stabilize the Cl_2 flow. When Cl radicals were used, the etching sequence is more complicated, as described as follows;

(i) *Etchant feed* : A Cl_2 gas pulse, synchronized with glow discharge plasma generation, is introduced into the etching chamber, and Cl radicals are generated. A shutter placed before the sample stage is closed, and the sample stage is biased to ground so that no ions reach the surface. Cl radicals generated in the etching chamber diffuse behind the shutter and adsorb to the GaAs surface.

(ii) *Etchant purge*: The Cl_2 introduction valve is closed and residual Cl in the etching chamber is purged. The Ar glow discharge is maintained during this step to desorb the Cl

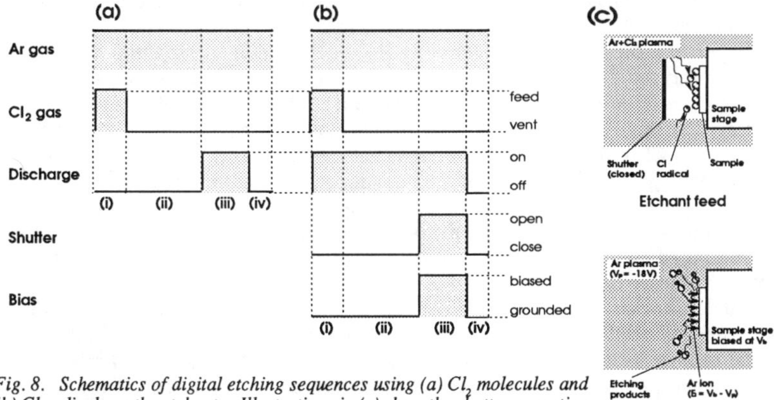

Fig. 8. Schematics of digital etching sequences using (a) Cl_2 molecules and (b) Cl radicals as the etchants. Illustrations in (c) show the shutter operation corresponding etching sequence (i) and (iii).

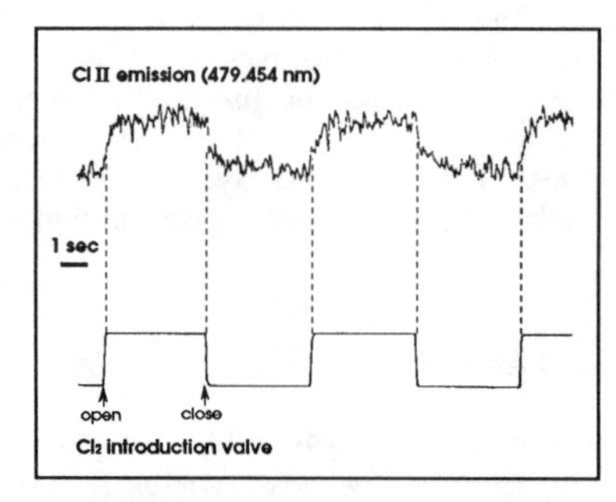

Fig. 9. Emission intensity change from Cl II in the etching chamber. Lower diagram shows the corresponding Cl_2 introduction valve operation.

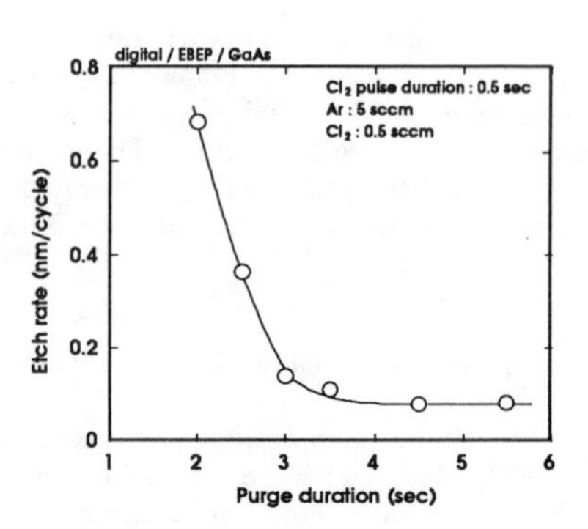

Fig. 10. Etch rate vs. purge time. Incident Ar^+ energy was fixed at 18 eV.

atoms adsorbed on chamber walls.

(iii) *Beam irradiation*: By applying an appropriate bias voltage to the sample followed by the immediate opening of the shutter, Ar ions bombard the surface. The impact energy of an Ar ion is given by the difference between plasma potential and applied bias.

(iv) *Products purge*: The discharge is turned off, the shutter is closed, and the sample is biased at ground again. During this step, all etching products are exhausted from the chamber.

3.2.2. *Etching characteristics*

When Cl_2 molecules are used as the etchant with Ar^+ irradiation of 22 eV, the etch rate shows saturation at around 1/3 ML/cycle.[22] The same behavior is observed in digital etching with sequential electron beam irradiation and Cl_2 molecule injection.[20] These results were interpreted as monolayer adsorption of Cl_2 molecules on the surface and generation of $GaCl_3$ and $AsCl_3$ as the etching products, as it was found that the etch rate was independent of incident electron current density. Thus, it is suggested that to realize ideal layer-by-layer controlled etching more Cl adsorption and/or generation of etching products, such as $GaCl_x$ or $AsCl_x$ ($x = 1$ or 2), is required. The digital etching characteristics using Cl radicals were also examined since they adsorb to GaAs more easily than Cl_2 molecules do.[35]

Figure 9 shows the emission intensity change from Cl II (479.454 nm) in the etching chamber and the Cl_2 introduction valve operation. When Cl radicals are used as the etchant, the residual Cl in the etching chamber can be observed from the emission because the glow discharge is maintained during the purging, as shown in Fig. 8(b). Emission intensity from Cl II decays to approximately zero after around 3.0 seconds. This result shows good agreement with the purge time dependence of etch rate shown in Fig. 10, which indicates that most residual Cl is exhausted after 3 .0 seconds purging. When the Cl_2 feed rate was fixed as 0.5 sccm, the partial pressure of Cl_2 during the Ar^+ irradiation was measured as 4×10^{-5} Torr. For example, when the Ar^+ irradiation time was set at 0.1 second, the decay time was measured as below 1.0 second, making a practical Cl exposure estimated of several langmuirs. A vapor phase reaction is probably involved, in addition to the surface reaction, when the purge time is set at 3.0 seconds. It was found that the digital etching technique using alternative exposures of Ar^+ and Cl radicals provides atomic-scale controllability in etched depth as indicated in Fig. 11. This figure shows the etch rate dependence on Cl_2 feed time and that little or no damage is induced during the

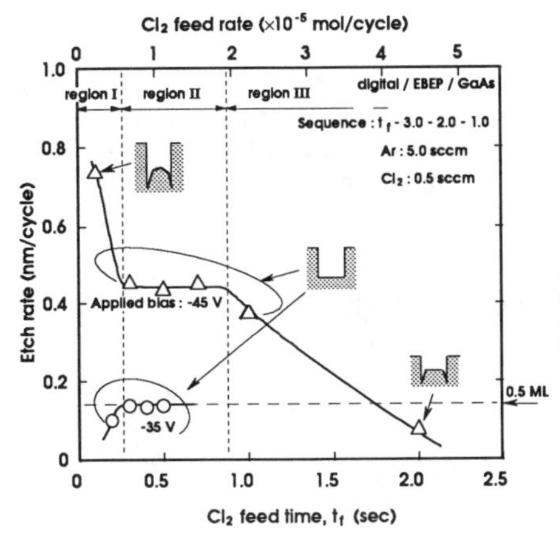

Fig. 11. Etch rate vs. Cl_2 feed time at 3.0 seconds of purge time. Inserted figures indicate cross-sectional etch profiles measured by surface profilometer.

etching procedure without any subtrench formation at the leading edge.[22] However, if the vapor phase reaction, owing to a short purge time, are involved in the etching procedure, the ideal self-limited etching reaction is not expected and the reaction scheme would become more complicated. Since Cl radicals introduced into the etching chamber are completely evacuated in purging within 3.0 seconds, vapor phase reaction can be ruled out by adopting a purge time greater than 3.0 seconds. The purge time was maintained at 10.0 seconds in the following experiments.

Figure 12 shows the variation of etch rate as a function of the Ar^+ incidence time for various Cl_2 feed time with 25 eV bombardment energy. As a whole, the etch rate increases with increasing Ar^+ incidence time until finally saturating. For the cases of 0.1 or 0.3 second Cl_2 feeding, etching arises without any onset incidence time and the etch rate at saturation is higher for the longer Cl_2 feed time. For Cl_2 feed time longer than 0.3 seconds, the etching starts after a certain lag time, with the onset of saturation taking longer for increased Cl_2 feed times, and the saturation level is independent of Ar^+ incidence time when the lag period is not considered.

3.2.2. *Plausible surface process*

Surface processes in digital etching are analyzed using a rate equation approach. The surface coverage of adsorbed Cl, *X*, is given by

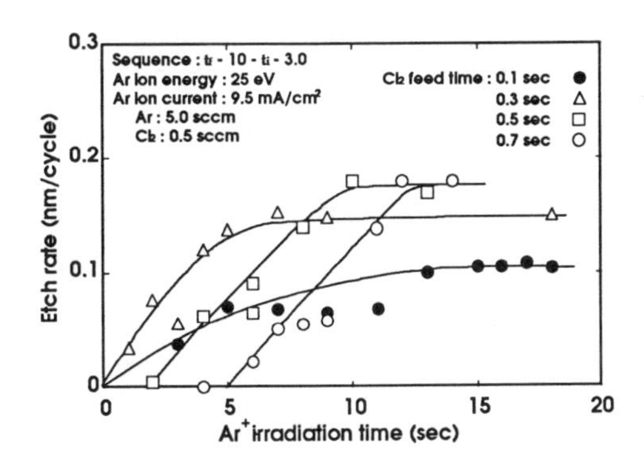

Fig. 12. Etch rate vs. Ar^+ irradiation time at various Cl_2 feed time.

$$dX/dt = -X/\tau_{reac} - X/\tau_{des}, \tag{1}$$

where τ_{reac} and τ_{des} are the time constants of surface reaction and desorption of Cl, respectively, and are assumed to be time-independent. The Cl radical is reported to accumulate on the GaAs surface at room temperature[36] and even physisorbed Cl layers are stable over several hours,[37] so that eq. (1) can be written in a simple form as

$$dX/dt = -X/\tau_{reac}. \tag{2}$$

The amount of Cl consumed by forming etch products is given by

$$dX_R/dt = (1-X_R)/\tau_{reac}, \tag{3}$$
$$1-X_R = X. \tag{4}$$

Considering the boundary condition of $X_R = 0$ at $t = 0$,

$$X_R \propto 1 - \exp(-t/\tau_{reac}). \tag{5}$$

By fitting eq. (5) with the experimental results shown in Fig. 12, τ_{reac} is determined to be ~ 2.5 second. This reaction time constant is estimated to be approximately 3 orders slower than the reaction time constant in ordinary RIBE,[38] suggesting that the surface reaction process in digital etching differs from conventional RIBE.

Figure 13 shows the plausible surface reaction processes in digital etching. We assumed that only Cl adsorbed within a critical thickness can contribute to etching and that excess Cl accumulation prevents the surface from etching. With thinner Cl adsorption than the critical thickness, etching occurs without any onset of the Ar^+ incidence time. The saturated etch rate becomes higher with larger Cl_2 feed time, since the saturated etch rate is decided by the initial Cl surface coverage. On the other hand, when the adsorbed Cl is thicker than the critical thickness, etching arises after the removal of excessive adsorption

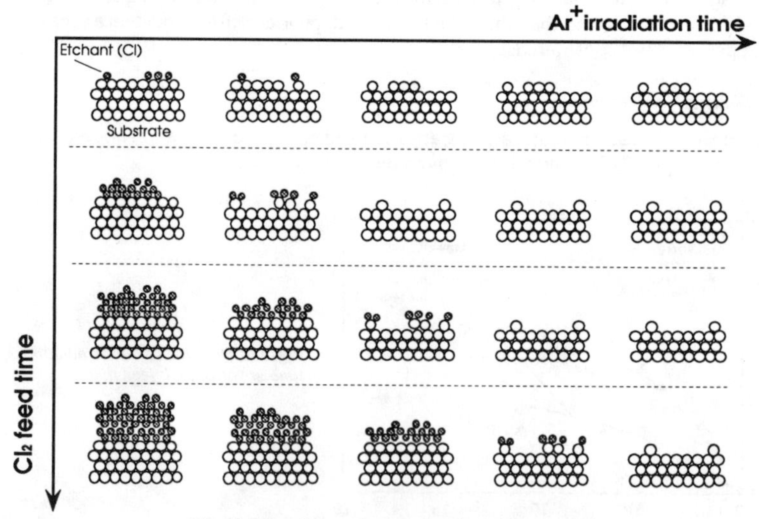

Fig. 13. Plausible surface processes in digital etching.

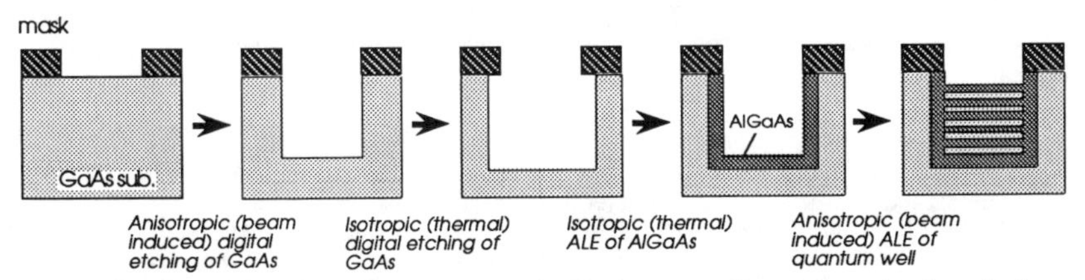

Fig. 14. Schematic procudure of fabrication of buried quantum well structure using layer-by-layer processes.

layer of Cl through the physical sputtering process. Etching in such a case proceeds only after the Ar^+ irradiation clears excess Cl. The saturated etch rate is independent of thickness of adsorbed Cl. In the saturated region, no Cl remains at the surface after one-cycle etching because Ar^+ irradiation is sufficient to remove Cl and the initial surface condition is retained in the digital etching procedure. Thus the etch rate is independent both of Cl_2 feed time and Ar^+ incidence time, and the self-limiting etching nature is obtained. At the present stage, the critical thickness in this digital etching procedure is not determined, but, in some cases, the critical thickness is considered to correspond with monolayer adsorption.

4. Summary

Laser-ALE and digital etching have been reviewed. Since the surface reactions involved in these layer-by-layer techniques described here are mainly induced by energetic beam irradiation, anisotropic surface reactions are expected. By the combination of isotropic reactions in thermal process, the beam induced anisotropic process can be utilized in the fabrication of various novel structures. To fabricate buried quantum well structure only by the layer-by-layer processes is one of the goals for its establishment. Figure 14 shows a schematic procedure of the layer-by-layer fabrication of the buried quantum well structure. First, a rectangular groove is fabricated by anisotropic digital etching and the wall and bottom of the groove are etched by the isotropic digital etching in what can be called a controlled undercut. Next, an AlGaAs layer is grown by isotropic ALE and then the quantum well structure is finally completed only at the bottom of the groove by anisotropic ALE. The layer-by-layer control offered by these process techniques allows precise control of structure dimensions in the quantum device, stressing ther importance. Unfortunately, many problems remain to be solved and extensive works are still required. This is especially true for digital etching, because of its short history and lack of basic understanding.

Acknowledgements

Part of this work was done with the colaboration of Dr. S. Iwai of the Laser Science Group, RIKEN and Mr. T. Arai of the Frontier Research System, RIKEN (Presently at Nippon Sanso, K. K.) in laser-ALE and with Mr. M. Ishii and Mr. H. Kodama of Hosei University in digital etching.

The authors thank Dr. Y. Iimura, Dr. K. Ishibashi ,Dr. J. Kusano and Dr. J. Simko of the Frontier Research System, RIKEN, and Dr. K. Ozasa of the Laser Science Group, RIKEN, for their helpful discussion, and also thank Mr. M. Mihara of the Laser Science Group, RIKEN, for his technical assistance.

This work was supported in part by a Grant-in-Aid for Scientific Research on Priority Area, "Electron Wave Interference Effects in Mesoscopic Structures" from the Ministry of Education, Science and Culture.

References

1) T. Suntola and J. Anston, Finnish patent no. 52359 (1974); US patent no. 4058430 (1977): T. Suntola, *Extended Abstracts of International Conference on Solid State Devices and Materials*, Kobe, 1984 (Publication Office Business Center for Academic Societies Japan, Tokyo, 1984) p. 647.
2) For example, *Proceedings of 1st International Symposium on Atomic Layer Epitaxy*, Helsinki, 1990, edited by L. Niinistö (Act. Polytechn. Scand. **Ch195** (1990)): *Atomic Layer Epitaxy*, edited by T. Suntola and M. Simpson, (Blackie, Glasgow, 1990): *Extended Abstracts of 22th (1990 International) Conference on Solid State Devices and Materials, Symposium D* (Business Center for Academic Societies Japan, Tokyo, 1990).
3) A. Doi, Y. Aoyagi and S. Namba, Appl. Phys. Lett. **48**, 1787 (1986).
4) A. Doi, Y. Aoyagi and S. Namba, Appl. Phys. Lett. **49**, 785 (1986).
5) Y. Aoyagi, A. Doi, S. Iwai and S. Namba, J. Vac. Sci. Technol. **B5**, 1460 (1987).
6) S. Iwai, A. Doi, Y. Aoyagi and S. Namba, Inst. Phys. Conf. Ser. **91**, 191 (1988).
7) A. Doi, S. Iwai, T. Meguro and S. Namba, Jpn. J. Appl. Phys. **27**, 795 (1988).
8) S. Iwai, T. Meguro, A. Doi, Y. Aoyagi and S. Namba, Thin Solid Films **163**, 405 (1988).
9) T. Meguro, T. Suzuki, K. Ozaki, Y. Okano, A. Hirata, Y. Yamamoto, S. Iwai, Y. Aoyagi and S. Namba, J. Cryst. Growth **93**, 190 (1988).
10) Y. Aoyagi, A. Doi, T. Meguro, S. Iwai, K. Nagata and S. Nonoyama, Chemtronics **4**, 117 (1989).
11) T. Meguro, S. Iwai, Y. Aoyagi, K. Ozaki, Y. Yamamoto, T. Suzuki, Y. Okano and A. Hirata, J. Cryst. Growth **99**, 540 (1990).
12) S. Iwai, T. Meguro and Y. Aoyagi, J. Cryst. Growth **107**, 136 (1991).
13) Y. Aoyagi, T. Meguro and S. Iwai, Act. Polytechn. Scand. **Ch195**, 55 (1990).
14) N. H. Karam, H. Lin, I. Yoshida, N. El-Masry and S. M. Bedair, Appl. Phys. Lett. **52**, 1114 (1988).
15) N. H. Karam, H. Lin, I. Yoshida, B. -L. Jiang and S. M. Bedair, J. Cryst. Growth **93**, 254 (1988).
16) N. H. Karam, H. Lin, I. Yoshida and S. M. Bedair, Appl. Phys. Lett. **53**, 767 (1988).
17) S. P. DenBaars and P. D. Dapkus, J. Cryst. Growth **98**, 195 (1989).
18) S. P. DenBaars, P. D. Dapkus, J. S. Osinski, M. Zandiand, C. A. Beyler and K. M. Dzurko, Inst. Phys. Conf. Ser. **96**, 89 (1989).
19) P. A. Maki and D. J. Ehrlich, Appl. Phys. Lett. **55** , 91 (1989).
20) T. Meguro, M. Hamagaki, S. Modaressi, T. Hara, Y. Aoyagi, M. Ishii and Y. Yamamoto, Appl. Phys. Lett. **56**, 1552 (1990).
21) T. Meguro, M. Ishii, M. Hamagaki, T. Hara, Y. Yamamoto and Y. Aoyagi, Act. Polytechn. Scand. **Ch195**, 163 (1990).
22) T. Meguro, M. Ishii, H. Kodama, M. Hamagaki, T. Hara, Y. Yamamoto and Y. Aoyagi, Jpn. J. Appl. Phys. **29**, 2216 (1990).
23) T. Meguro, M. Ishii, H. Kodama, M. Hamagaki, T. Hara, Y. Yamamoto and Y. Aoyagi, *Extended Abstracts of 22th (1990 International) Conference on Solid State Devices and Materials,* (Business Center for Academic Societies Japan, Tokyo, 1990), p. 893.
24) Y. Horiike, T. Tanaka, M. Nakano, S. Iseda, H. Sakaue, A. Nagata, H. Shindo, S. Miyazaki and M. Hirose, J. Vac. Sci. Technol. **A8**, 1844 (1990).
25) H. Sakaue, S. Iseda, K. Asami, J. Yamamoto, M. Hirose and Y. Horiike, Jpn. J. Appl. Phys. **29**, 2648 (1990).
26) Y. Aoyagi, M. Kanazawa, A. Doi, S. Iwai and S. Namba, J. Appl. Phys. **60**, 3131 (1986).
27) J. Kusano, Y. Segawa, S. Iwai, Y. Aoyagi and S. Namba, Appl. Phys. Lett. **52**, 67 (1988).
28) K. Nagata, Y. Iimura, Y. Aoyagi, S. Namba and S. Den, J. Cryst. Growth **95**, 142 (1989).
29) K. Kitahara, N. Ohtsuka and M. Ozeki, J. Vac. Sci. Technol. **B7**, 700 (1989).
30) M. Sasaki, Y. Kawakyu and M. Mashita, Jpn. J. Appl. Phys.,**28**, L131 (1989).
31) M. Tsuda, S. Oikawa, M. Morishita and M. Mashita, Jpn. J. Appl. Phys. **26**, L564 (1987).
32) T. Hara, M. Hamagaki, A. Sanda, Y. Aoyagi and S. Namba, Jpn. J. Appl. Phys. **25**, L252 (1986).
33) J. Z. Yu, T. Hara, M. Hamagaki, Y. Yoshinaga, Y. Aoyagi and S. Namba: J. Vac. Sci. Technol. **B6**, 1626 (1988).
34) J. Z. Yu, N. Masui, Y. Yuba, T. Hara, M. Hamagaki, Y. Aoyagi, K. Gamo and S. Namba, Jpn. J. Appl. Phys. **28**, 2391 (1989).
35) K. Asakawa and S. Sugata, in *Proceedings of Symposium of the International Engineering Congress ISIAT'83* (IEE of Japan, Tokyo, 1985) p. 759.
36) S. Sugata and K. Asakawa, IEE Jpn. Tech. Rep. **EFM-84-6** (IEE of Japan, Tokyo, 1984).
37) K. Shimmura, K. Kawasaki, T. Tanaka, I. Nakamoto, Y. Aoyagi, K. Gamo and S. Namba, *Extended Abstracts (The 38th Spring Metting, 1991); The Japan Society of Applied Physics and Related Societies*, p. 496.
38) M. Ishii and T. Meguro, unpublished data.

THE ROLE OF GAS PHASE DECOMPOSITION IN THE ALE GROWTH OF III-V COMPOUNDS

K.G. REID, H.M. URDIANYK, N.A. EL-MASRY, and S.M. BEDAIR

North Carolina State University, Electrical and Computer Engineering Department, Raleigh, NC

ABSTRACT

The effects of the growth temperature and exposure time to TMGa for ALE of gallium arsenide was studied using TMGa and AsH_3 in a modified, vertical, atmospheric, MOCVD reactor with a rotating susceptor. It was found that the temperature range for ALE growth could be extended from 450°C to 700°C by adjustment of the exposure time to TMGa. The maximum exposure time to TMGa was found to decrease as growth temperature increased with high temperature growth limited to exposures of only fractions of a second. Beyond a critical exposure time to TMGa, gallium droplets form on the surface. It is known that premature decomposition of TMGa in the heated gaseous boundary layer causes the formation of the gallium droplets and the consequent loss of ALE growth.

INTRODUCTION

Atomic layer epitaxy (ALE) is a thin film growth technique which gives monolayer per cycle growth over a range of operating conditions. A cycle for growth of GaAs using trimethylgallium (TMGa) and arsine (AsH_3) sources is one exposure of the substrate to AsH_3 followed by an exposure to TMGa. Ideally, ALE would be self limiting over a wide range of source fluxes, with the only requirement being a minimum flux determined by the amount needed for complete surface coverage. This condition is met by arsine where an increase above a lower limit of arsine flow does not affect the growth rate. However, the exposure to TMGa plays a crucial role in achieving ALE growth and in the autodoping of the GaAs film grown.

There are two general methods currently being used to achieve ALE in a MOCVD like reactor at near atmospheric pressure. One approach is based on exposure/purging of each reactant with a vent/run type manifold. Due to gases lagging in the reactor and valve switching times this method has a growth rate of about 0.1μm/h [1], and temperature range of 440°C - 560°C [2]. The second approach, developed in our lab [3], involves rotating the substrate between the different source gas streams that are continuously flowing through a specially designed vertical reactor. Growth rates in the range of 0.4 to 0.7 μm/h can be achieved with this approach for a temperature range of 450°C - 700°C. A schematic of the rotating susceptor is shown in Figure 1.

In this paper the effects of growth temperature and exposure time on growth rate and autodoping of ALE grown GaAs are presented. The results suggest the influence of gas phase decomposition in a boundary layer formed above the substrate. This results in the loss of the self limiting mechanism which causes conventional metalorganic chemical vapor deposition (MOCVD) or the formation of gallium droplets.

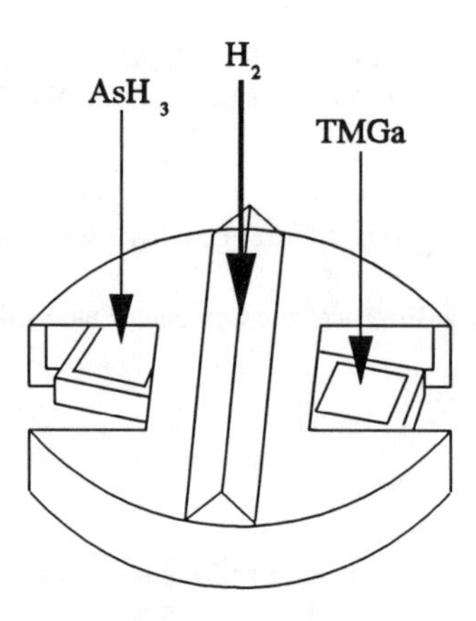

Figure 1 Schematic of the Rotating Susceptor

EXPERIMENTAL

The experiments were performed in a modified, atmospheric pressure, MOCVD, vertical style reactor shown in Figure 2. The quartz growth chamber has been modified to have three inlet tubes: one for the column III source (TMGa), one for the column V source (AsH_3), and a center line for a high flow of hydrogen. The hydrogen center line and a stationary graphite pyramid on top of the susceptor prevent mixing of the TMGa and AsH_3 which would impede the self limiting mechanism. The susceptor design has two crucial purposes to help achieve ALE: one is to minimize mixing of the column III and column V sources and the other is to minimize the thickness of the gaseous boundary layer above the substrate. To achieve these goals the graphite susceptor consists of a rotating sample holder with adjustable clearance under a stationary top plate (clearance for data presented in this paper $\approx$ 0.5 mm). The rotation speed and the stationary top plate help control the boundary layer for a given set of growth conditions. The windows cut out of the top plate allow controlled exposure of the substrate to the continuous precursor flows. The rotation is controlled by a stepper motor and computer. The exposure time is controlled by the rotation speed of the susceptor. The films were grown on (100) 2° off toward [110] GaAs substrates.

The effect of TMGa exposure on ALE growth rate and autodoping of the film were studied by varying the exposure time to TMGa while keeping the growth temperature constant. The minimum exposure time attempted (not necessarily the minimum required for complete surface coverage) was 0.2 seconds. Thus the experiments were performed to establish the upper limit of the exposure time to TMGa to maintain the self limiting mechanism.

The effect of the growth temperature on ALE growth rate and autodoping due to TMGa supplied carbon were also studied. For a given exposure time to TMGa and a TMGa flux of 0.08μmoles/sec we varied the growth temperature. Most films were grown to be 0.4μm thick. Thicknesses were measured by Nomarski microscopy, and the carrier concentration was measured by the Van der Pauw technique.

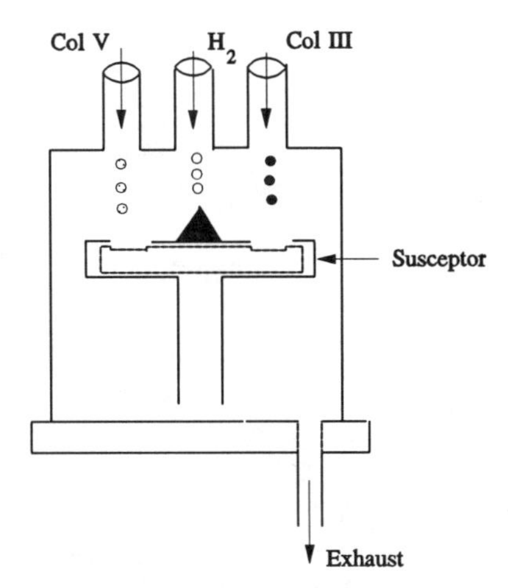

Figure 2 Schematic of the Growth Chamber

RESULTS AND DISCUSSION

The requirement for ALE growth is one monolayer per cycle with good film quality. As shown in Figure 3, we observed for a TMGa (0.08μmoles/sec) exposure time of 0.2 seconds ALE was obtained at 480°C - 700°C. For the same flux, but a one second TMGa exposure, ALE was achieved at 450°C - 600°C. These temperature ranges are much higher than reported by groups using the vent/run type reactor. Also, we found a decrease in the maximum exposure to TMGa decreased, as summarized in Figure 4. The shaded region is the ALE regime. The hatched area is a transition region where the monolayer per cycle growth is lost, but the surface has good morphology. Gallium droplets form on the surface in the dotted area.

Figure 5 shows the p-type background doping level dependence on the exposure time to TMGa for a given TMGa flux and growth temperature. For a given exposure time the doping decreased as growth temperature increased. And finally, for TMGa flux of 0.08μmoles/sec with a one second exposure, the doping showed a hill profile as a function of growth temperature, see Figure 6. The peak occurred at 500°C giving a doping concentration of 10^{19} cm^{-3}.

These results can be explained based on gas phase decomposition in the boundary layer. It is this layer just above the hot substrate where chemical reactions take place [4]. This chemical boundary layer is a coupling between the temperature dependence of the chemical reaction rate constants, a steep temperature gradient in the gas phase above the hot susceptor, and transient heating processes from the substrate to the impinging gases. The thickness of the boundary layer depends on: the gas flow rate, substrate rotation speed, substrate temperature, and the physical properties of the gases such as kinematic viscosity and thermal conductivity.

The ALE self limiting process requires surface reactions to dominate the growth process. There are two models that have been proposed to explain the self limiting mechanism. The selective adsorption model [5] which supports the formation of a monolayer of elemental gallium due to selective adsorption of TMGa onto As atom, but rarely on Ga atoms. The other mechanism is the adsorbate inhibition mechanism which proposes the adsorption of $Ga(CH_3)_n$ where n=1 or 2 which then prevents further adsorption once a monolayer is formed. The

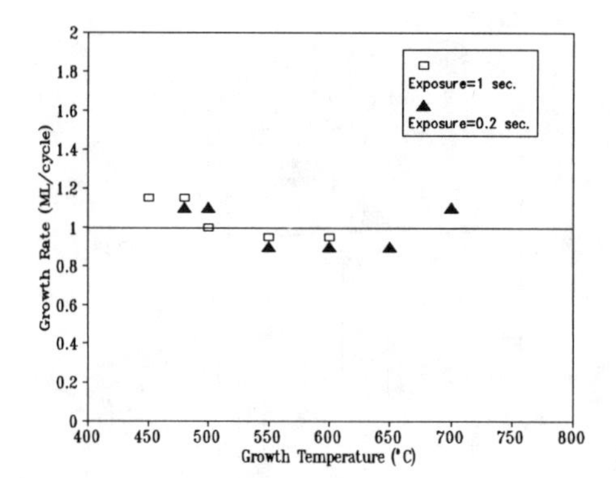

Figure 3 Growth rate as a function of temperature for both 0.2 and 1 second exposures to 0.08μmoles/sec TMGa. Thickness was measured using AB etch and Nomarski microscopy which led to scatter in the data.

methyl radicals are then displaced by arsine or react with hydrogen to form, for example, a methane molecule which leaves the surface.

For either theory, the self limiting mechanism requires minimum gas phase decomposition. For example, if TMGa molecules decompose in the gas phase to form Ga atoms, Ga droplets will be deposited on the surface and the self limiting process will not occur.

In our reactor, the substrate is exposed to heated gases only during the time the substrate cuts through the gas streams due to the shearing off of the gaseous boundary layer. The exposure time is not extended due to diffusion of the reacting gases away from the substrate as

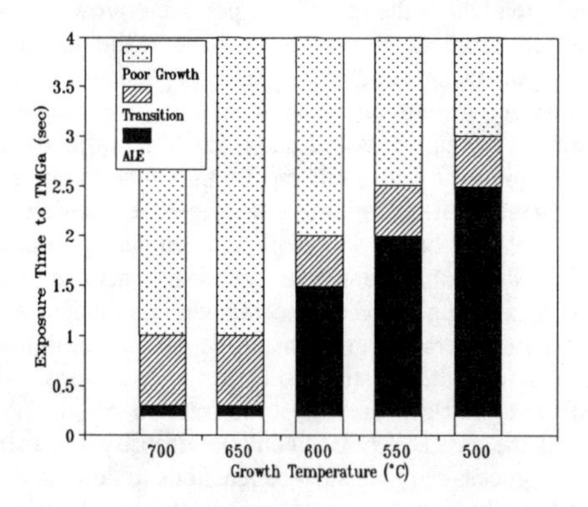

Figure 4 ALE growth regime established for our reactor.

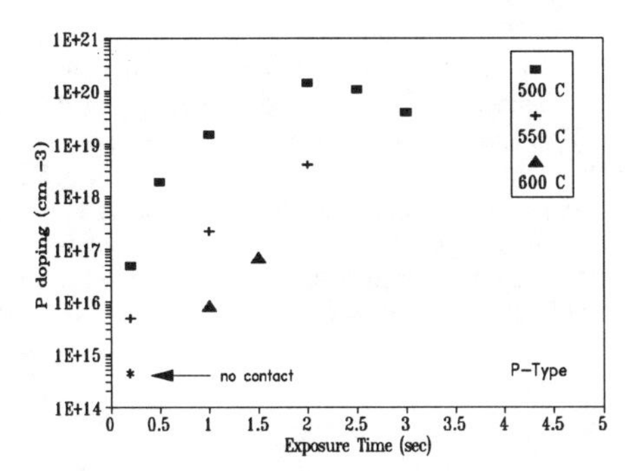

Figure 5 Dependence of the doping concentration on exposure time to TMGa (0.08μmoles/sec) for three growth temperatures: 500°C, 550°C, 600°C. At 600°C for a 0.2 sec exposure the material had high resistivity.

in the vent/run approach [6]. The results presented here are consistent with our earlier work on the exposure time during the growth of strained layer superlattices using the molecular stream epitaxy method [7].

It seems evident that an ALE reactor using TMGa should have several features not necessary for the MOCVD case. The ALE growth system should minimize heating of the gas phase, and allow short exposure time if high temperature operation and low carbon concentration are desired. The short exposure time requires sudden termination of the exposure period to

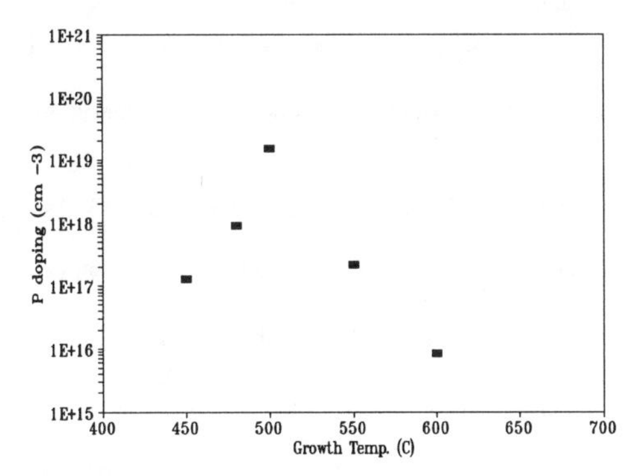

Figure 6 Temperature dependence of doping concentration for a 1 second exposure to TMGa (0.08μmoles/sec).

TMGa flux using means such as the mechanical approach, rather than relying on the out diffusion of the precursors away from the substrate surface.

The increased carbon incorporation with growth temperature in the low temperature range of 450°C to 500°C can be explained by the increased probability of decomposition of methyl radicals, such as $GaCH_3$. This leads to the release of more CH_3 radicals, and thus increases carbon incorporation due to increased number of collisions of the CH_3 molecules with the growing surface. This decomposition process of $Ga(CH_3)_3$ seems to be completed at about 500°C leading to maximum carbon doping [8]. Reduction of carbon incorporation with growth temperature above 500°C may be due to factors such as increased desorption rate of adsorbed CH_3 from the surface, more effective formation of CH_4 during the AsH_3 exposure, and others. An in-situ characterization tool is necessary to confirm the previous explanation.

SUMMARY

In a vertical, atmospheric pressure, reactor with a rotating susceptor the exposure time to TMGa plays a crucial role in achieving ALE growth and controlling the background doping. ALE was achieved over a wide temperature range of 450°C to 700°C due to the reactor design. By proper adjustment of the growth temperature and exposure time the p-type doping level was controlled from $10^{15}cm^{-3}$ to $10^{20}cm^{-3}$. These results suggest the influence of a boundary layer which inhibits the self limiting mechanism by allowing gas phase reactions to dominate substrate surface reactions.

This work is supported by ONR/SDIO, SERI and NSF. We would like to thank Brian McDermott for his assistance.

REFERENCES

1. K. Mochizuki, M. Ozeki, K. Kodama, and N. Ohtsuka, J. Cryst. Growth 93,557 (1988).

2. M. Ozeki, K. Mochizuki, N. Ohtsuka, and K. Kodama, Appl. Phys. Lett. 53, 1509 (1988).

3. S.M. Bedair, M.A. Tischler, T. Katsuyama, and N.A. El-Masry, Appl. Phys.Lett. 47, 51 (1985).

4. M.H.J.M. de Croon and L.J. Giling, J. Electrochem. Soc. 137, 2867 (1990).

5. K. Kodama, M. Ozeki, K. Mochizuki, and N. Ohtsuka, Appl. Phys. Lett. 54, 656 (1989).

6. E. Colas, R. Bhat, B.J. Skromme, and G.C. Nihous, Appl. Phys. Lett. 55, 2769 (1989).

7. T. Katsuyama and S.M. Bedair, J. Appl. Phys. 63, 5098 (1988).

8. Ming L. Yu, Ulrich Memmert, and Thomas Kuech, Appl. Phys. Lett.55 , 1011 (1989).

GROWTH RATE LIMITING AND CARBON REDUCTION PROCESSES FOR GaAs GROWN BY ALTERNATE GAS SUPPLY USING H_2 AND N_2 CARRIER GASES

M. SHINOHARA, Y. YOKOYAMA AND N. INOUE

NTT LSI Laboratories

3-1, Morinosato Wakamiya, Atsugi-shi, Kanagawa, Japan

Abstract

Processes limiting the growth of GaAs grown by an alternate gas supply are investigated by kinematical analysis. Based on these results, it is shown that the atomic layer epitaxial (ALE) window is expanded on the high temperature by the suppression of the decomposition of column III gas sources using a nitrogen carrier gas and on the low temperature side by the enhancement of their chemisorption to substrate surface atoms by a new method using a cracking tube. The latter enables us to achieve ALE of AlAs for the first time. Moreover, the carbon concentration is reduced by one order of magnitude by such a reaction control.

Introduction

Atomic layer epitaxy (ALE) is one of the most promising technique to grow atomically-controlled epilayers, and is expected to be applied to the fabrication of quantum effect devices. However, there are some serious problems that have prevented ALE from becoming a truly practical technique. One is the narrowness of the window for the substrate temperature called the ALE window. Another is the high level of carbon contamination in the epilayer.

Many efforts using laser irradiation,[1],[2] a fast gas stream from a gas nozzle,[3] and chloride column III gas sources [4] have been made to solve these problems. Although great progress has been made, the ALE window of AlAs, which is a promising material for hetero structures with GaAs, has not yet been obtained and growth margins for getting low carbon concentration are still narrow. These problems arise from the nature of the growth mechanism itself which has not yet been clarified.

In this paper, processes limiting growth at the high and low ALE window ends are examined. Based on these results, the expantion of the ALE window and achievment of ALE of AlAs are provided. In addition, the carbon reduction method is shown.

Experiments

Growth was carried out using trimethylgallium (TMG), trimethylaluminum (TMA), and 10 % AsH_3 diluted by hydrogen at a low pressure of 70 Torr. The epilayers were grown in the substrate temperature from 450°C to 560°C with input flows of TMG of 1.6×10^{-6} mol/cycle, TMA of 9.2×10^{-7} mol/cycle and arsine of from 6.8×10^{-6} to 3.4×10^{-5} mol/cycle. The substrate temperature was measured using a pyrometer calibrated by the melting points of Al and InSb. Either hydrogen or nitrogen were used as carrier gases, and their flow rates were fixed at 3 slm for all sequences. The gas was supplied for 1 sec and purged for 3 sec. In order to analyze the gas decomposition characteristics, a quadrupole mass spectrometer was used to sample gas through a capillary tube positioned 1 mm above the susceptor. The carbon concentration was determined by Hall effect measurement and secondary ion mass spectrometry.

Results

1. Processes limiting growth using hydrogen carrier gas

The growth rate dependence on the substrate temperature for GaAs is shown by the filled circles in Fig. 1. The reaction process limiting growth was determined by comparing the activation energy for growth to the known activation energies for several related reactions. The activation energy in the region of the growth rate above 1 monolayer/cycle is estimated to be 1.2 eV. According to the result measuring the decomposition characteristics of TMG in hydrogen,[5] the high ALE window end corresponds to the temperature where the TMG decomposition occurs. Moreover, Ga droplet formation on the epilayer surface shows that an excess Ga is supplied.[5] Since the Ga formation energy from TMG decomposition is much larger than the activation energy of the growth rate and Ga is excessively supplied, Ga supply does not limit the growth rate. The activation energy is close to the heterogeneous decomposition energy of arsine of 1 eV.[6] Therefore, the process limiting growth is caused by a lack of As to cover excess Ga, that is, by the decomposition of arsine.

The activation energy in the region of the growth rate below 1 monolayer/cycle is 1.9 eV corresponding to the decomposition energy of TMG on the As surface of 1.8 eV.[1],[7] Therefore, chemisorption of TMG with As on the surface is found to limit growth.

2. Expansion of the ALE window on the high temperature side by nitrogen carrier gas

From the process limiting growth at the high temperature region, the suppression of TMG decomposition in the vapor phase is found to be effective for expantion of ALE window on the high temperature side. Thus, nitrogen is substituted for hydrogen as the carrier gas. It is because hydrogen promotes the decomposition of TMG by reacting with methyl radicals, whereas TMG decomposes only thermally in nitrogen.[8] The growth rate dependence on the substrate temperature using nitrogen carrier gas is shown by triangles in Fig. 1. As expected, a 20°C expansion of the high ALE window end is achieved.

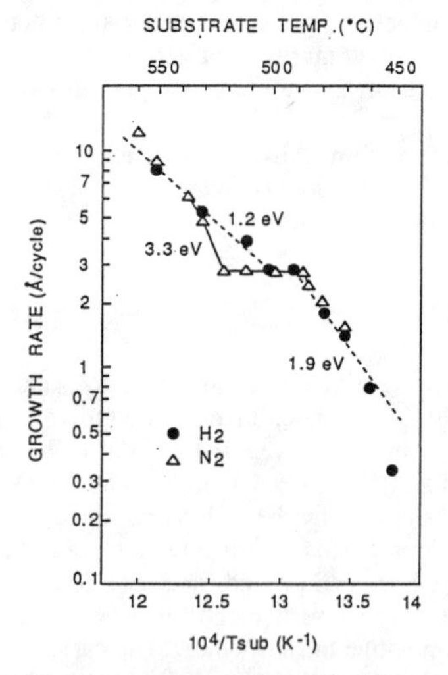

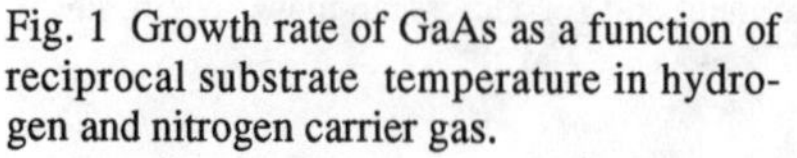

Fig. 1 Growth rate of GaAs as a function of reciprocal substrate temperature in hydrogen and nitrogen carrier gas.

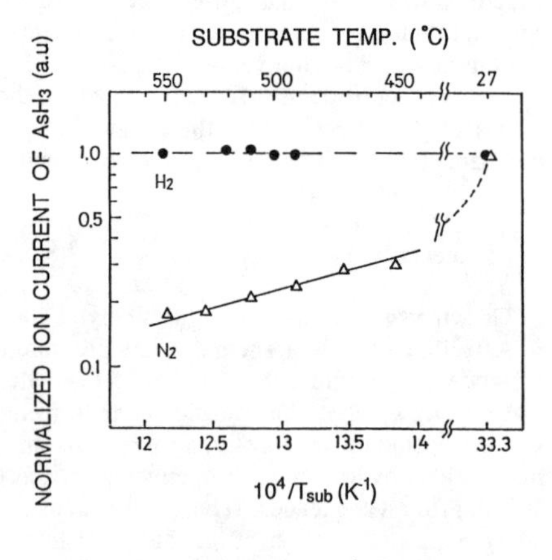

Fig. 2 Decomposition of arsine as a function of temperature in hydrogen and nitrogen. The ion current of arsine is normalized by the value at room temperature.

Compared to hydrogen, nitrogen has characteristics of lower thermal conductivity, higher density and dynamic viscosity. This makes the boundary layer thickness thinner. Accordingly, the effect of the thin boundary layer on the suppression of TMG decomposition would not be completely excluded. However, since a diffusion coefficient of TMG in nitrogen is lower than that in hydrogen, the time for TMG to stay within each boundary layer is nearly equal. Moreover, we have ascertained that the formation of methane with decomposition of TMG takes place only in hydrogen.[5] As a result, the suppression effect of TMG decomposition is thought to be mainly due to the difference in the reaction between TMG and the carrier gas.

3. Processes limiting growth using nitrogen carrier gas

The process limiting growth in the region above 1 monolayer/cycle is different from that in hydrogen. From the result that Ga droplets were not observed on this epilayer surface, [5] the supply of Ga atoms from the vapor phase is very low, and alkylgalliums are related to growth. In order that the growth rate exceeds 1 monolayer/cycle, the first 1 monolayer alkylgallium should decompose to Ga, and the following alkylgallium should adsorb to the Ga layer.

The activation energy of the process limiting the growth rate is a function of alkylgallium formation energy E_{for}, its decomposition energy to Ga E_{dec}, adsorption energy to Ga E_{ads} and desorption energy from Ga E_{des}. In this growth region their alkylgalliums are dimethylgallium (DMG) and monomethylgallium (MMG). Since these amounts forming in the vapor phase are thought to be much larger than those reaching the substrate surface, E_{for} is not necessary to be taken into consideration. Moreover, adsorption energies of DMA and MMA are thought to be very low because they are easily adsorbed through danling bonds. Accordingly, approximation of the growth rate R can be given by

$$R=n\cdot\exp(-E_{dec}/kT)-m\cdot\exp(-E_{des}/kT) \quad (1)$$

where, n and m are the constants. Although the decomposition energies (E_{dec}) of DMA and MMA have been estimated to be 4.8 eV and 3.4 eV, respectively,[9] their desorption energies are unknown. Therefore, it is difficult to determine what species contributes the growth. However, since the origin of carbon is deduced to be a methylradical dissociated from MMG as described later, it is liable that decomposition and desorption of MMG determine the growth.

The reason that the decomposition of arsine is not the limiting process for nitrogen carrier gas is found to be due to the enhancement of arsine decomposition. The temperature dependence of an arsine-ion current in hydrogen and nitrogen atmospheres as measured with a quadrupole mass spectrometer is shown in Fig. 2. The constant value for hydrogen means that it does not affect the arsine decomposition, whereas the steady decrease in nitrogen shows that arsine decomposes in the vapor phase. On the contrary, there is a report where arsine decomposition is independent of a carrier gas such as hydrogen, helium and nitrogen.[10] Although the model of decomposition enhancement is not clear, it is possible that convection may occur and promote heterogeneous decomposition of arsine on the reactor wall due to the high dynamic viscosity of nitrogen.

Where the growth rate is below 1 monolayer/cycle, the activation energy is the same as that for hydrogen carrier gas as shown in Fig. 1. This means the surface decomposition of TMG is independent of any kind of carrier gas.

4. Expansion of the ALE window on the low temperature side (application to ALE of AlAs)

To date, some efforts at ALE of AlAs have been made,[11],[12] but they have not yet succeeded. The most likely reason for their lack of success is that the decomposition energy of

TMA and TEA in the vapor phase is very close to the decomposition energy on the substrate surface. Actually, even in our experiment, ALE of AlAs is not achieved for conventional alternate gas supply growth as shown by the empty circles in Fig. 3.

The process limiting growth below the low ALE window end suggests either the promotion of surface decomposition of column III gas sources,[1],[2] or the supply of sources which are easily chemisorbed to As effectively expands the ALE window on the low substrate temperature side. Based on these premises, a new method with a cracking tube available for supplying partially decomposed column III gas has been developed and applied to the growth of AlAs.

The filled circles in Fig. 3 show the result for the new method attaching the cracking tube through which TMA is introduced to the reactor. Then, the temperature of the cracking tube was set high enough to decompose TMA into dimethylaluminum (DMA), monomethylaluminum (MMA), and Al.[13] It is noticed that the growth rate is 1 monolayer/cycle in the region from 460°C to 480°C. Achievement of ALE in this region was proved from the constant growth rate regardless of the TMA flow rate.[13]

Although the mechanism of ALE when using the cracking tube is not clear, one possible model is shown in Fig. 4. In the cracking tube, some TMA decompose into DMA, MMA and Al.[13] Almost all Al adhere to the inside wall of the cracking tube and don't come off of it. If the DMA and MMA reach the substrate, they would be easily chemisorbed to As bacause of having dangling bonds. However, it is unlikely that DMA and MMA radicals reach the substrate without reacting under short mean free path in a high pressure of 70 Torr because they are very active. So we suppose that a dimethylaluminum hydride (DMAH), which is a stable organoaluminum compound, is formed by the reaction of DMA with hydrogen carrier gas. The formation of DMAH is probable because the reaction of DMA with hydrogen at the heated substrate surface was found.[14] In the vapor phase monomethylaluminum hydride (MMAH) may be also formed. However, it would change to DMAH by the exchange reaction between hydrogen and methylradical because MMAH is very unstable. When DMAH reaches the substrate surface, hydrogen dissociates more easily than methylradicals due to

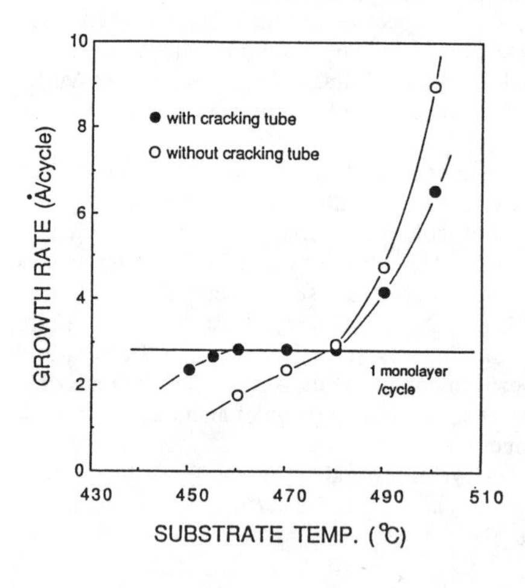

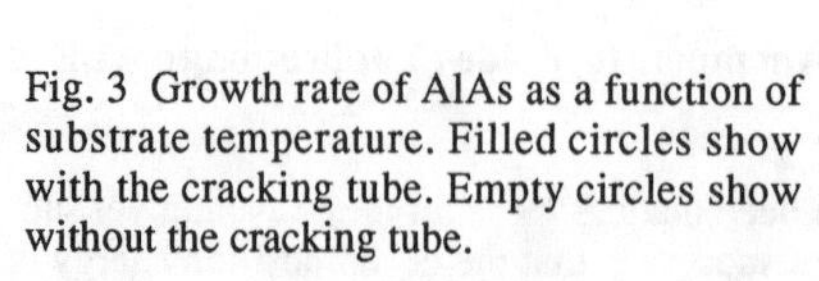
Fig. 3 Growth rate of AlAs as a function of substrate temperature. Filled circles show with the cracking tube. Empty circles show without the cracking tube.

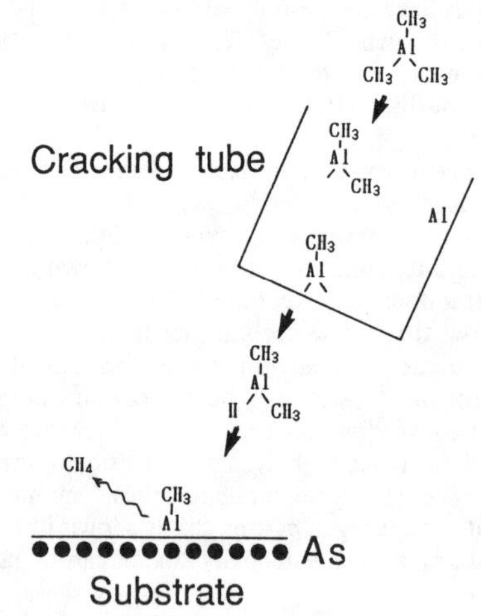

Fig. 4 One model of reactions of TMA using the cracking tube.

catalysis between Al and As. Thus, chemisorption to the substrate occurs at a relatively low temperature. The result is the extension of the ALE window on the low temperature side.

5. Carbon incorporation and reduction

The rate of carbon incorporation into GaAs as a function of the substrate temperature for the hydrogen and nitrogen carrier gas is shown in Fig. 5. The source of the carbon is generally said to be methyl radicals. In the ALE region, since the growth rate is constant, in other words, the number of methyl radicals are constant, the slight decrease in carbon incorporation with decreasing temperature suggests a carbon reduction process. The energy of carbon reduction is 0.33 eV. As the carbon reduction mechanism, we support the conventional view that methyl radicals react with hydrogen from the arsine to form methane.[15],[16] Although the activation energy of this reaction is not clear, the above energy is satisfactory because most actvation energies of reactions associated with alkyl radicals are about 0.4 eV.

In the region of the growth rate above 1 monolayer/cycle, the activation energy appears as the difference between the energies of the incorporation and reduction processes. The activation energy of the incorporation reaction is estimated to be 3.3 eV, which is close to the dissciation energy of methylradical from MMG (3.4 eV). This suggests that the source of the carbon is the single methyl radical that becomes dissociated from MMG.

The interesting point is that the carbon incorporation is one order of magnitude lower when using nitrogen carrier gas as shown in Fig. 5. Since the activation energy for the carbon reduction process is nearly equivalent to that in hydrogen carrier gas, the same reaction would occur. So, the noticeable carbon reduction is due to the increase in hydrogen radicals formed by the enhanced arsine decomposition.

In order to investigate the electrical quality, the hole mobility is measured. The hole mobility dependence on carrier concentration is shown in Fig. 6. The broken line indicates the calculated curve by Wiley[17] considering ionized impurity scattering. The filled circles represent the mobilities for using nitrogen carrier gas, whereas the empty ones represent the mobilities for hydrogen. Both data agree well with the calculation. This means that the carri-

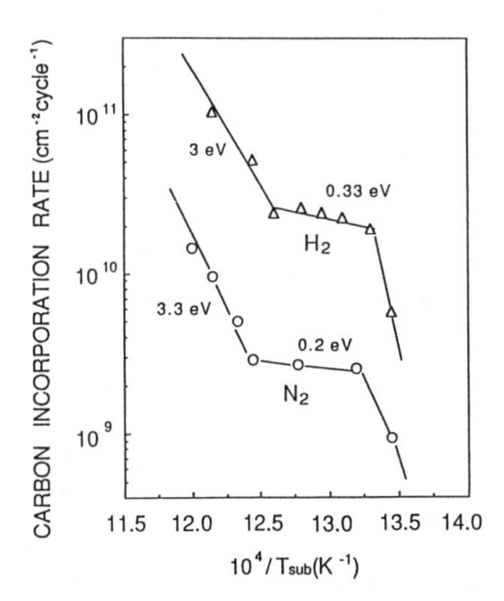

Fig. 5 Carbon incorporation rate for GaAs grown in hydrogen and nitrogen carrier gases as a function of substrate temperature.

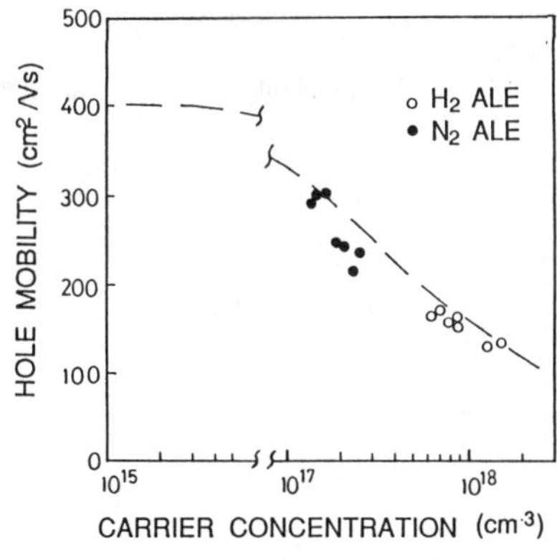

Fig. 6 Hole mobility dependence for GaAs on carrier concentration. The broken line is the calculated curve by Wiley considering ionized impurity scattering.

er compensation is very low in these epilayers.

Conclusion

The growth rate limiting processes in the low and high substrate temperature regions around ALE for hydrogen and nitrogen carrier gases are summarized in table 1. Based on these results, it is shown that the high ALE window end for GaAs is expanded by 20°C using nitrogen carrier gas by suppressing the TMG decomposition in the vapor phase. Moreover, the method to expand the low ALE window end is developed; by the enhancement of chemisorption of column III gases to the substrate surface using the cracking tube. By applying this method, ALE of AlAs is achieved. Nitrogen carrier gas also has the added benefit of cutting carbon concentration in GaAs from about 1 x $10^{18}/cm^3$ to about 1.5 x $10^{17}/cm^3$.

Table 1 Factors limiting the growth rate of GaAs

	limiting processes	
carrier gas	> 1 monolayer/cycle	< 1 monolayer/cycle
hydrogen	surface decomposition of arsine 1.2 eV	surface decomposition of TMG 1.9 eV
nitrogen	decomposition and desorption of DMG or MMG	

Acknowledgements

The authors would like to thank Takashi Kobayashi for helpful discussions and Kazuo Hirata for continous encouragement.

References

[1] Y.Aoyagi, A.Doi, S.Iwai and S.Namba, J. Vac. Sci. Technol. B5, 1460 (1987).
[2] Y.Kawakyu, H.Ishikawa, M.Sasaki and M.Mashita, Jpn. J. Appl. Phys. 28, L1439 (1989).
[3] M.Ozeki, K.Mochizuki, N.Otsuka and K.Kodama, Appl. Phys. Lett. 53, 1509 (1988).
[4] A.Usui and H.Sunakawa, Jpn. J. Appl. Phys. 25, L212 (1986).
[5] H.Yokoyama, M.Shinohara and N.Inoue, submitted to Appl. Phys. Lett.
[6] K.Tamaru, J. Phys. Chem. 59, 777 (1955).
[7] H.Ishikawa, Y.Kawakyu, M.Sasaki and M.Mashita, Jpn. J. Appl. Phys. 12, L2327 (1989).
[8] J.Nishizawa and T.Kurabayashi, J. Electrochem. Sci. 130, 413 (1983).
[9] M.G.Jacko and S.J.W.Price, Can. J. Chem. 41, 1560 (1963).
[10] R.Luckerath, P.Tommack, A.Hertling, H.J.Koss, P.Balk, K.F.Jensen and W.Richter, J. Crystal Growth 93, 151 (1988).
[11] T.Megro, S.Iwai, Y.Aoyagi, K.Ozaki, Y.Yamamoto, T.Suzuki, Y.Okano and A.Hirata, J. Crystal Growth 99, 540 (1990).
[12] M.Ozeki, K.Mochizuki, N.Ohtsuka and K.Kodama, J. Vac. Sci. Technol. B5(4), 1184 (1987).
[13] H.Yokoyama, M.Shinohara and N.Inoue, submitted to Appl. Phys .Lett.
[14] K.Masu, K.Tsubouchi, N.Shigeeda, T.Matano and N.Mikoshiba, Appl. Phys. Lett. 56, 1543 (1990).
[15] H.Mori and S.Takagishi, Jpn. J. Appl. Phys. 12 L877 (1984).
[16] T.F.Kuech and E.Veuhoff, J. Crystal Growth. 68, 148 (1984).
[17] J.D.Wiley, in Semiconductor and Semimetals 10, eds. R.K.Willardson and A.C.Beer (Academic, New York, 1975), P.91.

STUDY OF As AND P INCORPORATION BEHAVIOR IN GaAsP BY GAS-SOURCE MOLECULAR-BEAM EPITAXY

B. W. LIANG, H. Q. HOU, and C. W. TU
Department of Electrical and Computer Engineering, University of California at San Diego, La Jolla, CA 92093-0407

ABSTRACT

A simple kinetic model has been developed to explain the agreement between *in situ* and *ex situ* determination of phosphorus composition in $GaAs_{1-x}P_x$ ($x < 0.4$) epilayers grown on GaAs (001) by gas-source molecular-beam epitaxy (GSMBE). The *in situ* determination is by monitoring the intensity oscillations of reflection high-energy-electron diffraction during group-V-limited growth, and the *ex situ* determination is by x-ray rocking curve measurement of $GaAs_{1-x}P_x$/GaAs strained-layer superlattices grown under group-III-limited growth condition.

INTRODUCTION

Mixed group-V ternary and quaternary III-V compounds, such as GaAsP, InAsP, and InGaAsP, are important for optoelectronic applications. Hydride-, chloride-, and metalorganic vapor-phase epitaxy (MOVPE) [1-4], molecular-beam epitaxy (MBE) [5] and gas-source molecular-beam epitaxy (GSMBE) [6], including hydride-source MBE (HSMBE), metalorganic MBE (MOMBE) or chemical-beam epitaxy (CBE), have been used to grow these materials. One of the critical issues of epitaxial growth is controlling the composition of group-V elements in the compounds [7,8]. Most researchers in this area use post-growth characterization techniques, such as X-ray diffraction and photoluminescence spectroscopy, to correlate the composition with flow rates or fluxes during growth. Recently we have demonstrated an *in situ* determination of group-V composition in GSMBE (or HSMBE) of $GaAs_{1-x}P_x$ ($x < 0.4$), using group-V hydrides, arsine and phosphine, and elemental group-III sources [8]. The group-V composition of epilayers grown under normal MBE growth condition, where the growth rate is controlled by the group-III flux, can be obtained easily from the intensity oscillations of reflection high-energy-electron diffraction (RHEED) under the group-V-limited growth condition at the same flow rates and growth temperature. Obviously, there is a competition between As and P during growth of GaAsP [7] and understanding this competition is the key point to determination of group-V composition. Even though several models have been proposed in recent years for composition determination of mixed-group-V ternary or quaternary compounds grown by MOVPE and (GS)MBE, they can not explain the agreement between the *in situ* composition determination of mixed-group-V compounds by RHEED intensity and *ex situ* composition determination by X-ray rocking curves [2,3,5]. In this paper we take into account this competition process and propose a simple kinetic model for GSMBE of $GaAs_{1-x}P_x$ to understand the relationship between the group-III-limited and the group-V-limited growth modes.

A KINETIC MODEL

In GSMBE, the growth rate and composition may be controlled by four different kinetic processes, mass transfer (i.e., beam fluxes), adsorption, desorption, and surface reaction. For binary compounds like GaAs, under normal growth conditions (T_s in the 450 - 650°C range and R_g in the order of 1 monolayer/s), the growth rate is expected to be limited by mass transfer, which means that surface reaction rates are much faster than that of mass transfer. In this case the growth rate does not depend on the substrate temperature. On the other hand, in very low temperature or very high beam flux range, the growth rate may be controlled by surface reaction. The growth rate then will be sensitive to temperature variation. For mixed group-III ternary compounds such as AlGaAs, the situation is almost the same as that of GaAs. The growth rate and composition are controlled simply by Al and Ga beam fluxes. For mixed group-V ternary compounds, however, the situation is quite different. Even though the growth rate is still controlled by the group-III element beam flux, the composition is controlled by chemical reactions between group-III atoms and group-V species and thermal desorptions of group-V species on the growth front surface.

On a Ga-rich surface, the growth rate (R_g) of GaAsP is limited by the incorporation rates of As_2 and P_2 [8]. We can write

$$R_{g(GaAsP)} = R_{g(GaAs)} + R_{g(GaP)} \quad (1)$$

Because Ga atoms are abundant on the surface, it is reasonable to assume that As_2 or P_2 can easily react with two Ga atoms at one time. In other words, we assume that the following elementary chemical reactions occur on the surface during growth of GaAs and GaP,

$$V_2\,(g) \xrightarrow{(FS)_V} V_2\,(ad)$$

$$V_2\,(ad) + 2\,Ga\,(ad) \xrightarrow{K_V} 2\,GaV$$

$$V_2\,(ad) \xrightarrow{D_V} V_2\,(g)$$

where V stands for As or P; F, the beam flux; S, the sticking probability; K_V, reaction rate coefficient; and D_V, desorption rate coefficient. Then, the growth rates can be written in terms of reaction rate coefficients and surface activities a_i (i = Ga or As or P),

$$R_{g(GaAs)} = K_{As}\, a_{Ga}^2\, a_{As} \quad (2)$$

$$R_{g(GaP)} = K_P\, a_{Ga}^2\, a_P \quad (3)$$

$$R_{dV} = D_V\, a_V \quad (4)$$

where R_{dV} is desorption rate of group-V dimers. The total flux at growth front is then the sum of the growth rate and desorption rate.

$$(FS)_V = R_{gV} + R_{dV} \quad (5)$$

The phosphorus composition (x) in the epilayer can be written as

$$x = \frac{R_{g(GaAsP)} - R'_{g(GaAs)}}{R_{g(GaAsP)}}$$

where (') stands for the situation where only As_2 molecules are deposited. In principle, the As incorporation rate into GaAsP will change upon injection of P_2 due to displacement of As by P. Foxon *et al.* reported a slight decrease of As_4 sticking coefficient when P_4 was injected during MBE of GaAsP [7]. The situation is similar if dimers are used. In our case, however, excess Ga atoms are deposited on the surface before the group-V-limited RHEED oscillation measurements. Therefore, we can expect the displacement of As by P is negligible. In addition, As has much higher sticking coefficient than P, especially when both are present [7]. In short, for group-III-rich surface, the As incorporation rate is independent of whether P is present or not, at least for low x range. Therefore, $R_{g(GaAs)}$ and $R'_{g(GaAs)}$ conceal out with each other, i.e.,

$$x \approx \frac{\frac{(FS)_P}{A_P}}{\frac{(FS)_P}{A_P} + \frac{(FS)_{As}}{A_{As}}} \qquad (6)$$

where

$$A_V \equiv 1 + \frac{D_V}{K_V} = \left(1 + \frac{1}{\alpha}\right) \text{ for As and } \left(1 + \frac{1}{\beta}\right) \text{ for P}$$

Because of high vapor pressure and short surface lifetime of the group-V elements, if we assume that at the growth temperature considered, the desorption probability of group-V species is much greater than incorporation probability , i.e., $D_V/K_V >> 1$, then

$$x \approx \frac{1}{1 + \frac{F_{As}}{F_P}\frac{S_{As}}{S_P}\frac{\alpha}{\beta}} \qquad (7)$$

Since As_2 and P_2 beam fluxes are proportional to the AsH_3 and PH_3 flow rates, respectively [9], we can write

$$x \approx \frac{1}{1 + C\frac{f_{AsH_3}}{f_{PH_3}}} = \frac{1}{1 + C\frac{(1 - X_f)}{X_f}} \qquad (8)$$

where f's are hydride flow rates; X_f, the phosphine flow-rate fraction; and

$$C \equiv \frac{\alpha\, S_{As}}{\beta\, S_P}\sqrt{\frac{M_{P_2}}{M_{As_2}}}$$

Now we consider normal GSMBE growth condition (group-V rich), where not so many Ga atoms are available as in the Ga-rich condition. The following elementary surface reactions involving dimers are assumed,

$V_2\,(g) \rightarrow V_2\,(ad)$

$V_2\,(ad) + Ga\,(ad) \rightarrow GaV + V\,(ad)$

$V\,(ad) + Ga\,(ad) \rightarrow GaV$

$V\,(ad) + V\,(ad) \rightarrow V_2\,(ad)$

$V_2\,(ad) \rightarrow V_2\,(g)$

The growth-rate equation of GaAsP in this case is similar to equation (1), but it is limited by the Ga-beam flux.

$$R_{g(GaAsP)} = R_{g(GaAs)} + R_{g(GaP)}$$

$$= K_{As}\, a_{Ga}\, a_{As} + K_P\, a_{Ga}\, a_P = \lambda\, F_{Ga} \qquad (9)$$

where λ is a constant relating the Ga flux and the growth rate, $R_{g(GaAsP)} = \lambda F_{Ga}$. Therefore,

$$x = \frac{R_{g(GaP)}}{R_{g(GaAsP)}} = \frac{K_P\, a_{Ga}\, a_P}{\lambda\, F_{Ga}}$$

Using equation (4), we have

$$x = \frac{(FS)_P}{(FS)_P + (FS)_{As}\frac{\alpha}{\beta} + R_{g(GaAsP)}} \qquad (10)$$

$$x = \frac{1}{1 + \frac{F_{As}}{X_F F_{(As+P)}}\left(\frac{S_{As}}{S_P}\right)\left(\frac{\alpha}{\beta}\right) + \frac{\lambda F_{Ga}}{X_F S_P F_{(As+P)}}} \qquad (11)$$

where X_F is group-V beam flux fraction of P_2. Since V/III > 1 in term of beam flux and D_V/K_V >> 1, in low x range, $F_{As}S_{As} \gg \lambda F_{Ga}$. Also since As incorporation is more efficient than P incorporation at the growth temperature considered, i.e., $\alpha/\beta \geq 1$, we can have

$$x = \frac{1}{1 + \frac{F_{As}}{F_P}\frac{S_{As}}{S_P}\frac{\alpha}{\beta}} = \frac{1}{1 + C\frac{f_{AsH_3}}{f_{PH_3}}} = \frac{1}{1 + C\frac{(1 - X_f)}{X_f}} \qquad (12)$$

which is identical to equation (8).

COMPARISON WITH EXPERIMENTAL RESULTS AND DISCUSSION

Equations (8) and (12) reveal the relationship between phosphorus composition (x) in the epilayer and flow-rate fraction of PH_3 (X_f) as well as other parameters included in the parameter C. The interesting and important point is that under the physically reasonable assumptions we made above, they are applicable for two different growth modes, group-III-limited and group-V-limited, in the low x range. If we know these parameters, such as S, K and D, we can obtain the parameter C and calculate the growth rates under different conditions. Since K and D obey an Arrhenius relation, the phosphorus composition depends on the growth temperature. Unfortunately, experimental data of these parameters are not available. However, since in equation (12) x and X_f are determined experimentally at a given growth temperature, we can calculate C for different temperatures. For T_s = 500°C, the average of C equals 2.48. Then, we obtain the phosphorus composition as a function of phosphine flow rate for the same growth temperature, as shown in Fig.1.

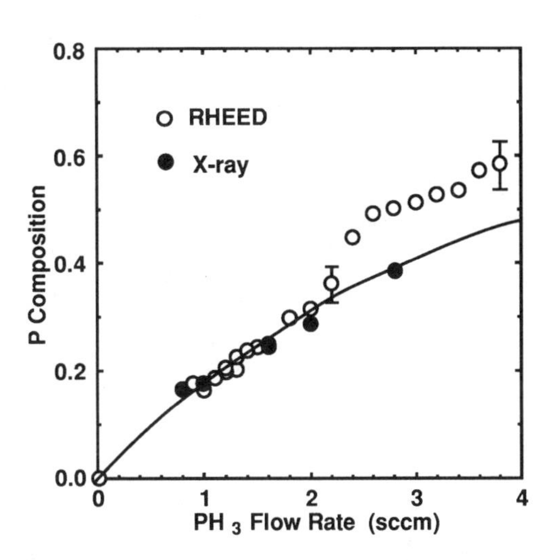

Fig.1 P composition in GaAsP on GaAs (001) as a function of PH_3 flow rate at AsH_3 flow rate of 1.6 sccm and growth temperature of 500°C. Closed circles are from X-ray diffraction study of GaAsP /GaAs SLS's. Open circles indicate RHEED oscillations results, and the line is calculation result based on equation (8), once C is determined from one data point.

Figure 1 also compares the P composition determined *in situ* by RHEED, shown by open circles. Obviously, the *in situ* and *ex situ* determinations agree well with each other in low x range ($x < 0.4$). For other growth temperatures, C= 2.73 for 420°C, 1.3 for 580°C and 0.96 for 700°C. According to equation (12) phosphorus composition in GaAsP epilayers is independent of V/III ratio in the temperature range considered now. This has been verified by our experimental results [8].

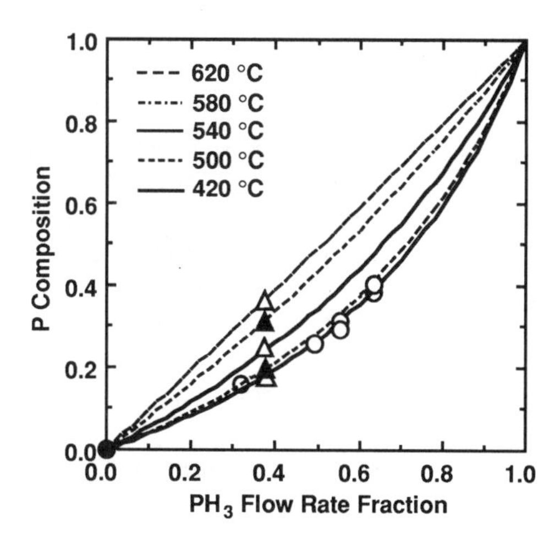

Fig. 2 P composition in GaAsP on GaAs (001) as a function of PH_3 flow rate fraction at different growth temperatures. The closed circles are from X-ray diffraction of GaAsP /GaAs SLS's grown at 500°C. Other symbols indicate SLS's grown at different substrate temperatures.The lines are calculations from the kinetic model.The triangles were used to get C.

In high x range, however, the results from the *in situ* determination does not agree with that from the *ex situ* ones. The reasons are believed to be (1) it is difficult to obtain reasonably good RHEED oscillation because of high strain; (2) the SLS structure may be partially relaxed because of strain; and (3) the assumptions in equation (6) that As and P incorporation being independent on a Ga-rich surface may not hold any more because of the existence of unnegligible amount of phosphorus on the growth front.

Figure 2 shows the P composition as a function of flow-rate fraction in gas phase at different growth temperatures. The lines come from our model. It is interesting to notice the similarity between results from GSMBE and those from MOVPE. It is well known that in MOVPE the growth rate are limited by diffusion of group-III species through the boundary layer above the susceptor under normal growth conditions. However, the group-V composition is controlled by competing reaction rates between group-V species and group-III's and desorptions of group-V species on the substrate surface. This is because that the group-V overpressure is very high and diffusion rates are much higher than surface reaction rates. In this aspect MOVPE is similar to the situation of GSMBE. Because GSMBE is carried out in high vaccum, the corresponding temperatures are lower than those in MOVPE. Also, it is interesting to notice that $C \approx 1$, when growth temperature is about 620°C. In this case the phosphorus composition is almost equal to the flow-rate fraction of PH_3. The composition control then is easier than at low temperatures.

CONCLUSIONS

A simple kinetic model has been proposed to explain the arsenic and phosphorus incorporation behavior in $GaAs_{1-x}P_x$ grown on GaAs (001) by GSMBE. For $GaAs_{1-x}P_x$ ($x < 0.4$) epilayers, one can use the *in situ* determination by RHEED to obtain the phosphorus composition. At a growth temperature of about 620°C, a simple relation exists between phosphorus composition in epilayers and flow-rate fraction.

ACKNOWLEDGEMENTS

We wish to thank T. P. Chin and M. C. Ho for assistance in GSMBE growth. This work is partially supported by the Office of Naval Research.

REFERENCES

[1] A. D. Huelsman, J. Electron. Mater. **18**, 91 (1989).

[2] L. Samuelson, P. Omling and G. Grimmeiss, J. Cryst. Growth **61**, 425 (1983).

[3] H. Seki and A. Koukitu, J. Cryst. Growth **74**, 172 (1986).

[4] G. B. Stringfellow, J. Cryst. Growth **70**, 133 (1984).

[5] T. Nomura, H.Ogasawara, M. Miyao and M. Hagino, J. Cryst. Growth (1991).

[6] M. B. Panish and S. Sumski, J. Appl. Phys. **55**, 3571 (1984).

[7] C. T. Foxon, B. A. Joyce and M. T. Norris, J. Cryst. Growth **49**, 132 (1980).

[8] H. Q. Hou, B. W. Liang, T. P. Chin and C. W. Tu, Appl. Phys. Lett. (1991).

[9] T. P. Chin, B. W. Liang, H. Q. Hou, M. C. Ho, C. E. Chang, and C. W. Tu, Appl. Phys. Lett. **58**, 254 (1991).

[10] K. Woodbridge, J. P. Gowers and B. A. Joyce, J. Cryst. Growth **60**, 21 (1982).

GROWTH OF $(GaAs)_{1-x}$ $(Si_2)_x$ METASTABLE ALLOYS USING MIGRATION-ENHANCED EPITAXY

T. Sudersena Rao and Y. Horikoshi
NTT Basic Research Laboratories, Tokyo 180, Japan.

ABSTRACT

Epitaxial $(GaAs)_{1-x}(Si_2)_x$ metastable alloys have been grown on GaAs (100) substrates using Migration-Enhanced Epitaxy in the composition range of $0<x<0.25$. The lattice constant a_0 of the alloys was found to decrease with increasing Si content from 0.56543nm at x=0 to 0.5601nm at x=0.25. Double-crystal x-ray diffraction rocking curve measurements and cross-sectional transmission electron microscopy studies made on a 10 period $(GaAs)_{1-x}(Si_2)_x$/GaAs strained layer superlattice indicated sharp and abrupt interfaces. High crystalline quality GaAs has been grown on Si substrates using $(GaAs)_{0.80}(Si_2)_{0.20}$/GaAs strained layer superlattices as buffer layers.

1. Introduction:

In recent years there has been increasing interest in the growth of metastable alloy systems of the type $(III\text{-}V)_{1-x}$ $(IV_2)_x$ due to their several interesting fundamental properties and potential for device applications [1]. Attempts to fabricate these alloys using conventional equilibrium melt cooling techniques have not been very successful due to the problem of phase separation [2]. Epitaxial layers of these alloys, however can be grown using -Metalorganic Chemical Vapor Deposition (MOCVD), Molecular Beam Epitaxy (MBE) and Sputtering techniques [1,3-5,7], which operate in a highly non-equilibrium regime. The present work is concerned with growth of a new $(III\text{-}V)_{1-x}(IV_2)_x$ system of GaAs-Si solid solutions. The lattice constant of these alloys can be tailored in between the values of 5.6534 Å for GaAs and 5.345 Å for Si, with a corresponding change in the band gap from direct 1.43 eV to indirect 1.1 eV respectively. Potential applicability of these alloys in semiconductor laser technology has also been recently demonstrated by Burnham et.al [5]. They have found that thin $(GaAs)_{1-x}(Si_2)_x$ alloy layers placed inside the quantum well of a AlGaAs/GaAs single quantum well lasers resulted in shifting the output emission to higher energy. Another technologically important area where these alloy layers can find immediate applications is as buffer layers during the growth of GaAs on Si. The deleterious effects of large lattice and thermal mismatch between GaAs and Si [6], may be reduced by employing a graded $(GaAs)_{1-x}(Si_2)_x$ or a $(GaAs)_{1-x}(Si_2)_x$ /GaAs Strained Layer Superlattice (SLS).

In this paper, we report, some of our recent results of structural studies made on $(GaAs)_{1-x}(Si_2)_x$ alloys epitaxially grown on GaAs (100) substrates using Migration-Enhanced Epitaxy (MEE). MEE is a modified MBE technique, which allows the growth of high quality GaAs/AlGaAs interfaces over a wide range of temperatures [9,10].

In the MEE mode growth of GaAs, the Group III and Group V atoms are alternately deposited on the substrate. As a result, the As concentration is reduced during the Ga cycle which is believed to enhance the surface migration of Ga atoms and to allow for the growth of high quality material [11].

We will also present some preliminary results concerning the growth of GaAs on Si using $(GaAs)_{1-x}(Si_2)_x$/GaAs SLS as buffer layers. The composition of epi-layers was characterized using Auger Electron Spectroscopy (AES) and Secondary Ion Mass Spectrometry (SIMS). Structural studies of the epi-layers were performed using cross-sectional Transmission Electron Microscopy (X-TEM) and Double -Crystal X-ray Diffraction rocking curve measurements (DCXRD).

2.Experimental:

A 3-chamber MBE system equipped with 30 kV RHEED gun for in-situ growth monitoring was employed to grow $(GaAs)_{1-x}(Si_2)_x$ alloys. Conventional PBN effusion cells were used for Si beam generation. During the GaAs growth in the MEE mode, the Ga and As_4 beam shutters are opened alternatively. The Ga, As_4 and Si cell shuttering sequence for the $(GaAs)_{1-x}(Si_2)_x$ alloy layer growth is shown as inset in Fig.1. Due to the low silicon beam flux and the relatively high growth temperature (>520°C) employed the As_4 beam shutter opening was coincided with Si deposition in order to suppress the re-evaporation of As_4. The alloy composition was varied by changing the shutter opening time for Si cell.

3. Results and Discussion:

In the present study, we have grown a large number of $(GaAs)_{1-x}(Si_2)_x$ layers with various composition x values of $0<x<0.25$ and at various growth temperatures in the range of 500-620°C. Initial growth experiments revealed that films grown with higher silicon content (x>0.10) and above 580°C exhibited very rough surface morphology and appeared milky to the naked eye. However, for substrate temperatures between 580°C and 530°C, films showing smooth, and shiny mirror like surfaces could routinely be obtained. Figure 1 shows the x-ray rocking curves for a 0.3μm thick $(GaAs)_{0.88}(Si_2)_{0.12}$ epi-layer on GaAs (100) substrates. Mei et al [8] have also recently reported X-ray diffraction scan for $(GaAs)_{1-x}(Si_2)_x$, x=0.11 alloys on GaAs (100) substrates grown by hybrid sputter/deposition technique. Higher order diffraction peaks of (200) and (600) were also observed in the in the x-ray diffraction patterns for alloy layers of $(GaAs)_{1-x}(Si_2)_x$ for $0<x<0.25$, indicating the zincblende structure [12]. The full width at half maximum (FWHM) x-ray intensity values of the alloy peaks depicted in Fig.1 is 144 arc-seconds. The value of 144 arc-seconds for 0.3μm thick $(GaAs)_{0.88}(Si_2)_{0.12}$ film, indicates high crystalline quality of the alloy epi-layer.

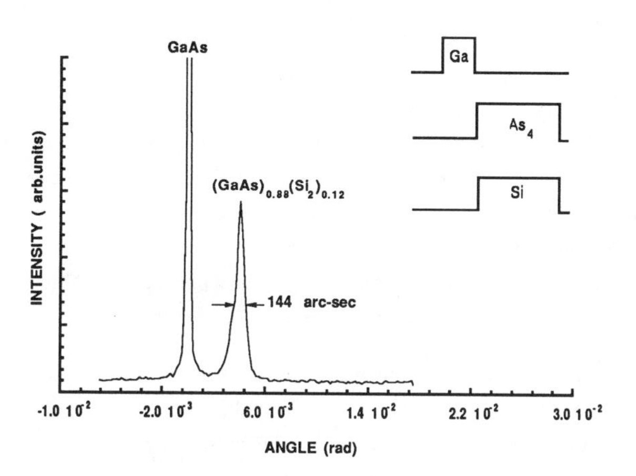

Fig. 1. DCXRD (400) rocking curves of a 0.3μm thick $(GaAs)_{0.88}(Si_2)_{0.12}$ alloy layer. The inset shows the effusion cell shuttering sequence during the $(GaAs)_{1-x}(Si_2)_x$ MEE

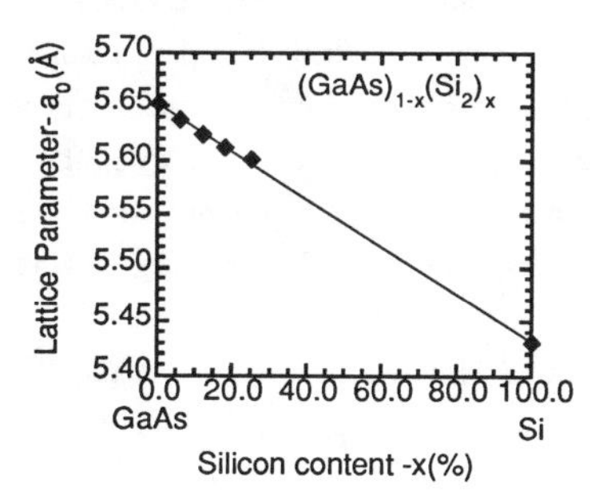

Fig.2: The variation of DCXRD (400) peak rocking curves FWHM value for $(GaAs)_{1-x}(Si_2)_x$ alloys as a function of Silicon content x.

Increasing the Si content in the alloy layers resulted in broader (400) x-ray peak FWHM values. The (400) X-ray peak broadening with increasing Si content may be due to defects generated in the epilayer as a result of the increasing lattice constant mismatch between the alloy layer and the substrate. The calculated lattice constant a_0 of the $(GaAs)_{1-x}(Si_2)_x$ alloys and the variation with Si content is shown in Fig. 2. The solid line in the inset corresponds to the expected lattice constant variation with increasing Si content based on Vegards rule. The alloy lattice constant decreases with increasing Si content from 0.56534nm at x=0 to 0.5601nm for x=0.25 in good agreement with Vegards rule. In order to test the metastability of these layers, we have performed Rapid Thermal Annealing (RTA) of these layers. No observable degradation in the structural properties could be detected for annealing temperatures upto 800^0C for 10secs. Currently RTA experiments at higher temperatures >800^0C are in progress.

In order to study the microstructure and defect generation, cross-sectional TEM studies were made on $(GaAs)_{1-x}(Si_2)_x$ alloy layers with different x values. For $(GaAs)_{1-x}(Si_2)_x$, x=0.10, the alloy layer were found to be almost featureless, with no evidence of dislocations or defects. For increasing Si content occasional stacking faults or a dislocation gliding through the thickness of the epilayer was observed. For higher Si content films (x>0.18), however, alloy layers were found to be heavily dislocated, with large areas of stacking faults and dislocations running through the thickness of the epi-layers.

One of the major objectives of the present study was to investigate the applicability of $(GaAs)_{1-x}(Si_2)_x$ alloy layers as lattice matching and thermal expansion coefficient matching layers for growth of GaAs on Si system. In order to achieve this, the crystalline quality of the $(GaAs)_{1-x}(Si_2)_x$ alloy layers for x>0.15, needed to be improved. This can be performed either by limiting the thickness of the alloy epi-layer below the critical thickness or a lattice matched template can be provided by using a properly designed $(GaAs)_{1-x}(Si_2)_x$ /GaAs strained-layer superlattices (SLS). Such a superlattice can also be very effective as dislocation barrier in GaAs on Si. Additionally, such superlattices can also provide a good tool for investigating the $(GaAs)_{1-x}(Si_2)_x$ interface abruptness. In view of this, we grew several $(GaAs)_{1-x}(Si_2)_x$ /GaAs SLS, and with varying period values. In Fig. 3, a typical double-crystal x-ray rocking curve for one such $(GaAs)_{1-x}(Si_2)_x$ /GaAs SLS (10 periods of 320Å, $d_{(GaAs)1-x(Si2)x}$= 160Å and d_{GaAs} =160Å). Apart from the main SLS peak at 33.332^0, satellite peaks up to third order are clearly visible in Fig.3.The main SLS peak is very sharp and its FWHM of 74 arc-secs indicates the high crystalline quality of the multilayer structure. The satellite peak separation can be used to calculate the period of the superlattice using standard kinematic assumptions and it comes out to be 317Å.

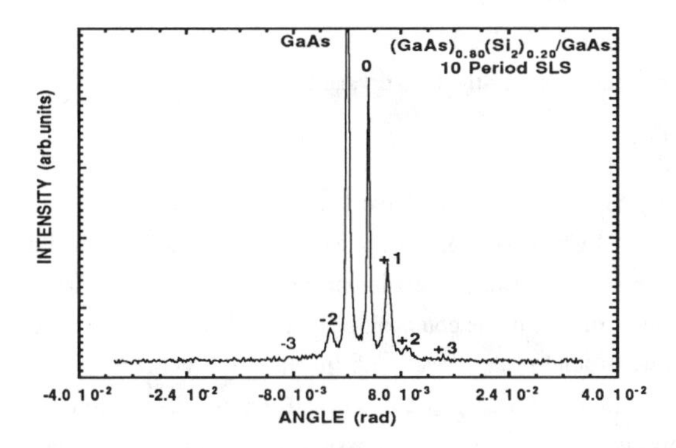

Fig. 3. DCXRD (400) rocking curves for a 10 period $(GaAs)_{0.80}(Si_2)_{0.20}/GaAs$ strained layer superlattice with a period of 320Å, $d_{(GaAs)1-x(Si2)x}$ = 160Å and d_{GaAs} =160Å.

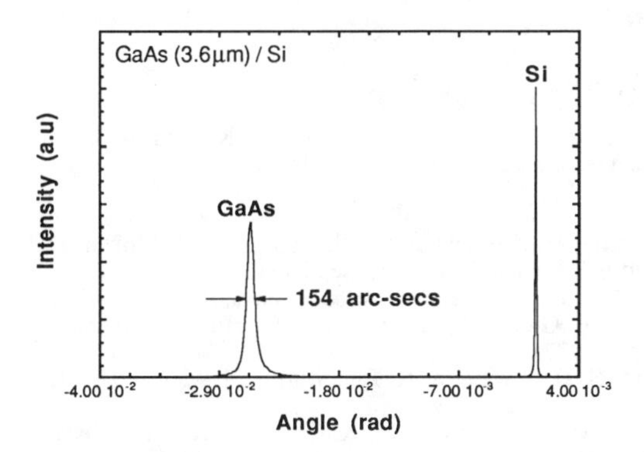

Fig4: DCXRD rocking curve for GaAs grown on Si with a 10 period$(GaAs)_{1-x}(Si_2)_x/GaAs$,x=0.20, SLS as buffer layer.

Currently, we are studying the applicability of $(GaAs)_{1-x}(Si_2)_x/GaAs$ SLS as buffer layers for improving the quality of GaAs grown on Si (100) substrates. We have grown several GaAs layers on Si (100) substrates using $(GaAs)_{1-x}(Si_2)_x/GaAs$ SLS as buffer layers using x=0.20. In Fig.4, the X-ray rocking curve for 3.5μm as grown GaAs grown on Si substrates, with a 10 period $(GaAs)_{1-x}(Si_2)_x/GaAs$ SLS as buffer layer is shown. A standard two step growth procedure has been employed for this sample, where after the customary oxide desorption at 1000^0 C, the sample is cooled to 300 ^{0}C and a initial nucleating layer is grown using MEE, followed by a 15 min annealing at 580^{0}C. Subsequently a 10 period $(GaAs)_{1-x}(Si_2)_x/GaAs$, x=0.20 has been grown at 530^{0}C using MEE. The final GaAs epi-layer of 3.0μm was grown at standard MBE conditions at 580^{0}C. The FWHM of the GaAs peak shown in Fig. 4 is 154 arc-secs, clearly demonstrates the high crystalline quality of the as grown sample.

4. Conclusions:

$(GaAs)_{1-x}(Si_2)_x$ for 0<x<0.25 have been successfully grown on GaAs (100) substrates using Migration-Enhanced Epitaxy. Structural and compositional analysis of the as grown $(GaAs)_{1-x}(Si_2)_x$ layers indicated single phase epitaxial single crystals with zincblende structure in the compositional range 0<x<0.25. The lattice constant a_0 of the alloys was found to decrease with increasing Si content from 0.56543nm at x=0 to 0.5601nm at x=0.25. Double-crystal x-ray diffraction rocking curve measurements and cross-sectional transmission electron microscopy studies made on a 10 period $(GaAs)_{1-x}(Si_2)_x$/GaAs strained layer superlattice indicate abrupt interfaces. High crystalline quality GaAs exhibiting a DCXRD FWHM of 154 arc-seconds was subsequently grown using a 10 period $(GaAs)_{1-x}(Si_2)_x$/GaAs SLS as 'Buffer' layer on Si substrates.

Acknowledgments:

The authors would like to thank Dr. T. Kimura for his keen interest and encouragement throughout this work.

References

1. Zh. I. Alferov, M. Z. Zhingarev,S. G. Konnikov,I. I. Mokan, V.P.Ulin, V.E. Umanskii and B.S.Yavich: Sov. Phys. Semicond.16 , 532 (1982).
2. C. Kolm, S. A. Kilin and B. L. Averbach: Phys. Rev 108 , 965 (1957).
3. S. I. Shah, J. E. Greene, L. L. Abels, Q. Yan and P. M. Raccah: J. Cryst. Growth 83 , 3 (1987).
4. R. J. Baird, H. Holloway, M. A. Tamor, M. D. Hurley and W. C. Vassell: J.Appl. Phys. 69 ,(1), 226 (1991).
5. R. D. Burnham, N. Holonyak. Jr., K. C. Hsich, R. W. Kaliski, D. W. Nain, R. L. Thornton and T. L. Paoli: Appl. Phys. Lett. 48 , 800 (1986).
6. W. Stolz, M. Naganuma and Y. Horikoshi: Jpn. J. Appl. Phys. 27 (1988) L283.
7. A. J. Noreika and M. H. Francombe: J. Appl. Phys. 48 , 3690 (1974).
8. D. H. Mei, Y. W. Kim, D. Lubben, I. M. Robertson and J. E. Greene: Appl. Phys. Lett. 55 , 2649 (1989).
9. Y. Horikoshi, M. Kawashima and H. Yamaguchi: Jpn. J. Appl. Phys. 25 , L868 (1986).
10. Y. Horikoshi, M. Kawashima and H. Yamaguchi: Jpn. J. Appl. Phys. 27 ,169 (1988).
11. Y. Horikoshi and M. Kawashima: Jpn. J. Appl. Phys. 28 , 200 (1989).
12. B.D.Cullity: Elements of X-ray Diffraction (Addesion-Wesley, Minto Prk, CA, 1978.)

MULTI-WAFER ATOMIC LAYER EPITAXY REACTOR FOR DEVICE QUALITY GaAs

P.C. Colter, S.A. Hussien, A. Dip, M.U. Erdogan and S.M. Bedair
North Carolina State University, Dept. of Electrical and Computer Engineering, Raleigh, NC 27695-7911

ABSTRACT

Reactor design considerations are discussed relevant to the two main problems, carbon contamination and low growth rate, facing Atomic Layer Epitaxy (ALE) of GaAs. A new reactor design addressing these problems is described. It utilizes the concept of rotating the substrate between streams of reactant gases. The growth chamber provides baffles and gas jets to shear off and sweep away the thermal boundary layer after exposure to the reactive gas streams. Construction is based on modification of a commercially available low pressure MOCVD reactor equipped with a load lock. The reactor is capable of processing three, two-inch wafers. A background carbon concentration of about $10^{15}cm^{-3}$ and a 77°K mobility of 30,000 cm^2/V-sec were achieved. Self limited growth was observed for a growth temperature as high as 600 °C. Controlled p- and n-type doping was accomplished by changing growth conditions and adding silane.

INTRODUCTION

Atomic layer epitaxy (ALE) possesses many advantages which make it an attractive choice as an MOCVD growth technique. Primarily, control over film thickness on the order of a single monolayer can be accomplished. Other benefits include the ability to easily scale up or down growth chamber dimensions without affecting growth rates or uniformity. Also, flexibility in the precise control of reactant gas flows and exposure times is available due to the self-limiting characteristic of ALE.

However, there have been several limitations with the ALE growth technique which have hindered its acceptance as a "next generation" growth technique for III-V semiconductors. Predominate among these limitations is the high level of carbon contamination from the source gases. Very low growth rates and restrictions on sample sizes have also posed serious barriers[1].

An MOCVD reactor design must incorporate several important features to achieve growth in the ALE regime. Among these are rapid switching of the reactant gases over the substrate and the minimal heating of the thermal boundary layer. If reactant gases such as trimethylgalluim (TMG) are allowed to dwell over the substrate for an extended period, decomposition of the TMG will result. This is unwanted as a growth rate greater than one monolayer per cycle or even the collection of gallium droplets may occur. In addition, excessive decomposition of TMG will cause a large incorporation of carbon into the film. This produces a heavily p-type semiconductor with loss of low level doping control[2].

In this paper we discuss the design of an MOCVD reactor that overcomes these limitations of the ALE growth technique and is simultaneously capable of producing device quality GaAs[3]. Most importantly for device applications, this system is capable of controlled doping by means of adjustment of the reactant flows (p-type) or by addition of silane (n-type) gas.

EXPERIMENTAL

The reactor used in this investigation is a modified Emcore 3200 system which was operated at 30 torr. It consists of three basic components: the gas delivery system, growth cabinet, and the system controller unit. The gas delivery system contains all of the equipment (mass flow controllers, tubing, etc.) to deliver the process gases at given flow and/or concentration into the growth chamber. Flows, temperatures, pressures and exposure times are all programmed into and executed by the system controller unit. The growth cabinet contains the growth chamber, load lock and pumps. A schematic of the principle gas streams and major system components is shown in Figure 1 .

In general, the principle distinguishing feature of growth systems is the growth chamber design and construction. The other components such as the gas delivery system tend to be standardized, since those designed for MOCVD use are adequate. In this reactor the gas delivery and system controller are those of a standard Emcore 3200 reactor. The growth chamber has been modified for ALE as we shall discuss shortly.

Previous growth chamber designs for use in ALE investigations have taken one of two approaches. First, consider the pulse jet epitaxy (PJE) which was reported by Mochizuki et al.[2,4]. In this method, the boundary layer is reduced by high speed injection of reactant gases onto the substrate. Gas switching between arsine and TMG is accomplished by the vent/run method. This precedes as follows. TMG is run into the chamber, then after some dwell time is switched back to the vent line. This is followed by a TMG purge cycle where hydrogen is run into the system. Following the purge cycle, arsine is then run into the chamber for some duration and switched back to the vent. Finally, another purge cycle occurs and the process repeats. Growth chamber design for this technique is straight forward, but material growth by this method has a very slow growth rate and is limited to temperatures below 550°C.

The second approach to the ALE chamber design was developed in our laboratory[5]. Here, the boundary layer is mechanically sheared off by a fixed obstruction. The two reactant gas streams (e.g. TMG and arsine) are run *simultaneously* through the chamber separated both by mechanical and gas barriers.

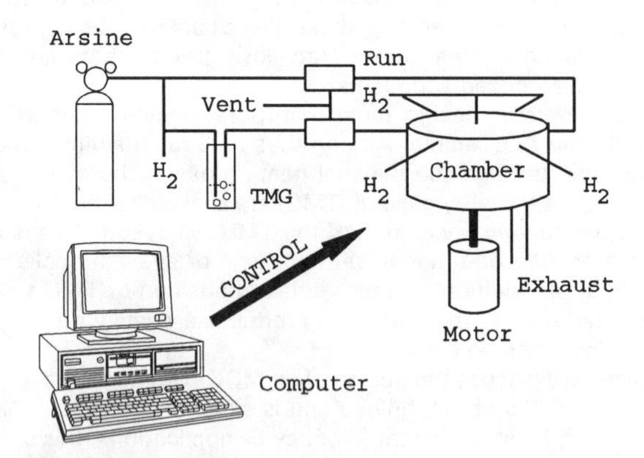

Figure 1 A schematic of the process gas flows and major system components.

The substrate is then rotated from the TMG to the arsine stream to complete one cycle. In the process of rotation, the boundary layer is sheared off by a mechanical barrier. The growth rate is primarily determined by the minimum dwell time under each gas stream for saturation of the surface by one monolayer of atoms. No purge cycle is required between the two streams. Growth by this method proceeds more quickly and has a greater upper temperature limit of about 700°C.

The growth chamber design discussed in this paper is a hybrid of the two designs outlined above. Gas flows are introduced into the chamber by means of six uniformly spaced coplanar injectors. The jets deliver the gases at a very high speed. In order to achieve high gas velocities, the reactor pressure was kept low and the outlets of the jets were kept small. Figure 2 shows a schematic of the growth chamber as viewed from above. The chamber is subdivided into six equally spaced compartments by the insertion of 0.01" thick molybdenum baffles. These partitions help to shear off the boundary layer above the substrate. Only two of the jets shown carry the reactant gases. The remaining four jets are used for hydrogen injection. These additional jets are used to purge reactant gases from the substrate as the substrate rotates.

In addition to these four hydrogen jets, the standard Emcore shroud flow dispersed by a perforated plate at the top of the chamber is present. It is thought that this background flow of hydrogen serves two purposes, both suppressing recirculation above the substrate and providing additional purging.

A growth cycle is composed of an exposure to TMG, flushing by H_2, an exposure to AsH_3, and flushing by H_2. This is done by rotating the substrate. Growth rates of 0.4 and 0.6 μm/h were achieved by rotational speeds of 24 and 36 rpm, respectively. Silane (n-type) and carbon (p-type) were used as dopants.

RESULTS

A growth rate curve for GaAs was generated by varying the TMG concentration at a temperature of 550 °C and is shown in Figure 3 . Points on this curve were generated by a series of two hour runs done at both 0.4 and 0.6 μm/h growth rates.

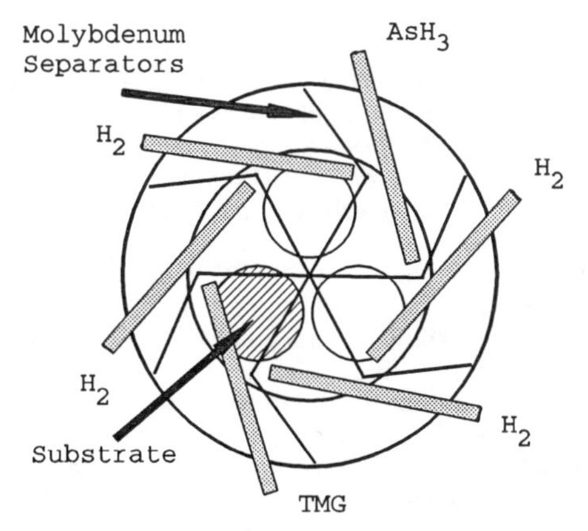

Figure 2 A drawing of the growth chamber as seen from above.

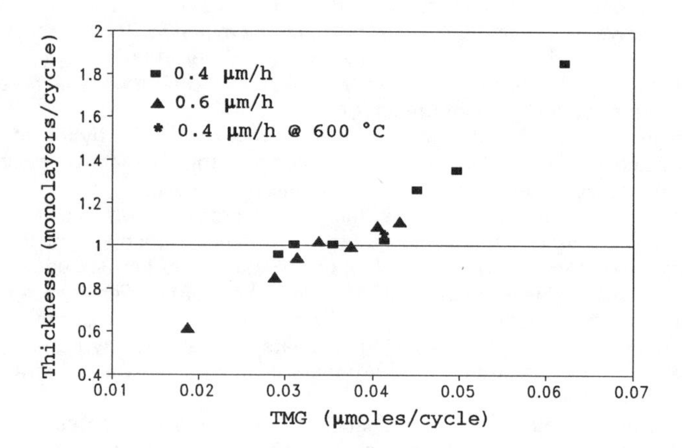

Figure 3 Growth curve for GaAs.

The most prominent feature of the growth curves is the plateau of saturated growth indicative of ALE. Thickness were evaluated by staining the samples with A-B etchant followed by measurements with an optical microscope. Optical thickness measurements were also cross checked by TEM. The interface regions of the deposited layers were so defect free that it was necessary to first grow an AlGaAs layer to delineate the GaAs epilayer from the substrate after the A-B etch.

Film uniformity across a two-inch substrate was observed to be good. Saturated growth at one monolayer/cycle for 75% of the wafer farthest from the center of rotation was observed. The 25% of the substrate closest to the central part of the susceptor had a growth rate which was always less than the rest of the substrate. This is due to the reduction in TMG flux at this location in the chamber.

By controlling the V/III ratio it was possible to control the doping of the ALE GaAs films with carbon. It was possible to control carbon doping over the range of 10^{15} to 10^{19} cm^{-3}. Figure 4 shows a doping curve obtained from varying the arsine concentration at 550 °C. Data points along this curve were made by Hall measurements. Increasing the V/III ratio (TMG flux was kept constant) was found to decrease the carbon doping level, which is in good agreement with results published by Mochizuki et al.[2] at Fujitsu.

At high V/III ratios, films grown at 580 °C were depleted and Hall data could not be obtained. We therefore assume that the carrier concentration was less than 10^{15} cm^{-3}. However, Figure 4 shows a data point for carrier concentration at 580 °C for a lower V/III ratio. The best liquid nitrogen Hall mobility obtained was about 30,000 cm^2/V-sec with a carrier concentration of low 10^{15} cm^{-3} and a room temperature mobility of about 5000 cm^2/V-sec. It is important to note that although high purity ALE material growth has been reported elsewhere[2,6], careful inspection will show that these materials were not grown in the self limiting range of ALE.

A series of experiments using a 1000 ppm silane source in H_2 was done to

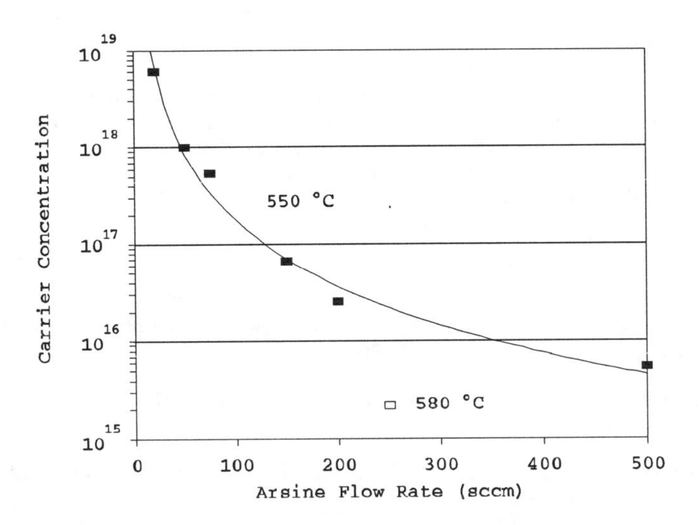

Figure 4 Carbon incorporation as a function of arsine flux.

obtain an n-type doping curve for ALE GaAs. Figure 5 shows a linear dependance on the doping level due to silane from 10^{16} to 10^{19} cm^{-3}. Points on this curve were obtained by Hall measurements on the same high quality ALE GaAs films discussed above.

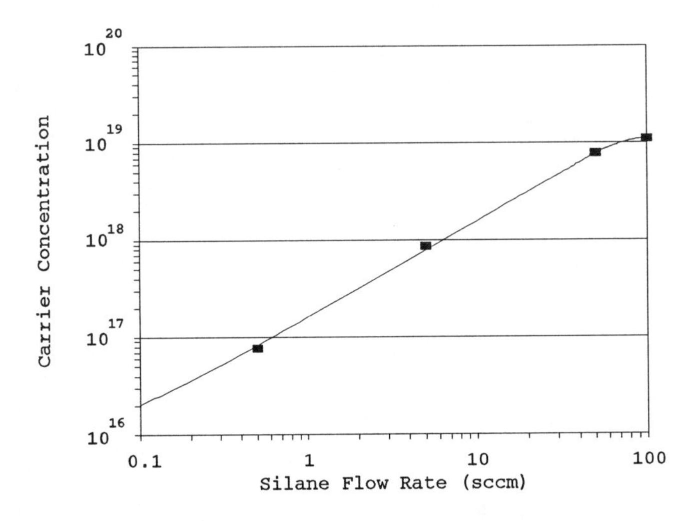

Figure 5 Carrier concentration as a function of silane flow rate.

We have demonstrated the growth by ALE of device quality GaAs at a technically useful growth rate as high as 0.6 μm/h on large area substrates. By changing the growth conditions and by using silane, the doping has been controlled from high resistivity (below 10^{15} cm^{-3}) to 10^{19} cm^{-3} for both n and p type doping. These results have been obtained in a low pressure, rotation switched (substrate rotates between isolated TMG and AsH_3 fluxes) ALE reactor.

This work is supported by the NSF, ONR/SDIO and SERI. We would like to acknowledge helpful discussions with Drs. J.R. Hauser and P.E. Norris.

REFERENCES

1. M.A. Tischler and S.M. Bedair, Atomic Layer Epitaxy, T. Suntola and M. Simpson eds., Blackie, Glascow and Linden, 1990.

2. K. Mochizuki, M. Ozeki, K. Kodama and N. Ohtsuka, J. Crystal Growth 93, 557 (1988).

3. P.C. Colter, S.A. Hussien, A. Dip, M.U. Erdogan, W.M. Duncan, and S.M. Bedair, APL, to be published (1991).

4. M. Ozeki, K. Mochizuki, N. Ohtsuka, and K. Kodama, Appl. Phys. Lett. 53 (16), 1509 (1988).

5. S.M. Bedair, M.A. Tischler, T. Katsuyama and N.A. El-Masry, Appl. Phys. Lett. 47, 51 (1985).

6. E. Colas, R. Bhat, B.J. Skronme and G.C. Nihous, Appl. Phys. Lett. 55, 2769 (1989).

HIGH-CONDUCTANCE GaAs TUNNEL DIODES BY OMVPE

R. VENKATASUBRAMANIAN, M.L. TIMMONS, and T.S. COLPITTS
Research Triangle Institute, Research Triangle Park, NC 27709

ABSTRACT

GaAs p^+-n^+ tunnel diodes have been grown by atmospheric-pressure organometallic vapor phase epitaxy (OMVPE) using zinc as the dopant for the p^+ regions and either Se or Si as the dopant for the n^+ regions. At a growth temperature of 700 °C, using a "cycled" growth for just the Zn-doped p^{++}-GaAs layer both the conductance and the peak current of the tunnel diode has been increased by a factor of ~ 65. The conductance of the tunnel diode, maximized at a growth temperature of 650 °C with the cycled growth, is comparable to the best reported values by MBE. Cycled growths for n^+ Se-doped regions reduce the tunnel-diode conductance by more than two orders of magnitude. However, the cycled growth for n^+-GaAs regions formed with Si doping shows no conductance degradation. A model for these observations is presented.

INTRODUCTION

Monolithic cascade solar cells, based on lattice-matched material systems such as AlGaAs/GaAs [1] and $GaInP_2$/GaAs [2], offer the promise of high-efficiency conversion for photovoltaic power systems. A key to these high efficiency multijunction solar cells is a highly conductive interconnect using a GaAs tunnel junction. A GaAs tunnel junction is also useful as an active load in GaAs FET logic circuits and can also be used as a memory element [3]. A discussion of the I-V characteristics of tunnel diodes is given elsewhere [4]. For the purpose of this paper, the conductance of the tunnel diode is defined as the ratio of the peak current density (J_p) to the voltage (V_p) across the diode at J_p.

Important to a tunnel junction is the excess current, a component over and above the tunneling and diffusion currents that exist in the forward-biased p^+-n^+ junction. The excess current can reduce the (J_p/J_v) ratio, reducing the negative differential region (NDR) in the I-V characteristic of the tunnel diode. While the lack of a well-pronounced NDR makes the tunnel diode unsuitable for digital applications, it is tolerable for interconnect application in multijunction solar cells. The excess current arises from band-tail tunneling between the p^+- and n^+-regions; the heavy doping invariably produces band tails in both the p^+- and n^+-type materials. Further, heavy doping can introduce additional deep states in the band gap that promote carrier tunneling between the p^+- and n^+-regions of the tunnel junction via the deep states.

In this paper we present results on conductance and the NDR behaviour of the GaAs p^+-n^+ tunnel junctions grown by atmospheric-pressure OMVPE with zinc as the dopant for the p^+ regions and either Se or Si as the dopant for the n^+ regions. An approach, called "cycled" growth, is examined and has been shown to improve the conductance of the tunnel diodes under certain doping conditions. The material properties of Zn:GaAs, Se:GaAs, and Si:GaAs using cycled-growth, is also described.

EXPERIMENTAL

The GaAs tunnel junctions were grown in a conventional atmospheric-pressure OMVPE growth system, using trimethylgallium (TMGa) and arsine (AsH_3). The growth temperature was varied between 650-700 °C. The AsH_3/TMGa ratio was typically 30 and the H_2 flow rate was 8 slm. Dimethylzinc (DMZn) was used for p-type doping. Silane (SiH_4) was used to obtain Si doping and hydrogen selenide (H_2Se) was used for Se doping. In the case of "cycled" growth, AsH_3 and the respective dopants were maintained in the growth ambient and the TMGa flow was introduced periodically. The TMGa on-time was about 7 sec and the off-time was about 13 seconds. The growth rate employed during the on-cycle was about 5 Å/sec, the same as during a continuous/regular growth. The full details of the tunnel junction device fabrication and ohmic-contact requirements will be discussed elsewhere [5].

RESULTS AND DISCUSSION

The Hall carrier data and the 300 K photoluminescence (PL) peak energies of Zn:GaAs, using regular and cycled growths, are indicated in Table 1. Here the DMZn partial pressure and the growth temperature were kept constant at about 0.4 Torr and 700 °C, respectively, for both the growth cases. The hole concentration level falls by more than a factor of five for the cycled growth. This decrease was initially unexpected. Also, the lower PL peak energy in the cycled-growth sample, inspite of the lower hole concentration, is noteworthy.

We believe that during the off-cycle of the GaAs:Zn growth, the simultaneous presence of AsH_3 and DMZn over the GaAs surface results in the preferential adsorption of Zn onto interstial sites. Interstitial absorption of dopant atoms has been considered elsewhere [6]. Following the off-cycle, when growth is initiated, the Zn adatoms are probably trapped and incorporated onto the interstitial sites. The interstitial Zn (Zn_i) is expected to be a donor in GaAs [7] and therefore, can cause n-type compensation of the p^+-type material, resulting in a lower hole concentration. Further, the energies of Zn_i donors near the valence-band edge [7] can cause the overlap of the interband transition and the conduction band to Zn_i donor transition, causing the PL peak energy to shift to lower energies.

We are presently conducting a secondary ion mass spectrometry (SIMS) analysis of GaAs:Zn, both the regularly-grown and the cycled-grown, to compare the total amount of zinc to confirm the above speculation. However, in GaAs:Zn grown by

OMVPE, it has been shown that Zn incorporation increases (as measured by hole concentration) with growth rate of GaAs. This behaviour has been explained by a similar trapping mechanism [8].

The material characteristics of GaAs:Se, using regular and cycled growths, are indicated in Table 1. Here the H_2Se partial pressure and the growth temperature were kept fixed at about 0.015 Torr and 700 ° C, respectively. The electron concentration is very similar in both cases. The 300 K PL peak energy of the regularly-grown GaAs:Se, with a slightly higher free-carrier concentration, is higher than that of the cycled-grown sample. This is expected because of the increased band-filling effects of the higher free-electron concentration, causing the PL peak energy to shift to higher values. However, the PL intensity of the cycled-grown GaAs:Se was about 4-5 times higher than that of the regularly-grown GaAs:Se. This suggests that the non-radiative processes are less significant in the cycled-grown GaAs:Se, and it is likely that V_{Ga}-Se_{As} deep level complexes, which are known to occur in heavily doped n-GaAs [9,10], are thermally annealed out during the off-cycle of the cycled-growth. The thermal annealing of such deep levels and the consequent effect on lifetime (in turn on PL intensity) has been previously noted [11]. In effect, we observe that regularly-grown GaAs:Se is likely to have a larger concentration of V_{Ga}-Se_{As} deep levels than the cycled-grown GaAs:Se.

The Hall carrier data and the 300 K PL characteristics of GaAs:Si, using regular and cycled growths, are indicated in Table 1. The net free-carrier concentration in either growth are almost the same. However, the mobility of the cycled-grown GaAs:Si is significantly lower than that of the regularly-grown GaAs:Si, suggesting increased compensation with cycled-growth. Further, the 300 K PL peak energy of the cycled-grown sample is significantly lower than that of the regularly-grown sample in spite of very similar net free-carrier concentration. Both these observations are well explained by increased Si_{As}-acceptor compensation in the cycled-grown GaAs:Si. Si incorporation onto As-sites during the off-cycle is probably a result of less-than-unity As-coverage on the GaAs surface. It has been shown that Si_{As} is the dominant acceptor compensator in heavily-doped GaAs:Si grown by OMVPE [12] and that V_{Ga}-Si_{Ga} acceptor complexes play a negligible role. Thus, the thermal annealing is not expected to introduce any effect on near bandedge PL intensity.

The characteristics of GaAs tunnel diodes, using regular growth for GaAs:Se and either regular or cycled growth for GaAs:Zn, are shown in Figs. 1a and 1b. The peak current density (J_p) of the tunnel junction with the cycled GaAs:Zn, is higher by about 65, and the specific resistivity is correspondingly smaller by the same ratio than that of the junction with regular GaAs:Zn. The increase in conductance (Fig. 1b) is also accompanied by the disappearance of a well-marked NDR (as in Fig.1a). By plotting the J_p values versus effective doping concentration (as discussed in Ref.13) for both cases, we find that J_p in Fig.1a is smaller than the theoretical value by a factor of two,

	REGULAR GROWTH	CYCLED GROWTH
GaAs : Zn Hole Concentration (cm^{-3})	4.9×10^{19}	9.3×10^{18}
Hole Mobility (cm^2/Vs)	56	66
300 K PL Peak Energy (eV)	1.414	1.397
GaAs : Se Electron Concentration (cm^{-3})	6.6×10^{18}	6.1×10^{18}
Electron Mobility (cm^2/Vs)	1034	915
300 K PL Peak Energy (eV)	1.445	1.434
GaAs : Si Electron Concentration (cm^{-3})	4.5×10^{18}	4.3×10^{18}
Electron Mobility (cm^2/Vs)	1237	922
300 K PL Peak Energy (eV)	1.484	1.426

Table 1. Characteristics of GaAs:Zn, GaAs:Se, and GaAs:Si using regular and cycled growths.

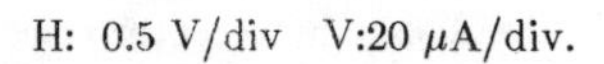

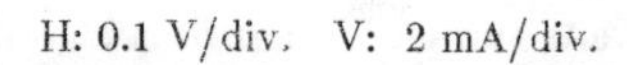

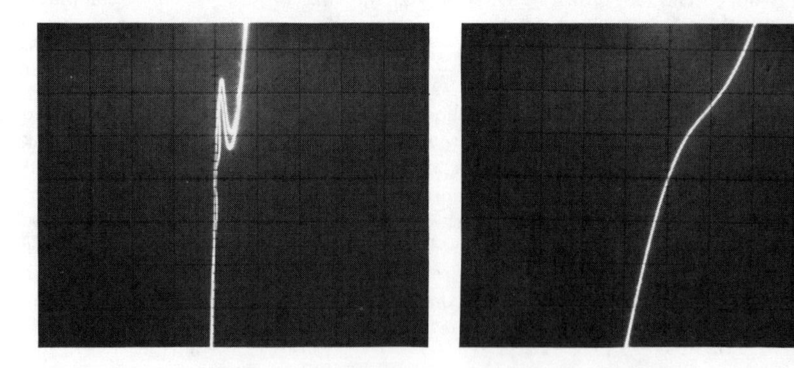

Figure 1. GaAs tunnel diode characteristic with a) regular growth and b) cycled growth for GaAs:Zn.

while the J_p in Fig.1b, is higher than the theoretical value by almost two orders of magnitude. We have also seen that a GaAs tunnel junction grown with both cycled GaAs:Zn and GaAs:Se regions, re-acquires the NDR as well as indicates a J_p value within a factor of two of the theoretical value.

Similar improvements in J_p and the conductance of the tunnel junctions with cycled GaAs:Zn regions were observed at a growth temperature of 650 ° C. However, the increase in J_p was ~50 compared to ~65 seen at a growth temperature of 700 ° C. At a growth temperature of 650 ° C, we have obtained a J_p of 12.8 A/cm^2 and a specific contact resistance of 3.3 x 10^{-3} ohm-cm^2. This is comparable to the lowest-reported value of 3.0 x 10^{-3} ohm-cm^2, for a GaAs tunnel junction grown by MBE at 520 ° C [14].

The only effect of cycled GaAs:Zn layer in GaAs tunnel junctions, grown using Si as the dopant for the n^+-region, was to reduce the J_p and proportionately increase the specific-resistance, mainly as a result of a fall in hole concentration in the cycled GaAs:Zn, as seen in Table 1. The J_p values were within about 20-30% of the theoretical values for the respective doping levels. The cycled growth of GaAs:Si did not affect the J_p and the conductance values, whether cycled or regular GaAs:Zn layer was used to make the tunnel junction. The I-V data and other characteristics such as thermal stability vis-a-vis tunnel diodes with GaAs:Se will be presented elsewhere [5].

We can explain the tunnel junction results in terms of the material properties discussed earlier. In the case of tunnel diodes with GaAs:Se (regular) and GaAs:Zn (cycled) doping, the presence of V_{Ga}-Se_{As} defects in the n^+-GaAs region and Zn_i in the p^+-GaAs region assist in excess carrier tunneling. Thus the J_p is about two orders of magnitude higher than the theoretical value, and the NDR disappears as a result of this excess current. However, with cycled GaAs:Se layer for n^+ GaAs, the concentration of V_{Ga}-Se_{As} centers is small due to thermal-annealing and even though Zn_i defects are in p^+-GaAs, there no accompanying states in n^+-GaAs to assist in carrier tunneling. Thus GaAs tunnel junction with cycled GaAs:Se and GaAs:Zn layers, show NDR as well as J_p close to the theoretical value for the doping level.

In the case of GaAs:Si and GaAs:Zn system, the negligible presence of Si_{Ga}-V_{Ga} complexes in Si:GaAs probably does not result in insignificant excess carrier tunneling even though Zn_i defects are present in cycled GaAs:Zn region. Hence, under all conditions of growth of GaAs tunnel junctions with GaAs:Si doping for the n^+-GaAs region, we obtain near-ideal J_p, consistent with the doping levels. We have also observed, in general higher J_p/J_v ratios (maximum of ~6) with GaAs tunnel diodes with Si doping for the n^+ GaAs [5].

SUMMARY

In summary we have examined an approach denoted as "cycled" growth in conventional OMVPE. The cycled growth apparently results in a significant fall in hole concentration in GaAs:Zn compared to regular growth. In the case of GaAs:Se, the free-electron concentration is unaffected with cycled growth while the 300 K photoluminescence intensity improves considerably. Neither the free carrier level nor the 300 K bandedge photoluminescence intensity is affected in cycled-grown GaAs:Si. The usefulness of the cycled-growth in the growth of high-conductance GaAs tunnel junctions is indicated.

ACKNOWLEDGMENT

This work was supported by Solar Energy Research Institute under Subcontract No. XM-0-18110-2, with Dr. R.L. Mitchell and Mr. T.S. Basso as the technical monitors.

REFERENCES

1. S.M. Bedair, J.A. Hutchby, J.P.C. Chiang, M. Simons, and J.R. Hauser, Proc. of the 15th IEEE Photovoltaic Specialists Conf. (IEEE, New York, 1982), p.692.
2. J.M. Olson, S.R. Kurtz, A.E. Kibbler, and P. Faine, Appl. Phys. Lett. 56, 623 (1990).
3. J.V. DiLorenzo and D.D. Khandelwal, "GaAs FET Principles and Technology", Artech House, Dedham, Massaschsetts, 646 (1982).
4. S.M. Sze, "Physics of Semiconductor Devices", 2nd Edition, John Wiley, New York (1981).
5. R. Venkatasubramanian, M.L. Timmons, and T.S. Colpitts, To be Published.
6. R.J. Field and S.K. Ghandhi, J. Cryst. Growth, 74, 551 (1986).
7. A. Luque, J. Martin, and G.L. Arujo, J. Electrochem. Soc., 123, 249 (1976).
8. T.F. Kuech, M.A. Tischler, R. Potemski, F. Cardone, and G. Scilla, J. Cryst. Growth, 98, 174 (1989).
9. D.T.J. Hurle, J. Phys. Chem. Solids, 40 (1979) 627.
10. R. Venkatasubramanian, S.K. Ghandhi, T.F. Kuech, J. Cryst. Growth, 97, 827 (1989).
11. G. Mathur, Ph.D Thesis, Rensselaer Polytechnic Institute, Troy, NY (1985).
12. R. Venkatasubramanian, K. Patel and S.K. Ghandhi, J. Cryst. Growth, 94, 34 (1989).
13. S.K. Ghandhi, R.T. Huang, and J.M. Borrego, Appl. Phys. Lett., 48, 415 (1986).
14. D.L. Miller, S.W. Zehr, and J.S. Harris, Jr., J. Appl. Phys., 53, 744 (1982).

A POSSIBILITY OF ALE GROWTH OF InN BY USING $InCl_3$

KAZUHITO HIGUCHI, AKIO UNNO and TADASHI SHIRAISHI
Tokai University, Dept. of Communications, 1117 Kitakaname
Hiratsuka 259-12, Japan
Fax. 81-463-58-1812

ABSTRACT

A possibility of the Atomic Layer Epitaxy, ALE, for InN was demonstrated by using $InCl_3/N_2$ and NH_3/N_2. The $InCl_3$ is a solid at room temperature and can be supplied in the reactor by heating with N_2 carrier gas. When the solid $InCl_3$ is heated in an inert gas, InCl, $InCl_3$ and In_2Cl_6 gases are formed. It was clear that the In_2Cl_6 which is the largest molecule of the three results in solid structural defects. The ALE growth temperature was from 440°C to 505°C. The fact that the ALE was performed at the temperature range from 440°C to 505°C indicates that In was supplied as $InCl_3$, suggesting the possibility of InN ALE by using $InCl_3$ and NH_3.

INTRODUCTION

In recent years, there has been growing interest in the production and applications of the nitrogen compound films such as SiN, SbN, BN and AlN. In particular, we are interested in InN semiconductors, due to its direct band gap of 1.9 eV which may be a promising material for optoelectronic applications.

In the past decade, many researchers are working on InN films in order to obtain high quality samples for electrical and optical applications[1-5]. Tansley et al[5] have reported 3000 cm^2/vs of Hall mobility and $5x10^{16}$ cm^{-3} of carrier concentration at 300 K in sputtered poly-crystalline film. The films, however, had a lot of antisite defects due to big nitrogen vacancy concentration. These defects dominant the properties of InN sputtered films. We attempted to make InN films by plasma assisted Chemical Vapor Deposition, CVD[6]. The films, however, were polycrystallines possessing a high defect concentration. These native defects are considered to result from the ion bombardment accelerated by electric field.

We have focused an Atomic layer Epitaxy (ALE) method which can be done without any field and in low cost apparatus.

In this paper, we will report results on the ALE growth InN films using $InCl_3$ and NH_3 gases as source materials.

EXPERIMENTAL

The nitrogen dissociation pressure of InN is 1 atmospheric pressure at 580 °C[7]. This gives us the high equilibrium vapour pressure of N_2 over InN even at relatively low temperatures.
Further, the dissociation energy of N_2 is 110.8 kcal/mol., and therefore N_2 is very stable at the temperature below 580°C.

This shows that it is difficult to thermally form InN by using In metal and N_2 gas. N_2 gas is used, however, as the carrier gas of $InCl_3$ and NH_3. Figure 1 shows a schematic drawing of the apparatus used in this experiment. This is hot wall reactor with a resistive furnace. The gas transport systems were heated the desired temperature. The substrate is set in one pipe of the reactor and is alternately exposed to $InCl_3/N_2$ and NH_3/N_2 by rotating the substrate holder.

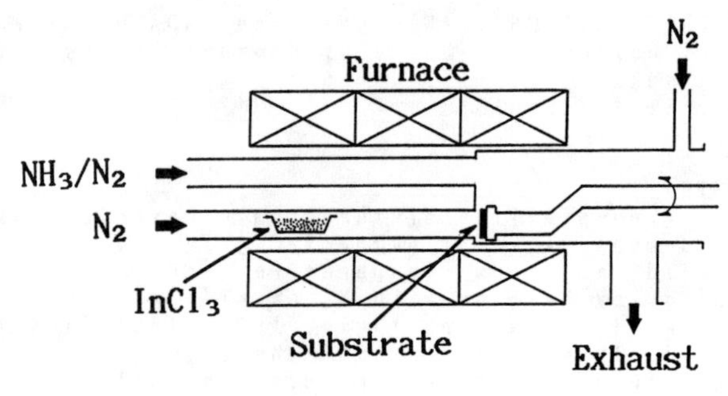

Fig. 1. Aparatus used in the experiment.

In preliminary experiments, we have chosen InCl as a source material having relatively high equilibrium vapor pressure by analogy of diethylgalliumchloride, GaCl, for ALE growth GaAs[8]. The analogous compound for InN, however, could not obtained at growth temperature (Tg) range from 400°C to 600°C. The reasons are not clear at this stage.

Next, solid $InCl_3$ was chosen as a source material by analogy of $GaCl_3$ for ALE GaAs[8] This was expected the partial pressure of about 7.1×10^{-3}atm. at 520°C. The assumed growth reaction is given by

$$InCl_3 + NH_3 \leftrightarrows InN + 3HCl. \quad 1)$$

The growth of InN films were observed on quartz substrate in Tg range from 460°C to 500°C. The X-ray diffraction patterns are shown

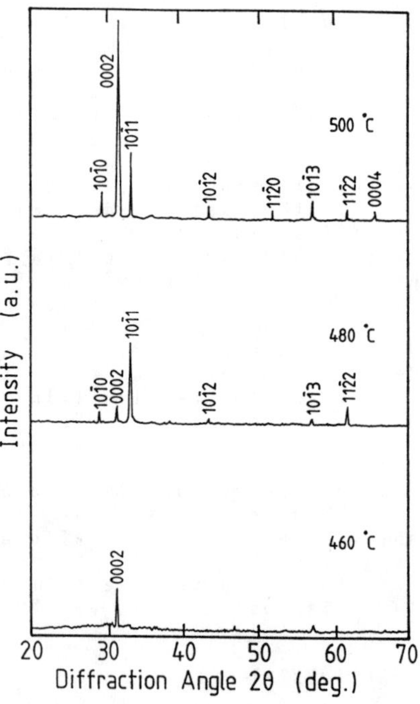

Fig. 2. X-ray diffraction patterns in InN films. (see text)

in Fig. 2 and the films are polycrystalline.

When solid $InCl_3$ was heated in an inert gas, InCl, $InCl_3$ and In_2Cl_6 gases were formed.

Figure 3 shows the calculated partial pressure as a function of temperature. This shows that the In_2Cl_6 molecules predominant in the gas phase below 460°C. These large In_2Cl_6 molecules predicted to be an obstacle to ALE growth, because of it can not structurally formd monolayer. Therefore, in this experiment the growth temperature, Tg may be limited in the range from 460°C to 580°C. In this temperature range, In_2Cl_6 still partially remains in the gas phase.

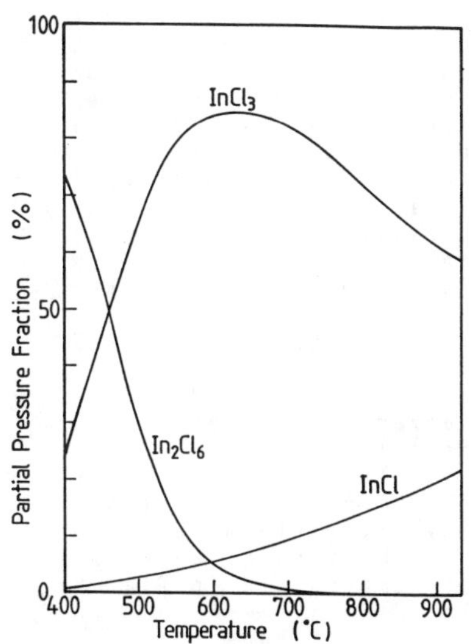

Fig. 3. Partial pressure fraction as a function of temperature. (see text)

ATOMIC LAYER EPITAXY

Table 1 shows ALE growth condition.

Table 1. ALE growth condition

Substrate	quartz and α -Al_2O_3 c-plane	
Growth temperature	420°C-520°C	
$InCl_3$ partial pressures	$\sim 3.0 \times 10^{-2}$	atm
NH_3 partial pressure	0.5	atm
Total flow rate	1580	cc/min
NH_3 purging gas velocity	15.5	cm/sec
Reactor pressure	1.0	atm

Although Si (111) plane has less lattice mismatch (-7.8%) than that of α-Al_2O_3 (0001) (sapphire) c-plane, 28.9%, no growth of InN films were observed in the desired Tg range, eventhough the InN films grew on SiO_2. The polished single crystal sapphire wafers chosen as substrate.

Figure 4 shows the time sequence of gas flows. This

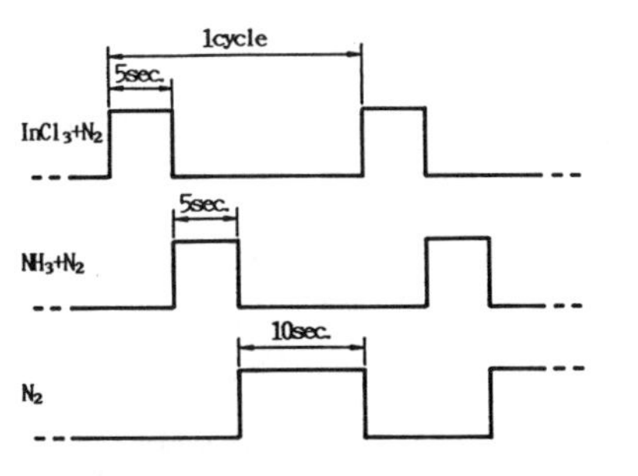

Fig. 4. Gas flow time sequences for the InN ALE growth.

sequence was a procedure optimal at determined by our cumulative experience In Fig. 3, we have present the partial pressures of InCl, $InCl_3$ and In_2Cl_6 in inert gas. InCl has the lowest pressure of the three and this did not form InN films as mentioned above. Both of $InCl_3$ and In_2Cl_6 take part in ALE growth. Therefore, we have examined the temperature dependence of films growth.

The FWHMs obtained from (0002) plane of X-ray diffraction patterns are shown as a function of temperature in Fig. 5. In Fig. 5, ω-mode and $2\theta-\theta$ mode were obtained from rocking and 2θ curves, respectively. These measurement give us the information on the degree of the dispersion of lattice direction and lattice constant, respectively. Both show the minimum FWHMs at 470°C, which is, the optimum ALE growth condition in these experiments. Grown surfaces, however, were observed to have an island structure in SEM images.

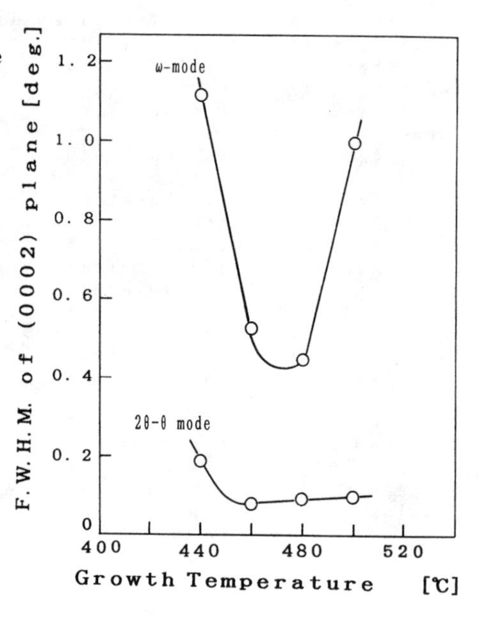

Fig. 5. The FWHMs obtained by X-ray diffraction patterns. (see text)

Figure 6 shows film thickness per cycle as a function of Tg. Unfortunately, completely selective growth was not

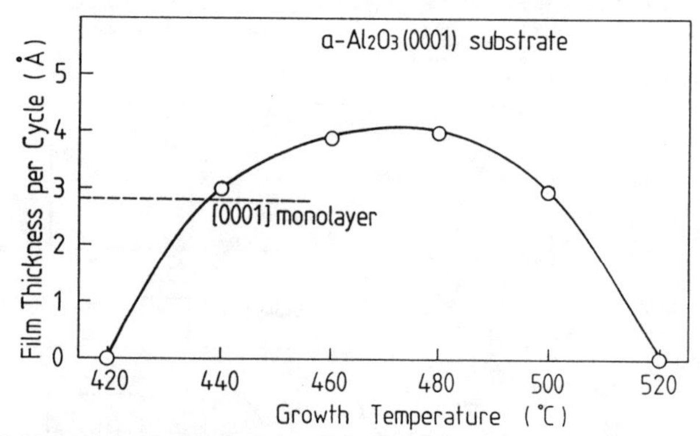

Fig. 6. Film thickness per cycle as a function of growth temperature. The partial pressure of $InCl_3$ and In_2Cl_6 are changed with growth temperature from 440°C to 505°C (see text).

obtained in the Tg range from 440°C to 505°C. The Fig. 6 shows no self limited characteristic which is expected the saturation at 2.8 Å of InN c-axis lattice constant. The over saturation of thickness per cycle is understandable that even in 470°C, still remaines the In_2Cl_6 of about 40 % in a partial pressure, as shown in Fig. 3. An In_2Cl_6 adsorbed on the surface makes a solid structure defects and could not be removed for gas purging of 10 sec or more This was considered an obstacle of AlE growth. If the vaporized indium chloride gases are partially heated above 700°C, the In_2Cl_6 can be removed to more than 99 % extent, and completely selective growth may be expected and successfully achieved.

CONCLUSION

A possibility of the ALE growth of InN films was researched by using solid $InCl_3$ as a source material. It is clear that there are several problems such as no lattice matched substrates for InN and adsorbed molecules which can be formed solid structural defects. If these problems were overcome, the successful ALE growth of InN will be achieved.

REFERENCES

1) H.J.Hovel and J.J.Cuomo; Appl. Phys. Letter, 20, 71 (1972)
2) C.P.Foley and T.L.Tansley; Appl. Surf. Science 14, 254 (1984)
3) N.Puychevrier and M.Menoret; Thin Solid Films 36, 141 (1976)
4) B.R.Natarajan, A.H.Eltoukhy, J.E.Greene and T.L.Barr; ibid, 69, 201 (1980)
5) T.L.Tansley and C.P.Foley; Electronics Letters, 20, 1066 (1984)
6) T.Shiraishi and T.Matsui; Proc. of 8th Int. Sympo. Plasma Chemi., 4, 2355 (1987)
7) J.B.MacChesney, P.M.Bridenbaugh and P.B.O'Connor; Mater. Res. Bull., 5,783 (1970)
8) K.Mori, M.Yoshida, A.Usui and H.Terao; Appl. Phys. Letters, 52, 27 (1988)
9) Y.Jin, R.Kobayashi, K.Fujii and F.Hasegawa; Japan. J. Appl. Physics, 29, 1350 (1990)

PART IV

Group IV Semiconductor Studies

UV PHOTOSTIMULATED Si ATOMIC-LAYER EPITAXY

D. Lubben, R. Tsu, T. R. Bramblett, and J. E. Greene
Materials Science Department, the Coordinated Science Laboratory, and the Materials Research Laboratory, University of Illinois, 1101 W. Springfield Avenue, Urbana, IL 61801

ABSTRACT

Single-crystal Si films have been grown on Si(001)2x1 substrates by UV-photostimulated atomic-layer epitaxy (ALE) from Si_2H_6. The ALE deposition rate R per growth cycle remains constant at 0.4 monolayers (ML) over a wide range of deposition parameters: growth temperature (T_s = 180-400 °C), Si_2H_6 exposure (peak pressure during gas pulse = 0.1-5 mTorr), UV laser energy density ($\mathscr{E}$ = 250-450 mJ cm^{-2} where $\mathscr{E}_{max}$ is determined by T_s), and number of UV laser pulses per cycle. A film growth model, based upon the results of the present deposition experiments and Monte Carlo simulations, together with our previous adsorption/desorption measurements, is used to describe the reaction pathway for the process. The H-terminated silylene-saturated surface formed by adsorption and desorption of disilene is thermally stable and passive to further Si_2H_6 exposure. ArF or KrF laser pulses (≈ 20 ns) are used to desorb H, following a Si_2H_6 exposure, and the growth cycle is repeated until the desired film thickness is obtained. At T_s < 180 °C, the growth process becomes rate limited by the surface dissociation step and R decreases exponentially as a function of $1/T_s$ with an activation energy of ≈ 0.5 eV. At T_s > 400 °C, H is thermally desorbed and pyrolytic growth competes with ALE. Transmission electron micrographs together with selected-area electron diffraction patterns show that the ALE films are epitaxial layers with no observed extended defects or strain.

INTRODUCTION

Atomic-layer epitaxy (ALE) is a vapor-phase epitaxial film growth technique incorporating self-limiting kinetic processes which result in the deposition of Θ monolayers (ML) per growth cycle where $\Theta \leq 1$ ML. For III-V and II-VI compounds and alloys, ALE has been accomplished by sequential introduction of gas-phase molecular precursors whose heterogeneous reactivities depend upon the surface termination.[1,2]

Although there is a large and growing literature on ALE of II-VI and III-V compounds,[2] very little has been published regarding ALE of group IV elements and compounds. In two previous papers, we outlined a proposed mechanism for achieving ALE growth of Si films.[3,4] The mechanism was based upon the results of electron energy loss spectroscopy (EELS) and reflection high energy electron diffraction (RHEED) studies of the adsorption, surface dissociation, and subsequent desorption (both thermal [3] and UV-laser induced[4]) reaction paths of Si_2H_6 on Si(001)2x1 surfaces. Based upon our results, and adsorption studies of previous workers,[5,6] dissociative chemisorption of Si_2H_6 appears to be initiated via scission of the Si-Si bond allowing adsorption of silyl (SiH_3) radicals on adjacent dangling bonds. The silyl radicals then further decompose to silylene (SiH_2) and H to form a mixed (2x1):H monohydride and (1x1)::2H dihydride surface. With further Si_2H_6 exposure, a saturated passive (1x1)::2H surface is obtained. Surface reactions in the saturated dihydride surface were studied as a function of temperature between ambient and 680 °C by following the intensities and positions of EELS peaks associated with surface dangling bonds, backbond states, and Si-H bonds. However, the important result for ALE was that annealing dihydride-saturated surfaces at 500-600 °C for 15 s led to the evolution of H leaving an epitaxial Si overlayer. Hydrogen desorption was also obtained with a pulsed excimer laser (ArF, 193 nm, 6.4 eV, 17 ns pulses) at photon energy densities (≤ 120 mJ cm^{-2}) well below the Si melting threshold.[7] The mechanism of photo-induced H desorption was shown to be photothermal rather than photolytic with peak temperatures of 527 °C at 120 mJ-cm^{-2} and photothermal decay times of the order of 100 ns to 1 µs.[4]

In this paper, we present the results of an investigation of the mechanism and kinetics of self-limited UV-photostimulated ALE growth of Si from disilane. Epitaxial Si films were deposited, using sequential gas dosing and KrF laser irradiation, as a function of the steady-state substrate temperature T_s, peak pressure P during Si_2H_6 exposure, and laser energy density $\mathscr{E}$. KrF (248 nm) irradiation was used instead of ArF (193 nm), as in our previous experiments, in order to minimize Si_2H_6 gas-phase photodecomposition. Self-limited film growth kinetics, due to self-limited disilane adsorption, were obtained over the following range in deposition conditions: 180 °C < T_s < 400 °C, 0.1 < P < 5 mTorr, and 250 < $\mathscr{E}$ < 470 mJ cm^{-2}, where $\mathscr{E}_{max}$ is determined by T_s. The self-limited ALE deposition rate R was found to be ≈ 0.4 ML per growth cycle. This is in agreement with Monte Carlo simulations which show that R is self-limited due to site blocking reactions during the adsorption and surface dissociation steps. A film growth model, based upon experimental observations and the Monte Carlo simulations, is proposed to explain the overall reaction path including adsorption, surface dissociation reactions, site blocking, and desorption.

EXPERIMENTAL PROCEDURE

The film-growth apparatus used in the deposition experiments consists of a stainless-steel ultra-high-vacuum (UHV) system which is shown schematically in Figure 1. The sample introduction chamber is evacuated with a 330 l-s^{-1} turbomolecular pump to pressures less than 10^{-8} Torr while the growth chamber is separately pumped to base pressures of ≤ $2x10^{-10}$ Torr using both ion and Ti sublimation pumps. During disilane exposure and film growth experiments, the deposition chamber pumps were valved off and the chamber was continuously evacuated through the transfer chamber with the turbomolecular pump. RHEED was used for in-situ

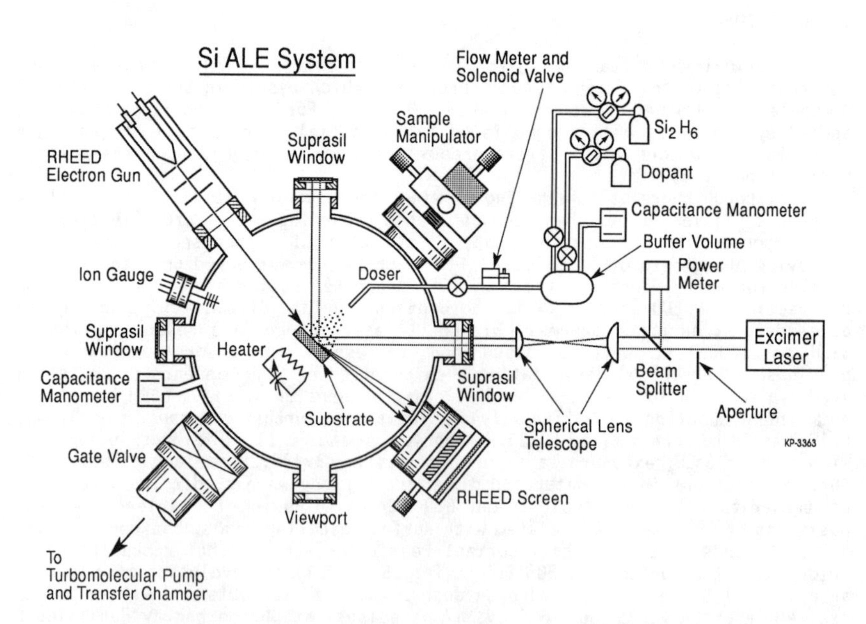

Figure 1. Schematic diagram of the Si atomic-layer epitaxy system.

structural analyses of the substrate surface and as-deposited layers. The electron accelerating voltage was 17 kV and the beam was adjusted to intercept the Si substrate at an angle of approximately 1.5°.

UV laser irradiation was provided by a Questek model 2820 excimer laser operated with KrF (248 nm, 5.0 eV, 22 ns pulses). The beam was focused and collimated to 0.5 cm^2 with a spherical lens telescope and coupled to the samples via a high-purity fused-silica window mounted on the UHV chamber. Energy densities $\mathscr{E}$ at the Si surface ranged from 0 to 500 mJ cm^{-2}.

The Si(001) substrates used in these experiments were 7 mm x 20 mm plates cleaved from 0.4 mm thick p-type (B doped, resistivity = 3-4 Ω-cm) wafers. Initial cleaning consisted of degreasing by successive rinses in trichloroethane, acetone, methanol, and distilled water. The substrates were then blown dry in dry N_2, exposed to UV irradiation from a low-pressure Hg lamp (15 mW-cm^{-2}) for 40 minutes in air to remove C-containing species[8], and introduced into the deposition system through the sample-exchange chamber. The wafers were degassed at 600 °C for 1 h, and rapidly heated to 900 °C for 5 min to desorb the oxide overlayer. Following the in-situ cleaning procedure, the substrates exhibited sharp 2x1 surface reconstruction patterns typical of clean Si(100) surfaces while AES spectra, obtained in the analytical chamber, showed no indication of C or O. After cleaning, a 50-nm-thick buffer layer was grown at a temperature of 700 °C by disilane gas-source molecular-beam epitaxy. T_s was then reduced to the desired ALE growth temperature.

Disilane flow was regulated via a computer-controlled solenoid valve and introduced into the chamber through a gas doser directed at the Si surface. The pressure rise P in the chamber, which was continuously evacuated by the turbomolecular pump, was monitored with a capacitance manometer. The dosing cycle was 1 s and P was varied from 0.15 to 5 mTorr. The total amount of gas introduced per cycle was 3.5×10^{17} - 1.7×10^{19} molecules as determined by measuring the pressure rise in the sytem with the turbomolecular pump valved off. After the gas pulse and prior to laser irradiation, a programmable delay (typically 1-10 s) was used to allow the growth chamber to be evacuated to a pressure $\leq 1\times10^{-6}$ Torr in order to minimize gas-phase photolytic reactions. The sample surface was then irradi laser irradiation was used rather than ArF (193 nm, 6.4 eV, 17 ns), as in the previous EELS experiments,[4] in order to further reduce the possibility of gas-phase photolysis since the absorption cross section for disilane at 248 nm is several orders of magnitude lower than at 193 nm ($< 1 \times 10^{-19}$ vs 3×10^{-18} cm^2 [9]). The growth cycle was repeated until the desired film thickness was achieved. Deposited film thicknesses were measured using a microstylus profilometer. Plan-view TEM and cross-sectional TEM (XTEM) together with selected-area electron diffraction were used to examine the microstructure of the film. In addition, XTEM served to confirm the microstylus thickness measurements. Both TEM and XTEM specimens were prepared by a combination of grinding, dimpling, and Ar^+-ion milling as described previously.[10] The samples were viewed in either an EM 420 STEM or a Phillips CM12 operated at 120 keV.

EXPERIMENTAL RESULTS

Figure 2 shows the film growth rate R as a function of the steady-state substrate temperature T_s with and without pulsed laser irradiation during each deposition cycle. The peak pressure during the 1 s Si_2H_6 exposures was 1.5 mTorr, sufficient to reach surface saturation, as shown below, in all experiments. The maximum KrF laser energy density $\mathscr{E}_{max}$ at the sample surface in the irradiation experiments was decreased from 470 to 250 mJ cm^{-2} as T_s was increased from 130 to 550 °C. The energy density window $\Delta\mathscr{E}$ over which R was found to be constant for a given T_s was ≈ 150 mJ-cm^{-2}. In all cases, $\mathscr{E}(T_s)$ was well below the melting threshold.

R in Figure 2 is plotted in units of ML per growth cycle where one monolayer in the [001] direction is taken to be 6.8×10^{14} cm^{-2} corresponding to a thickness of 0.136 nm. For the unirradiated films, R varied exponentially with $-1/T_s$ at $T_s \geq 550$ °C yielding an apparent activation energy of 1.2 eV. At lower

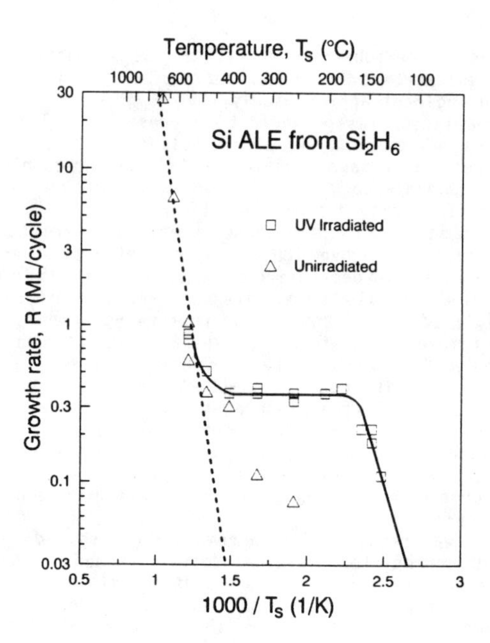

Figure 2. Si growth rate R as a function of substrate temperature T_s.

temperatures the slope is reduced, and no measurable growth occured at T_s below 200 °C. Films deposited with $T_s \geq 625$ °C were smooth, high-quality epitaxial layers, as judged by RHEED, while films deposited at lower temperatures exhibited 3-dimensional RHEED patterns with increased diffuse background intensity indicative of residual disorder.

In contrast, all films deposited at low temperatures with pulsed UV laser irradiation were smooth and well ordered, as judged by both RHEED and TEM, and exhibited much higher deposition rates at $T_s \leq 400$ °C. In fact, at $T_s = 250$ °C, R increased by more than an order of magnitude in the presence of laser irradiation. Moreover, there is a range in growth temperatures between ≈ 180 and 400 °C over which R is independent of T_s and deposition proceeds by a kinetically self-limited ALE mechanism with no apparent activation barrier. Varying the number of laser pulses per cycle from 10 to 50 also had no effect on R which remained constant at ≈ 0.4 ML/cycle. For $T_s > 400$ °C, the growth rate of the laser irradiated films increases with increasing T_s as deposition due to thermal pyrolysis begins to compete with ALE. At $T_s < 180$ °C, R decreases with an apparent activation energy of ≈ 0.5 eV.

The film growth rate as a function of the peak Si_2H_6 pressure P during a 1 s exposure with $T_s = 250$ °C, $\mathcal{E} = 400$ mJ-cm^{-2}, and $n_p = 10$ pulses is shown in Figure 3. Measureable R values were obtained for $P \geq 0.1$ mTorr and saturation was observed at $P_{sat} \approx 1.5$ mTorr with R ≈ 0.4 ML/cycle. These results are further evidence of a self-limited kinetic growth mechanism. The Si_2H_6 saturation exposure in Figure 3 is in good agreement with the exposure required to obtain a saturated (1x1)::2H dihydride surface during Si_2H_6 adsorption experiments at room temperature.[3] However, the incident flux, and hence the time of exposure, in the latter case was approximately 4 orders of magnitude lower than that employed in the present growth kinetics experiments.

The ALE film growth rate was also found to be constant for a wide range in incident UV energy densities. While $\Delta\mathcal{E}$ remained approximately 150 mJ-cm^{-2}, $\mathcal{E}_{max}$

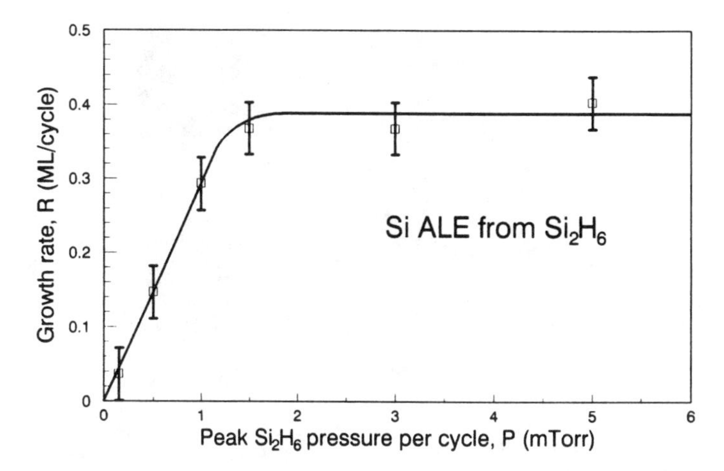

Figure 3. Si growth rate R as a function of the maximum chamber pressure P during the Si_2H_6 gas pulse. The film growth temperature T_s was 250 °C, the KrF irradiation energy density $\mathscr{E}$ was 400 mJ-cm^{-2}, and the number of laser pulses per gas pulse n_p was 10.

had to be decreased as the substrate temperature increased in order to achieve the same peak temperatures. In all cases, the laser-induced temperature was sufficient to cause complete removal of the hydrogen but less than that required for melting. One-dimensional heat-flow calculations, based upon optical absorption and incorporating temperature-dependent thermal parameters,[7] were used to estimate the melting threshold as a function of T_s and $\mathscr{E}$. For 22 ns KrF irradiation, the 1/e absorption depth is ≈ 5.6 nm and the threshold energy density for melting was found to vary from 480 to 410 mJ cm^{-2} as T_s was increased from 250 to 400 °C. At T_s = 250 °C, ALE growth with R ≈ 0.4 ML per growth cycle was obtained for 300 < $\mathscr{E}$ < 470 mJ-cm^{-2}, while for T_s = 400 °C the measured useable $\mathscr{E}$ range was 220 to 365 mJ-cm^{-2}.

A typical plan-view TEM micrograph and corresponding diffraction pattern, from an ALE film (T_s = 250 °C, $\mathscr{E}$ = 400 mJ cm^{-2}, n_p = 50) is shown in Figure 4. The (004) bright-field plan-view micrograph is featureless with no evidence of extended defects and the diffraction pattern is consistent with a high-quality single crystal. No dislocations were observed throughout the entire sample, indicating that the dislocation density was < 10^4 cm^{-2}. The {111} lattice fringes in the high-resolution XTEM micrograph are continuous across the buffer-layer/film interface with no evidence of disorder. Film thicknesses determined from XTEM micrographs were equal to that predicted from the product of the number of growth cycles and the saturated growth rate, R = 0.4 ML/cycles.

DISCUSSION

The results presented in section III demonstrate that single crystals of Si can be grown on Si(001) by photostimulated ALE from Si_2H_6 at T_s between 180 and 400 °C. The growth rate, ≈ 0.4 ML/cycle, remains constant over a wide range in deposition parameters. We have shown previously using EELS that Si_2H_6 is dissociatively chemisorbed on Si(001)2x1 surfaces.[3] There is evidence from thermally programmed desorption (TPD)[5,6] and multiple internal reflection spectroscopy (MIRS)[11] experiments that chemisorption on Si(111)7x7 occurs by

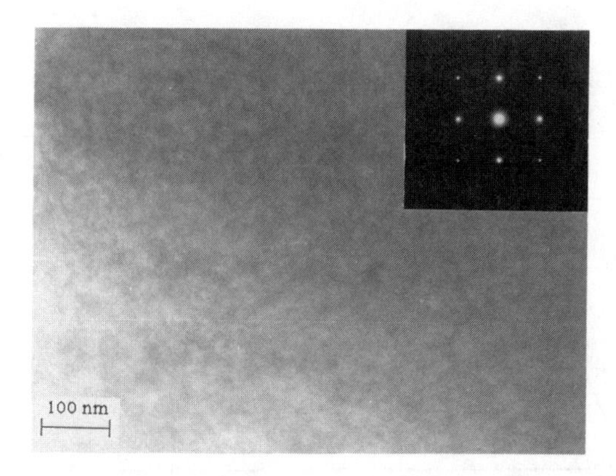

Figure 4. Bright-field (004) plan-view TEM micrograph of a Si film grown by ALE with T_s = 250 °C, $\mathcal{E}$ = 400 mJ-cm^{-2}, and n_p = 50. The inset shows a selected-area diffraction pattern.

Si-Si bond scission with a small negative apparent activation energy via a molecular precursor state. This is reasonable based upon the lower bond energy for Si-Si (3.3 eV) than for Si-H (3.8 eV) in Si_2H_6.[12] From the MIRS results[11] together with indirect evidence from TPD[5,6] and high-resolution EELS,[6] the adsorption products on Si(111)7x7 at temperatures < 200 K are SiH_3 which decompose at higher temperatures to form SiH_2 and SiH. Greenlief et al.[13] recently showed, using static secondary ion mass spectrometry (SSIMS), that SiD_3 is still detectable (≈ 0.04 functional groups per Si) on Si(111)7x7 surfaces exposed at 353 K to atomic D at saturation coverage. With temperature-programmed SSIMS, the SiD_3 signal decreased to zero at ≈ 700 K. The activation energy for the decomposition of adsorbed SiD_3 was estimated to be ≈ 1.3 eV.

It is reasonable to expect that the initial adsorption products on Si(001)-2x1 surfaces are also silyl radicals. Si dimers on the 2x1 surface, with a Si-Si separation of 2.40 Å[14], provide facile sites for dissociative adsorption of Si_2H_6 (Si-Si separation = 2.33 Å) onto dangling bonds. Another possible pair of adsorption sites are adjacent dimer atoms in the same row which have a Si-Si separation of 3.84 Å. Adsorption on neighboring dimers in different rows is less likely due to the increased Si-Si separation, 5.28 Å, and, more importantly, the fact that this reaction inserts two Si atoms in the same bulk-terminated 1x1 site. In fact, the placing of a silyl radical on the dangling bonds of one Si dimer effectively blocks, through steric hindrance, adsorption on the neighboring Si atoms in the adjacent dimer rows. While other reaction site pairs are possible, e.g. adsorption on dimer atoms in adjacent dimer rows separated by one lattice spacing, we regard those shown in Figures 5a to be the most probable and have labeled them configuration <u>D</u> (adsorption on a single dimer) and configuration <u>A</u> (adsortion on neighboring dimer atoms in adjacent rows) for purposes of discussion.

The self-limited Si growth kinetics exhibited in the ALE deposition mode (see Figure 2) are due directly to the self-limited adsorption kinetics in which available dangling bond sites are rapidly filled at deposition temperatures for which multilayer adsorption cannot occur. EELS studies showed that a dose of ≈ 10^{15} cm^{-2} was sufficient to remove the dangling bond peak in spectra acquired at 300 K.[3] The adsorption of SiH_3 and subsequent surface dissociation reactions give rise to steric hinderances on the surface which prohibit otherwise available adsorption sites from capturing Si-containing species. This, in turn, limits R to less than one ML per growth cycle.

Overall SiH_3 surface dissociation pathways are illustrated schematically

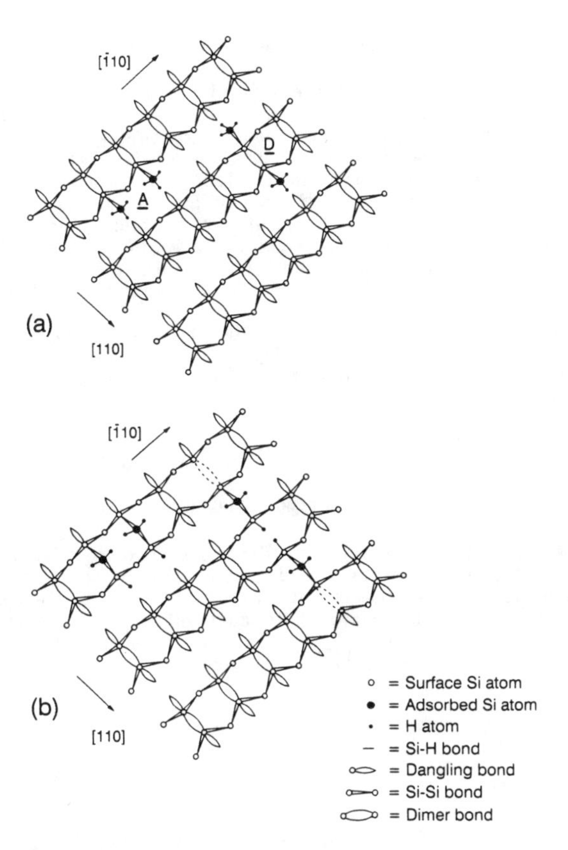

Figure 5. Schematic illustration of the Si(001) surface after (a) dissociative chemisorption of Si_2H_6 onto a single dimer (D) and onto adjacent dimers (A), and (b) dissociation of the adsorbed SiH_3 radicals to form SiH_2 and H.

in Figure 5. The upper panel shows silyl radicals obtained from dissociatively adsorbed Si_2H_6 in configurations D and A. SiH_3 dissociation reactions, involving minimal atomic motion and bond breaking, are illustrated in Figure 5b. For configuration D, the H atoms released by the reaction $2SiH_3(ad) \rightarrow 2SiH_2(ad) + 2H(ad)$ attack the dimer bond, break it, and attach to the resulting dangling bonds. The unsaturated SiH_2 groups attach to dangling bonds in adjacent dimer rows placing deposited Si atoms in tetrahedrally coordinated 1x1 positions with H termination. The dimer bonds associated with the Si atoms to which the silylene groups attach may be weakened or broken (indicated by dashed lines in Figure 5b). Nevertheless, the overall reaction is energetically favorable since dimer bond breaking requires $\approx$ 2.3 eV[15] while H termination of dangling bonds releases $\approx$ 3.6 eV.[16] The reaction, however, effectively blocks further Si_2H_6 adsorption on the adjacent dimers to which the SiH_2 groups bonded since one of the two dangling bonds associated with each dimer is now occupied. In addition, the Si atoms which formed the initial dimer adsorption site are now tetrahedrally coordinated with no dangling bonds and further reactions at these sites with nascent Si_2H_6 are blocked as well.

Dissociation of SiH_3 adsorbed onto adjacent dangling bonds in the same dimer row, configuration A (Figure 5b), is assumed to take place via insertion of SiH_2 into a dimer bond and attachment of the dissociated H to the dangling bonds. (The reaction could also proceed across dimer rows but this is energetically less favorable since it would result in strained or broken dimer bonds). The dangling bonds on the non-H-terminated sides of the dimers involved in the adsorption reaction are now blocked from further adsorption/dissociation reactions. Although chemisorption can take place, dissociation of SiH_3 is not possible through the processes discussed above due to lack of further reaction sites. In any case, SiH_3 remaining on the surface following the gas pulse would be more weakly bound than SiH_2 since it has only one backbond and hence, as discussed below, is likely to be desorbed during the laser irradiation step and thus not contribute to net Si deposition.

The adsorption process can be described by the following set of coupled equations. The time dependence of the silyl population density n_{SiH3} is just the adsorption rate less the desorption and dissociation rates,

$$\frac{dn_{SiH_3}}{dt} = 2J_{Si_2H_6}S_{Si_2H_6}(T_s)\left[1-N_{oc}-N_{bl}\right] - n_{SiH_3}k_{d,SiH_3} - (n_{SiH_3})^2k_{a,Si_2H_6} - n_{SiH_3}k_{dis,SiH_3}. \tag{1}$$

J_{Si2H6} in equation (1) is the incident disilane flux, N_{oc} is the fraction of occupied dimers, N_{bl} is the fraction of dimers blocked by steric hindrance, S_{Si2H6} is the disilane dissociative chemisorption probability as a function of temperature, $k_{d,SiH3}$ is the SiH_3 desorption rate constant, $k_{a,Si2H6}$ is the associative desorption rate constant, and $k_{dis,SiH3}$ is rate constant of the dissociation reaction.The net rate at which silylene is produced by the dissociation reaction is

$$\frac{dn_{SiH_2}}{dt} = n_{SiH_3}(t)k_{dis,SiH_3} - n_{SiH_2}k_{d,SiH_2} - n_{SiH_2}k_{dis,SiH_2} \tag{2}$$

where $k_{d,SiH2}$ is the silylene desorption rate constant, $k_{dis,SiH2}$ is the rate constant of the SiH_2 dissociation reaction and we have ignored associative back reactions. Tetrahedrally bonded H-terminated SiH_2 is inert to further reactions with Si_2H_6.[3]

The silyl dissociation rate constant in equations (1) and (2) is given by

$$k_{dis,SiH_3} = \nu e^{\left(-E_{dis}/kT_s\right)} \tag{3}$$

where ν is the frequency factor and E_{dis} (eV) is the activation barrier for decomposition.

The simultaneous solution of equations (1)-(3) for $n_{SiH3}(t)$ and $n_{SiH2}(t)$ requires an additional equation relating N_{bl} and N_{oc}. This was obtained using Monte Carlo simulations of the adsorption/dissociation processes illustrated in Figure 5. Dissociative adsorption is assumed to proceed with probability S_{Si2H6} on unoccupied and unblocked dimers and zero probability on H-terminated or blocked sites. Figure 6 shows the simulation results for Θ vs $S_{Si2H6} \times J_{Si2H6}$ using either configuration D or A as the primary adsorption geometry, where Θ is the coverage of SiH_x (x = 2,3) adspecies. The saturation coverage $n_{sat} = 2.92 \times 10^{14}$ cm^{-2} = 0.43 ML, in good agreement with the experimental results of Figure 2 for the ALE kinetically-limited deposition rate per growth cycle at temperatures between 180 and 400 °C. If we assume that neither D nor A dominates, but that both occur with equal probability, the increase in the number of possible adsorption configurations leads to a slight increase in the saturation coverage to n_{sat}

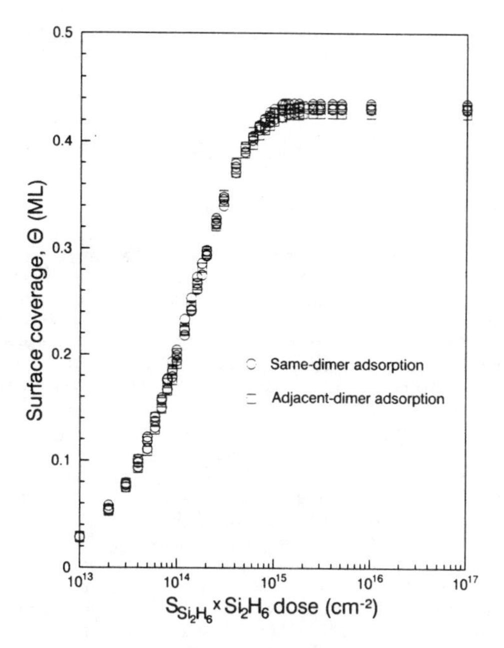

Figure 6. Monte Carlo simulation results showing surface coverage Θ of SiH_x (x=2,3) as a function of the product of the Si_2H_6 dose and the sticking probability $S_{Si2H6}(T_s)$.

≈ 0.52. However, based upon geometric and energetic considerations we expect that configuration D will dominate.

The observed experimental results for photostimulated Si ALE from disilane can be explained based upon equations (1)-(3) together with the results of the Monte Carlo simulation. When the sample is dosed to saturation at low deposition temperatures, Ts ≤ 130 °C, $k_{d,SiH3}$ and $k_{dis,SiH3}$ are negligible, as all sites are either occupied by SiH_3 or blocked by steric hindrance, and $n_{SiH3} \rightarrow n_{sat} \approx 0.4$ ML. We assume, from our previous studies[3,4] and the results on Si(111)7x7 discussed above, that the rate limiting step for the production of SiH_2 is the dissociation reaction, equation (2). Thus, at sufficiently low deposition temperatures and short delay times Δt between Si_2H_6 gas dosing and laser irradiation, the surface dissociation reaction will not be allowed to go to completion and the surface, prior to the laser irradiation step, will contain both SiH_3 and SiH_2 species. Since $k_{dis,SiH2}$ is also negligible at low temperatures, the number density of SiH_2 immediately prior to the laser pulse can be expressed approximately as

$$n^*_{SiH_2} \approx 0.4 n_s \nu(\Delta t) e^{(-E_{dis}/kT_s)}. \quad (4)$$

E_{dis} and ν can be estimated from the slope and intercept, respectively, of the low temperature region of the solid curve in Figure 2 as 0.57±0.15 eV and 2×10^5 s^{-1}. Thus, the number density of n^*_{SiH3} remaining prior to irradiation is

$$n^*_{SiH_3} \approx (0.4 n_s - n_{SiH_2}) \quad (5)$$

n_s is the number density of surface atoms. If the laser pulse desorbs both the remaining silyl radicals and hydrogen, then the film growth rate per cycle is just

$$R = n^*_{Si} = n^*_{SiH_2} \tag{6}$$

where n^*_{Si} is the number density of deposited Si atoms. Thus, as T_s is decreased below 130 °C (where n^*_{SiH2} from equation (4) is ≤ 0.06 ML for Δt = 10 s), $R \rightarrow 0$ as observed experimentally. At higher deposition temperatures, up to 180 °C, R increases exponentially with $1/T_s$ as the film deposition rate is limited by the extent of the SiH_3 to SiH_2 dissociation reaction (equation 2) whose rate is determined by equation (3). $n^*_{SiH2} \rightarrow n_{sat}$ at $180 < T_s < 250$ °C, so R remains constant and equal to ≈ 0.4 ML per deposition cycle and growth proceeds by ALE with self-limiting kinetics. For temperatures > 250 °C, SiH_2 dissociation accompanied by H desorption becomes significant during the delay between the gas and laser pulses which opens up sites for adsorption during the next gas pulse. Although this does not affect the ALE growth rate (since all H is desorbed by the laser in any case) it does result in an increase in the rate of thermal deposition (See Figure 2). At T_s = 400 °C and above, H desorption (and subsequent additional SiH_3 adsorption) occuring during the gas pulse becomes significant and the thermal growth rate exceeds that of the ALE process. For $T \geq 550$ °C thermal growth dominates and proceeds with an activation energy of 1.2 eV.

Note from equation (4) that n^*_{SiH2} also depends upon Δt and thus R for the ALE process can be increased at low temperatures, and a larger ALE temperature range achieved, by increasing the delay time between Si_2H_6 exposure and the laser pulse. We have found experimentally that films obtained at T_s = 160 °C and Δt = 10 s had a growth rate of ≈ 0.25 ML per cycle while those grown under the same conditions except that Δt was increased to 20 s achieved the saturated deposition rate of R = 0.4 ML per cycle. It is thus possible to obtain deposition even at room temperature as demonstrated by the EELS and RHEED experiments in reference (4). This involves a trade-off, however, in which the real-time deposition rate decreases exponentially.

ACKNOWLEDGEMENTS

The authors gratefully acknowledge the financial support of the office of Naval Research, through contract number N00014-90-J-1241, and the Semiconductor Research Corporation during the course of this research. We would also like to thank Y. W. Kim for performing the TEM analyses.

REFERENCES

1. C. H. L. Goodman and M. V. Pessa, J. Appl. Phys. 60, R65 (1986).
2. T. Suntola, Mater. Sci. Reports 4, 261 (1989).
3. Y. Suda, D. Lubben, T. Motooka and J. E. Greene, J. Vac. Sci. Technol. A8, 61 (1990).
4. Y. Suda, D. Lubben, T. Motooka and J. E. Greene, J. Vac. Sci. Technol. B7, 1171 (1989).
5. S. M. Gates, Surf. Sci. 195, 307 (1988).
6. R. Imbihl, J. E. Demuth, S. M. Gates, and B. A. Scott, Phys. Rev. B 39, 5222 (1989).
7. K. Suzuki, D. Lubben, and J. E. Greene, J. Appl. Phys. 58, 979 (1985).
8. J.P. Noël, J. E. Greene, N. L. Rowell, S. Kechang, and D. C. Houghton, Appl. Phys. Lett. 55, 1525 (1989).
9. U. Itoh, Y. Toyoshima, H. Onuki, N. Washida, and T. Ibuki, J. Chem. Phys. 85, 4867 (1986).
10. J. P. Noël, N. Hirashita, L. C. Markert, Y.-W. Kim, J. E. Greene, J. Knall, W.-X. Ni, M. A. Hasan, and J. E. Sundgren, J. Appl. Phys. 65, 1189 (1989).

11. K. J. Uram and U. Jansson, J. Vac. Sci. Technol. B7, 1176 (1989).
12. Paras. M. Agrawal, Donald L. Thompson, and Lionel M. Raff, J. Chem. Phys. 92, 1069 (1990).
13. C. M. Greenlief, S. M. Gates, and P. A. Holbert, Chem. Phys. Lett. 159, 202 (1989).
14. M. Kitabatake, P. Fons, and J. E. Greene, J. Vac. Sci. Technol. A8, 3726 (1990) and references therein.
15. Inder P. Batra, Phys. Rev. B41, 5048 (1990).
16. Douglas C. Allan, J. D. Joannopoulos, and William B. Pollard, Phys. Rev. B25, 1065 (1982).

GROWTH MECHANISMS DURING Si/Ge DEPOSITION

JOHN E. CROWELL, GUANGQUAN LU, AND BOB M.H. NING

Department of Chemistry, University of California at San Diego, La Jolla, CA 92093-0314

ABSTRACT

The adsorption and decomposition behavior of disilane and digermane are quite similar on the Ge(111) surface. Both precursors are weakly bound at low temperatures, but dissociatively adsorb at temperatures above 150K. Trihydride species are produced and stable at low temperatures, but decompose to di- and monohydride species at slightly higher temperatures. The desorption of hydrogen from the resulting layer is strongly dependent on the Si and Ge composition of the layer.

INTRODUCTION

Semiconductor superlattices and heterostructures are of interest for both technological device applications and for exploration of their fundamental physical properties. Of significant importance from both points of view is the construction of atomic scale structures with control of the electronic and optical properties at the quantum level. The silicon / germanium system is receiving considerable attention because high speed electronic devices have been successfully fabricated from these materials [1,2], and because these materials take advantage of the technological base already established for Si. In addition, these Group IV semiconductors can form uniform alloys with abrupt interface structure, where the strain caused by the lattice mismatch can be used to engineer the band structure of the device [3,4]. There is also evidence [4] suggesting that direct bandgap behavior can be obtained by carefully controlling the monolayer thickness of superlattices consisting of alternating Si and Ge layers. It has been shown that atomic layer growth [5] and device quality structures [1,2,4] of Si/Ge and $Si_{1-x}Ge_x$/Si can be produced using chemical vapor deposition (CVD). Despite the broad interest in these materials, little is known about the fundamental surface processes occurring during deposition and growth from typical CVD precursor gases.

We have used surface-specific spectroscopies to study the deposition of Si and Ge monolayers on Ge and Si surfaces. The methodologies used include multiple internal reflection infrared spectroscopy (MIRIRS), Auger electron spectroscopy (AES), and temperature programmed desorption mass spectrometry (TPD). Using these methods, we have delineated the reaction mechanisms operable during Si and Ge deposition from disilane and digermane, respectively. We have also studied the desorption properties of hydrogen from these surfaces.

EXPERIMENTAL METHODS

The MIRIRS experiments were performed using a Nicolet 60 SXB FT-IR spectrometer coupled to an ultrahigh vacuum chamber. The Ge(111) single crystal [6] is trapezoidally shaped, 50 mm in length and 1 mm thick, permitting 50 internal reflections of the IR radiation. All spectra shown are the average of 5000 scans recorded at 4 cm^{-1} resolution, and are ratioed to the clean surface spectrum at that same temperature. Both (111) faces of the crystal were cleaned using Ar ion sputtering ($\approx$ 5-8 μA, 500 eV) at 300 K, followed by annealing to 875 K for $\approx$ 5 minutes. The disilane was used as purchased without purification (Matheson, 99.99 % purity). All exposures cited are background exposures (1L = 10^{-6} torr-sec), uncorrected for ion gauge sensitivity.

The TPD studies cited were performed in a separate UHV system equipped with AES, TPD, and XPS capabilities. The Si(100) crystals used were B doped at 5-10 mΩ-cm and heated resistively using a linear temperature ramp. The mass spectrometer was multiplexed to a computer. Various coverages of Ge on the Si(100) crystal surface were achieved using exposure to digermane (Voltaix, 99.5% purity) at 300K. The resulting surface Ge hydrides were thermally decomposed by flashing the crystal to 940K. Subsequent AES analysis at 300K was used to determine the surface Ge coverage. A white hot tungsten spiral filament, approximately 5 cm from the crystal surface, was used to dissociate D_2 (Linde, 99.5%) to affect D atom exposure of the Ge/Si(100) surface. All exposures cited (in langmuirs) are background dosing of the molecular source gas, and are also uncorrected for ion gauge sensitivity.

RESULTS AND DISCUSSION

MIRIRS Studies of Si and Ge Deposition on Ge(111)

Delineation of the reaction mechanism leading to deposition of Si and Ge from Si_2H_6 and Ge_2H_6, respectively, has been possible using MIRIRS. The MIRIRS technique enables high resolution IR spectroscopy to be used to study adsorption and reaction processes on semiconductor surfaces. MIRIRS can provide a direct probe of the chemical nature of a semiconductor surface, and hence used to determine the identity and geometry of adsorbates present on the surface. Furthermore, the high resolution of the method, together with the good surface sensitivity, allows distinction between species with similar vibrational frequencies. By studying adsorption using MIRIRS on germanium surfaces, we are able to detect the low frequency deformation modes, permitting determination of the extent of Si-Si or Ge-Ge bond scission. By comparing the adsorption of hydrogen, digermane, and disilane, we are able to conclusively identify the surface species produced. The choice of reactants is made based on the significantly higher adsorption probability expected for disilane and digermane vs. that of silane and germane [7].

The high resolution IR spectra obtained after disilane adsorption on Ge(111) at 120 and 150K are shown in Fig. 1. The high frequency bands (due to symmetric and asymmetric stretching vibrations) are shown in the left panel, while those bands due to bending vibrations are shown in the right panel. Spectrum 1a is characteristic of molecularly adsorbed Si_2H_6 on Ge(111) and is similar to the IR spectrum of solid disilane [8]. The additional mode just visible in curve 1a at 855 cm^{-1} is due to the formation of a silyl species, SiH_3, upon dissociative adsorption of disilane on the Ge surface. This silyl symmetric deformation mode is not visible for disilane adsorption at 104K (not shown), is more intense for adsorption at 120K (curve 1a), and dominates the spectrum for adsorption at 150K (curve 1b).

The absence of an 811 cm^{-1} feature for exposure at 150K indicates that Si_2H_6 adsorbs only weakly on the Ge(111) surface. Temperature dependent studies have shown that most of the disilane desorbs upon heating to 135K. The 811 cm^{-1} feature is thus characteristic of molecularly adsorbed Si_2H_6, and the 855 cm^{-1} feature is characteristic of SiH_3 formation, and a measure of the extent of Si-Si bond scission. The vibrational spectrum of SiH_3 is distinct from that of the parent molecular species, perturbed due to scission of the Si-Si bond and formation of a Si-Ge bond. Molecular adsorption dominates at low temperatures (≤ 120K), but no longer occurs by 150K. Dissociative adsorption of Si_2H_6 to produce adsorbed SiH_3 is first detected at 110K, and the only process observed for adsorption at 150K. Curve 1b is representative of adsorbed silyl, SiH_3.

Our results suggest that a competition exists between molecular and dissociative adsorption at temperatures of about 110-140 K. The decrease in sticking probability observed with increasing temperature is indicative of a negative apparent activation energy. In essence, disilane molecularly adsorbs in a weak precursor state at low temperatures on Ge(111), and can dissociate upon achieving the proper adsorption site and adsorption geometry. The residence time in the weakly bound precursor state is reduced with increasing temperature, increasingly favoring molecular desorption over dissociation.

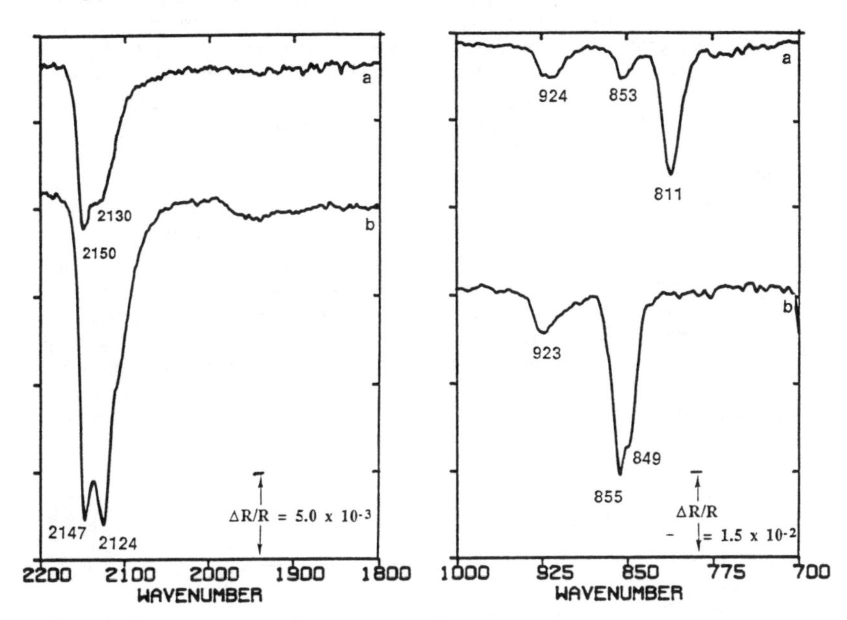

Fig. 1. IR spectra of Ge(111) after exposure to Si_2H_6 at (a) 120 and (b) 150K. The Si-H and Ge-H stretching region is shown on the left-hand side, and the deformation region is shown on the right-hand side. The spectra have been offset from each other for clarity. All spectra were recorded at the same temperature at which dosing occurred.

The species produced upon adsorption at higher temperatures involve scission of both the Si-Si and Si-H bonds. On warming a silyl saturated Ge surface (produced by Si_2H_6 exposure at 150K) to higher temperatures (200-400K), one observes a significant reduction in the intensity of the silyl bands, and the development of additional features due to generation of SiH_2, SiH and GeH [9]. The Ge monohydride band observed at 1950 - 1959 cm^{-1} is evidence that (i) Si-H bond scission occurs at temperatures as low as 150K to a minor extent, and (ii) significant Si-H bond breaking occurs at 300 - 400K. The Ge monohydride produced exists in a number of different environments with different geometric orientations (i.e. tilted and normal bonding orientations), as evident from the broad nature of the observed band.

The results indicate that Si_2H_6 weakly bonds to the Ge(111) surface, desorbing by ca. 135K. Dissociative adsorption of disilane occurs above 110K to produce adsorbed SiH_3. The decomposition of the adsorbed silyl leads to GeH formation, and the production of the mono- and dihydrides of Si. The SiH_2 is most abundant near 300 K, while the SiH species is more prevalent at higher temperatures. Comparison with studies of digermane and H atom adsorption on Ge(111) shows that the di- and trihydrides of Ge are quite distinct spectroscopically from the monohydride of Ge. Hence, only GeH is produced when the silyl species decomposes (or upon disilane exposure at elevated temperatures).

The adsorption and decomposition behavior of Ge_2H_6 on Ge(111) is quite similar to that discussed above for disilane adsorption [10]. Our IR studies show that digermane molecularly adsorbs on Ge(111) at low temperatures in a weak precursor state. Dissociation can also occur near these temperatures, producing adsorbed GeH_3. Molecular desorption and dissociation to adsorbed germyl are competitive processes at these temperatures (110-150K). Ge-H bond scission occurs at temperatures above 150 K, giving rise to the production of GeH_2 and GeH. Complete removal of hydrogen occurs by 600K. The vibrational bands observed for Ge_2H_6 on Ge(111) display the same behavior with exposure and temperature as observed for Si_2H_6, except that all modes are at frequencies that are ca. 100 cm^{-1} lower. The similarities add further support for the surface processes delineated using these methods.

Comparison of the H exposed surface with the digermane exposed surface shows that similar surface hydride species are produced at a given temperature [10]. However, the H atom exposed surface is significantly more heterogeneous than the Ge_2H_6 derived surface, suggesting that etching or reconstruction of the Ge surface occurs upon saturated hydrogen exposure. The adsorption of digermane results in surface species that are adsorbed on top of the Ge(111) surface, as the GeH_3 and GeH_2 species contain Ge derived from the gas phase precursor. In comparison, exposure of hydrogen results in the degradation of the Ge surface itself, as Ge-Ge bonds must be broken in order to form surface GeH_2 and GeH_3 species. The MIRIRS method is quite sensitive to this difference, as evident from the large differences in band width.

Desorption of Deuterium from Ge/Si(100) Surfaces

TPD has been used to probe the structure and composition of the deposited epilayer following chemisorption of a probe gas such as hydrogen. For example, we have deposited Ge onto Si(100) at submonolayer concentrations through repeated exposure / anneal cycles to digermane. The resulting Ge/Si(100) surface was subsequently dosed with deuterium atoms at 300K, and a TPD performed. The presence of Ge on the Si surface strongly alters the TPD profile of deuterium (and similarly hydrogen) from the surface, and provides a convenient nondestructive probe of the surface structure and composition. Representative TPD spectra following a 400L D atom exposure are shown in Fig. 2.

The desorption of D_2 from clean Ge(111) results in a single desorption state centered at ca. 610K [11]. The desorption of D_2 from clean Si(100) results in two desorption states at 690 and 790K. As shown in Fig. 2, the presence of Ge on Si(100) results in a significant shift in the D_2 desorption to lower temperatures. The general trend is to shift from that typical of D_2 desorption from Si, to that typical of D_2 desorption from Ge surfaces. However, at the low Ge coverages shown in Fig. 2, it is apparent that the Ge/Si(100) surface displays desorption features that are intermediate and distinct. In particular, studies performed as a function of D_2 exposure [12] for the Ge coverages shown in Fig. 2 display multiple desorption features intermediate between the extremes of pure Si or pure Ge. Also, comparison of D_2 desorption at low exposures (25L) at various Ge coverages shows a single desorption feature at high temperatures which shifts gradually with Ge coverage (of 0 to 32%) from ca. 790K to 715K. These latter observations indicate that the desorption of D_2 is not simply a combination of desorption from Si and Ge, but that electronic interactions in the Ge/Si adlayer results in intermediate affects distinct from either pure surface.

These and other TPD results [12,13] suggest that Ge strongly perturbs the desorption of hydrogen (or deuterium) from Ge/Si or $Si_{1-x}Ge_x$/Si surfaces. The presence of Ge in submonolayer quantities on Si, or its presence in a $Si_{1-x}Ge_x$ alloy results in a lowering of the activation energy for desorption [13]. The magnitude of this effect is strongly dependent upon the surface Ge coverage. A consequence of this perturbation in the desorption energy has been observed by a number of workers [14-16] as a growth rate change during CVD growth of $Si_{1-x}Ge_x$ alloys at moderate temperatures. The introduction of Ge precursors during Si deposition results in significant growth rate enhancements, an enhancement that is strongly dependent upon the Ge content of the deposited layer [14]. Presumeably, Ge acts to lower the rate-limiting desorption [17] of hydrogen at these temperatures, giving rise to open sites for decomposition of the precursor gases. Up to a point, (a point that is strongly dependent on the Ge coverage and the deposition temperature), the film growth rate increases with increasing Ge content of the deposited epilayer.

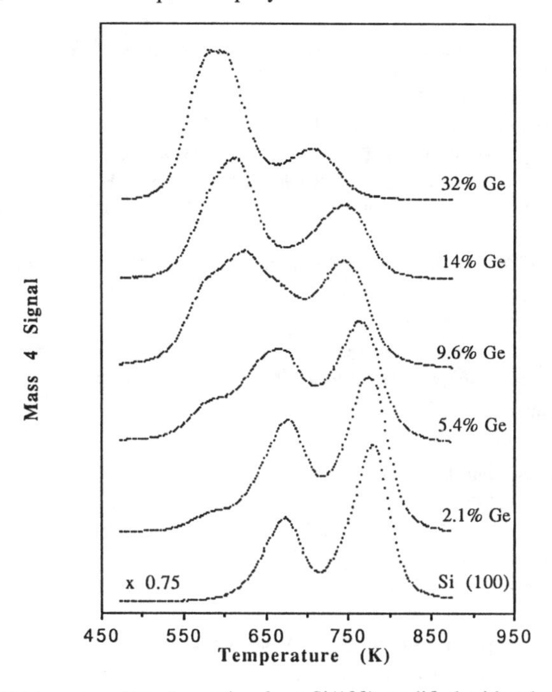

Fig. 2. TPD spectra of D_2 desorption from Si(100) modified with submonolayer concentrations of Ge. The Ge/Si(100) layer was prepared by exposure to digermane at 300K followed by a flash to 940K. A constant 400L D atom exposure was used in all spectra shown. The heating rate was 4.8 K/sec.

CONCLUSIONS

Our IR studies have shown that disilane and digermane molecularly adsorb on Ge(111) at low temperatures in a weak precursor state. Dissociation adsorption to the corresponding trihydride begins as low as 120K, and dominates near 150K. Dissociation of the trihydride occurs at temperatures above 150K, giving rise to the production of mono- and dihydrides, and surface hydrogen. The hydrogen primarily desorbs from these Ge-rich layers near 600K.

Our TPD studies have shown that desorption of hydrogen from Si-rich layers modified by Ge results in a lowering of the desorption energy, affecting $Si_{1-x}Ge_x$ growth rates at certain temperatures.

ACKNOWLEDGEMENTS

This research is supported by the Office of Naval Research. We gratefully acknowledge fruitful discussions with Kevin J. Uram.

REFERENCES

1. G. L. Patton, J. H. Comfort, B. S. Meyerson, E. F. Crabbe, G. J. Scilla, E. D. Fresart, J. M. C. Stork, J. Y. C. Sun, D. L. Harame, and J. N. Burghartz, *IEEE Electron Device Lett.* **11**, 171 (1990); S. S. Iyer, G. L. Patton, D. L. Harame, J. M. C. Stork, E. F. Crabbe, and B. S. Meyerson, *Thin Solid Films* **184**, 153 (1990).
2. C. A. King, J. L. Hoyt, and J. F. Gibbons, *IEEE Trans. Electronic Devices* **36**, 2093 (1989); J. L. Hoyt, C. A. King, D. B. Noble, C. M. Gronet, J. F. Gibbons, M. P. Scott, S. S. Laderman, S. J. Rosner, K. Nauka, J. Turner and T. I. Kamins, *Thin Solid Films* **184**, 93 (1990).
3. T.P. Pearsall, *CRC Crit. Rev. in Solid State and Mat. Sci.* **15**, 551 (1989).
4. S. S. Iyer, G. L. Patton, J. M. C. Stork, B. S. Meyerson, and D. L. Harame, *IEEE Trans. Electron Devices* **36**, 2043 (1989).
5. Y. Takahashi, H. Ishi, K. Fujinaga, *J. Electrochem. Soc.* **136**, 1826 (1989); Y. Takahashi, Y. Sese, and T. Urisu, *Jpn. J. Appl. Phys.* **28**, 2387 (1989).
6. Harrick Scientific Corp., Ossining, NY.
7. S. M. Gates, *Surface Sci.* **195**, 307 (1988).
8. J. Durig and J.S. Church, *J. Chem. Phys.* **73**, 4784 (1980).
9. J.E. Crowell and G.Q. Lu, "Epitaxial Heterostructures", *Mat. Res. Soc. Symp. Proc.* **198**, 533 (1990); G.Q. Lu and J.E. Crowell, "Symposium on Superlattice Structures and Devices", *Proc. Electrochem. Soc.* **90-15**, 450 (1990).
10. J.E. Crowell and G.Q. Lu, *J. Electron Spectrosc. and Relat. Phenom.*, **54/55**, 1045 (1990).
11. L. Surnev and M. Tikhov, *Surface Sci.* **138**, 40 (1984).
12. B.M.H. Ning and J.E. Crowell, to be submitted.
13. B.M.H. Ning and J.E. Crowell, submitted to *Appl. Phys. Lett.*
14. B.S. Meyerson, K.J. Uram and F.K. LeGoues, *Appl. Phys. Lett.* **53**, 2555 (1988); K.J. Uram and B.S. Meyerson, *Mat. Res. Soc. Symp. Proc. Vol.* **102**, 307 (1988).
15. P. M. Garone, J. C. Sturm, P. V. Schwartz, S. A. Schwarz, and B. J. Wilkens, *Appl. Phys. Lett.* **56**, 1275 (1990).
16. T. I. Kamins and D. J. Meyer, *Appl. Phys. Lett.*, in press.
17. M. Liehr, C. M. Greenlief, S. R. Kasi, and M. Offenberg, *Appl. Phys. Lett.* **56**, 629 (1990).

DIGITAL PROCESS FOR ADVANCED VLSI'S AND SURFACE REACTION STUDY

H. Sakaue, K. Asami, T. Ichihara, S. Ishizuka, K. Kawamura and Y. Horiike
Department of Electrical Engineering, Hiroshima University,
Saijo, Higashi-hiroshima, Japan

Abstract

Digital etching was carried out by repeating the fundamental reaction cycles of adsorption, reaction and desorption for fluorine(F) or chlorine(Cl)/Si systems. In the F/Si case, atomic layer etching of Si(100) was achieved by adsorption of F atoms produced by a remote discharge of F_2/99.8%He on the cooled Si surface and subsequent Ar^+ ion (≃20eV) irradiation. The digital method revealed that the cryogenic etching occurred by ion bombardment on physiosorbed F atoms on the cooled Si surface. Adsorption of Cl atoms on Si at room temperature allowed self-limiting reaction with etch rate of 0.4 Å/cycle. The etching increased rapidly over 40 V of substrate voltage. Secondly, reaction of TES (triethylsilane) with hydrogen(H) atoms was also found to lead to conformal CVD (Chemical Vapor Deposition) of Si film involving organic species. Then Si oxide and nitride films were formed by digital CVD which repeated a cycle of first deposition of this film and subsequent its oxidation and nitridation. The electrically excellent multilayer stacked oxide and nitride film was filled in to deep trench. In-situ FTIR-ATR spectroscopy demonstrated that the surface reaction was predominant for the TES/H process.

1. Introduction

As minimum feature size in VLSI is reduced to submicron dimensions, device structures with a high aspect ratio (depth, height/linewidth) is increasingly required for capacitors and multilayer metallization. However the high energy processes such as reactive ion etching (RIE), causes a wide variety of radiation damage not only on the bottom surface, but also on the sidewalls leading to electrical defects around 0.1 μm wires. The fundamental etching reaction occurs through three steps: (1)reactive gas, such as fluorine(F) or chlorine(Cl), adsorption or physiosorption on the surface, (2)energetic (e.g. plasma, ion, photon, electron) beam induced reaction between adsorbates and the surface, and (3)the desorption of reaction products. This surface reaction has to be studied in order to minimize the damage and enhance reaction rate. Hence layer-by-layer etching, called the digital etching, has been studied[1]. This is carried out by repeating these three steps at atomic level with a minimum energy[1,2]. The separation of the reaction steps by using the digital method also allows us to elucidate the mechanism of the reactive ion etching(RIE) in which the radical adsorption and ion-induced desorption processes proceed simultaneously[3~6].

On the other hand, conformal CVD (Chemical Vapor Deposition) technology, which is dominated by a surface reaction, has been developed to deposit a Si oxide film into a deep trench by employing an organic silane such as TEOS (tetraethylorthosilicate)/ozone reaction[7]. In future multilevel metallization schemes, the breakdown between wires on a same plane is more serious than that between the upper and the lower lines. Thus high integrity is necessary for quality of the insulator film filled into narrow width trench. To this end, a multilayer insulator such as stacked SiO_2/Si_3N_4 is desirable. TEOS which originally involves oxygen atoms can provide only SiO_2 film. Conformal CVD of organic Si films employing TES (triethylsilane; $SiH(C_2H_5)_3$) without oxygen bonds and hydrogen(H) reaction has been studied[8]. The digital CVD of the multilayer stacked Si oxide and nitride films has been studied by repeating a process of first depositing the Si film involving organic species and subsequent oxidation and nitridation of the film, layer by layer[9].

In this paper, the atomic layer etching of Si is reported, and the ion induced etching mechanism of F /Si and Cl/Si systems in this digital etching method is discussed. Secondly, the film characteristics obtained by the TES/H reaction and study on the reaction process with in-situ FTIR-ATR are discussed and electrical results of the multilayer stacked film are reported.

2. Digital Etching of Si and Discussion of F/Si and Cl/Si Reaction

2.1. Experimental

The experimental setup to achieve the digital etching is shown in Fig.1. Si etching with F was carried out through cycles of mono- or a few layer removal by employing the two separate reaction steps consisting of fluorine adsorption or physiosorption and reaction products desorption as a result of Ar^{+} ion irradiation induced surface reaction. A Si(100) wafer was attached on a rotatable disk in a vacuum chamber through a loadlock system. The wafer temperature was controlled from room temperature to -180 °C. Fluorine radicals produced by a remote microwave discharge of F_2+95~99.8% He in an Al_2O_3 tube were effused onto the Si surface held at temperatures of -120 ~ -170 °C which can stop the spontaneous etching of Si by fluorine. The exposure time was fixed for 1 sec. An adsorption chamber in Fig.1 was always differentially exhausted to prevent fluorine flowing into other chambers. Immediately after fluorine exposure on Si, the Si wafer was transferred to another chamber by rotating the substrate table. Next, the surface was irradiated with Ar^{+} ions produced by a downstream ECR plasma at a pressure of $1x10^{-3}$ Torr to promote the desorption of reaction products. The Ar^{+} ion energy was estimated to be 20 eV by the probe measurement, and the measured physical sputtering rate of Si was $9.5x10^{-3}$ Å/sec. At the beginning of this study it was not clear whether the F/Si reaction occurs immediately after fluorine adsorption or physiosorption on the Si surface or the reaction is promoted only by Ar^{+} ion bombardment of the surface.

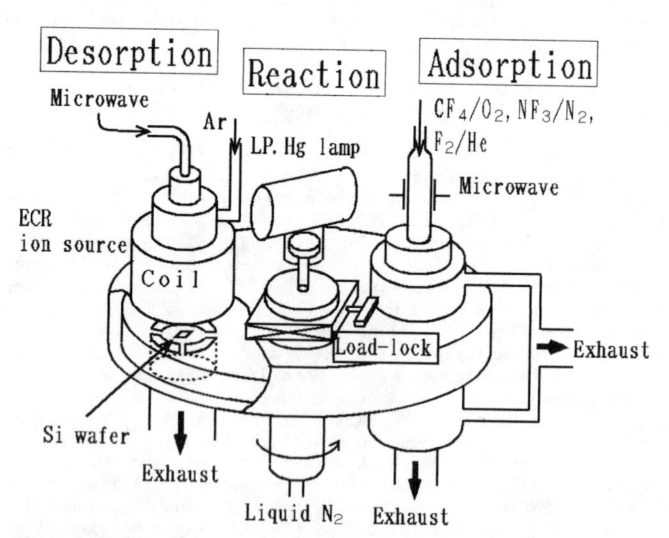

Fig.1 Schematic illustration of the digital etching method.

2.2. Results and Discussion

2.2.1. Atomic Layer Etching with F_2/He

At first, the temperature dependence of the Si spontaneous etch rate in the F_2/95%He system was measured in the adsorption chamber, as shown in Fig.2. The etch rate first decreased gradually, then dropped rapidly around -70 °C. The similar result was reported in an ECR etching with SF_6 [10]. The digital etch rates of Si by discharge of a F_2 diluted by 95%He and 99.8%He are shown in Fig.3 as a function of Ar^+ ion irradiation time, where the digital etch rate is defined as the total etched depth divided by the number of cycles. The etch rate first increases and then reaches the plateau region with an increase of Ar^+ ion irradiation time. The fluorinated surface layer thickness increases with irradiation time in the linearally increasing region, while the fluorinated surface layer produced by ion irradiation is completely removed in the plateau region. The saturated etch rate of 5.7 Å/cycle for the F_2+95%He at a pressure of 1.5×10^{-2} Torr. It should be noted that the exact atomic layer etching of Si(100) proceeds in the discharge of F_2+99.8% He and that adsorption of a trace amount of F atoms (0.2%) can lead to its monolayer reaction with the Si surface atoms. However this results insists that self-limiting characteristic in which the etch rate is independent on parameters of ion energy and flow rate is not realized in the case of fluorine chemistry.

2.3.2. Low Temperature Etching Mechanism

A drawback of the digital etching is a very low throughput. In order to enhance the etch rate, gradient of the digital etch rate by Ar^+ irradiation time as shown in Fig.3 should be increased. This means that the F/Si

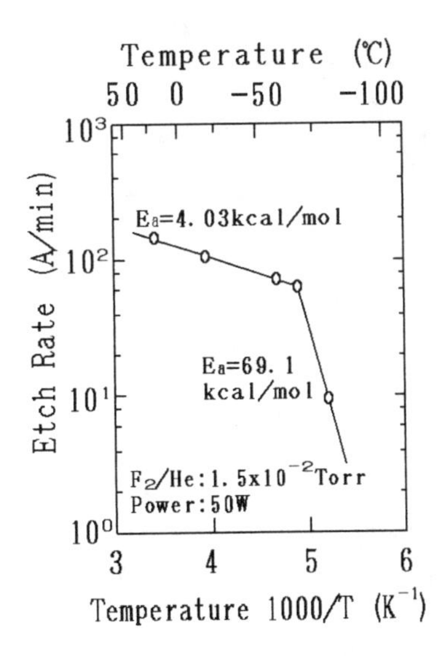

Fig.2 Si spontaneous etch rate with F atoms generated by a remote plasma of F_2/95%He in an alumina tube as a function of substrate temperature.

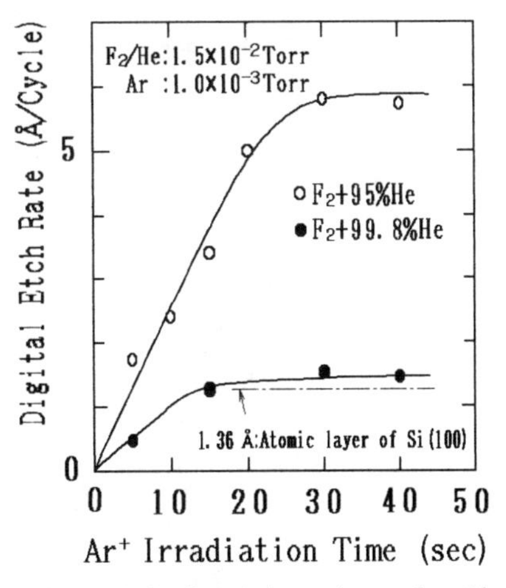

Fig.3 Digital etch rate of Si versus Ar^+ ion irradiation time at a pressure of 1.5×10^{-2} Torr of F_2+He down-stream plasmas. The substrate temperature was -110°C.

reaction should be promoted. It is known that in F/Si RIE the fluorinated Si surface consists of SiF_x (X=1-4) bonds. Changes of surface products to more volatile species such as SiF_2 and SiF_4 can improve the etch rate. Hence, a low pressure mercury (Hg) lamp was irradiated to the reaction chamber shown in Fig. 1. Nevertheless, the digital etch rate decreases with Hg lamp irradiation time as shown in Fig.4. The 253.7 nm (about 5 eV) photon from the lamp is considered to promote the fluorine desorption from the Si surface. Thermal desorption due to photo irradiation also may take place. For deeper insight into this reaction, pure F_2 diluted by 95%He without discharge which does not lead to an appreciable spontaneous etch rate at room temperature was investigated as an etching gas instead of F atoms. Surprisingly, the etching occurred by Ar^+ ions irradiation onto the molecular fluorine adsorbed Si surface cooled at -150°C as shown in Fig.4. In addition, the Hg lamp irradiation retarded the etch rate as well.

The F/Si reaction in the low temperature was also investigated employing in-situ XPS (x-ray photoelectron spectroscopy). Variations of F_{1s} and Si_{2p} peaks for F/Si samples reacted at a room temperature (R.T.) and -110 °C were measured. At R.T., silicon fluoride (SiF_1~SiF_4) layer was detected besides fluorine and metallic Si peaks. However, when the sample was cooled to -110 °C, any Si fluoride peak was not observed except for high fluorine coverage and low Si peaks[12].

These results demonstrate that fluorine is not in chemisorbed state but in merely physiosorbed state on the cooled Si surface. It is assumed that, as soon as F atoms adsorbed on the Si surface, they recombined through three body process to produce condensed molecular fluorine. The cryogenic etching[10] occurs only when low energy ions impinge onto the physiosorbed fluorine on Si through ion-induced F/Si reactions. Based on this mechanism, the results of Fig.2 and Fig.3 can be explained: The fluorinated surface is formed in the higher temperature region but F atoms/molecules are only accumulated on the surface in the lower temperature region, producing a very high activation energy of spontaneous etch rate.

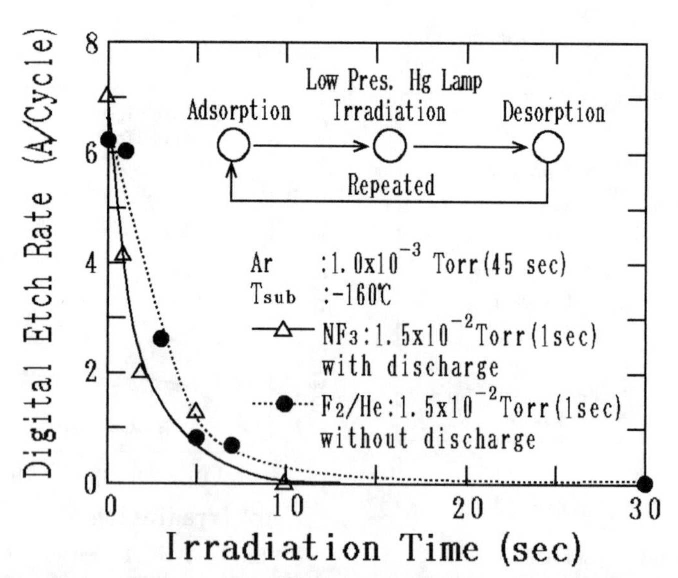

Fig.4 Si digital etch rate employing F radicals or molecules obtained with or without discharge of F_2/95%He versus a low pressure Hg lamp irradiation time.

2.2.3. Ion Energy Dependence of F or Cl/Si Reaction System

The Cl/Si system was studied to realize self-limiting atomic Si etching, because one Cl atom is reported to adsorb on each dangling bond site of Si in the Cl/Si reaction system[13]. Figure 5 shows the digital etch rate as a function of Ar^+ ions irradiation time. Pure Cl_2 was adsorbed at substrate temperature of R.T. The digital etch rate of 0.4 Å/cycle was obtained in spite of change in the chlorine suppling rate. The value of 0.4 corresponds to about one third of the atomic layer of Si(100). Although a self-limiting etching was achieved in the Cl/Si system, an origin of this value in the crystallography and the surface chemistry can not be understood.

It is important to investigate a threshold energy which initiates the ion induced etching reaction. In RIE with F/Si reaction at room temperature (R.T.), SiF_x layers in about one nanometer thickness formed on a Si surface is known to be sputtered physically by ion bombardment. To contrast with this R.T. reaction, we can study a significant ion induced F/Si reaction by utilizing the present physiosorbed fluorine on a cooled Si surface. Ar^+ ion energy dependences of digital etch rate in both F/Si and Cl/Si systems as a function of negative bias voltage were measured as shown in Fig.6(a,b). The bias was supplied at the substrate in the desorption stage in Fig.1. It should be noted that about 15 volt of plasma potential is added even at an applied bias of 0 volt. The Ar^+ ions irradiation time was set to 30sec in which the digital etch rate reached the saturated value. In the F/Si system (Fig.6(a)), the digital etch rate increased with the bias voltage without a threshold feature. This result demonstrates that some amount of SiF_x layer produced by ion bombardment to the F/Si surface desorbs, with the desorption rate increasing with ion energy.

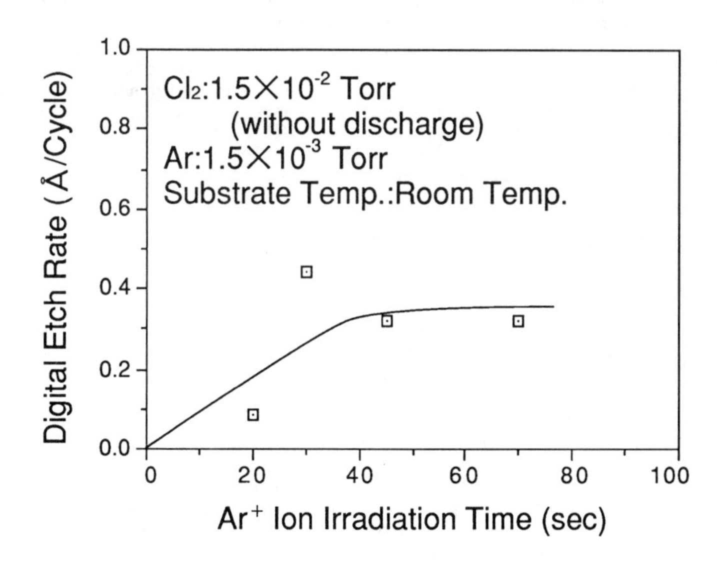

Fig.5 Digital etch rate of Si versus Ar^+ ion irradiation time at a pressure of 1.5×10^{-2} Torr of pure Cl_2 down-stream. The wafer temperature was room pemperature.

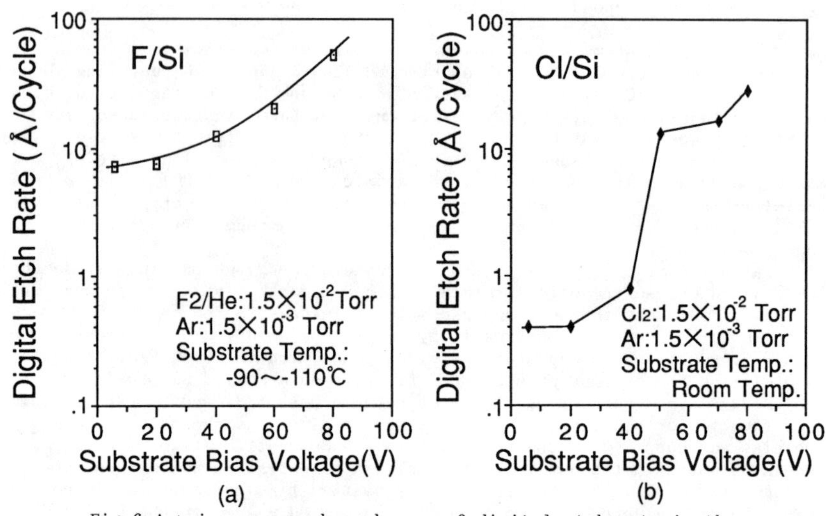

Fig.6 Ar^+ ion energy dependences of digital etch rate in the F/Si(a) and Cl/Si(b) system as a function of negative bias voltage. The substrate temperature was -110°C in F/Si case, and room temperature in Cl/Si case.

In the Cl/Si system, first the digital etch rate of about 0.4 Å/cycle was maintained until 40 volt, then it increased rapidly with bias voltage. An in-situ XPS measurment of Cl/Si reaction surface subjected to irradiation of Ar^+ ions accelerated at 15V and 60V indicated that the amount of higher order products as the $SiCl_4$, $SiCl_3$ increased and the low order products as the SiCl decreased with ion energy. These results demonstrate that Cl/Si reaction does not proceed and desorb sufficiently below about 40 V (55 V in the case of inclusion of plasma potential). More efforts should be made for deeper understanding of these interesting results for the Cl/Si reaction and of surface chemistry in order to achieve thick and higher order $SiCl_x$ products with a low energy beam irradiation.

3. Digital CVD

3.1. TES/H_2 reaction system

At first, TES and hydrogen radicals are introduced simultaneously to an Si wafer set on a substrate temperature of 250 °C in a reactor through separate Al_2O_3 tubes (see Fig.9). Total pressure was kept at 1.0 Torr. Figure 7 shows the deposition rate of films and step coverage as a function of H_2 concentration in TES. No film was formed with TES alone, while the deposition occurred rapidly with increasing H_2 in TES, and also, the value of the B/A ratio of step coverage became almost unity. Here A and B represent thickness at the plane of the films and lateral thickness at the upper edge in the trench, respectively. After the maximum deposition rate of 300Å/min at 60% H_2 concentration, both the deposition rate and B/A ratio decreased rapidly for the more H_2 addition.

In the FTIR (Fourier Transfer Infrared Spectroscopy) measurements of the films obtained by varying H_2 concentration in TES, the absorbance for $Si-C_2H_5$ (1250cm^{-1}), CH_2 (2926cm^{-1}) and CH_3 (2962, 2872cm^{-1})[14] bonds was observed at 40% and 60% H_2 concentrations, while these organic species were

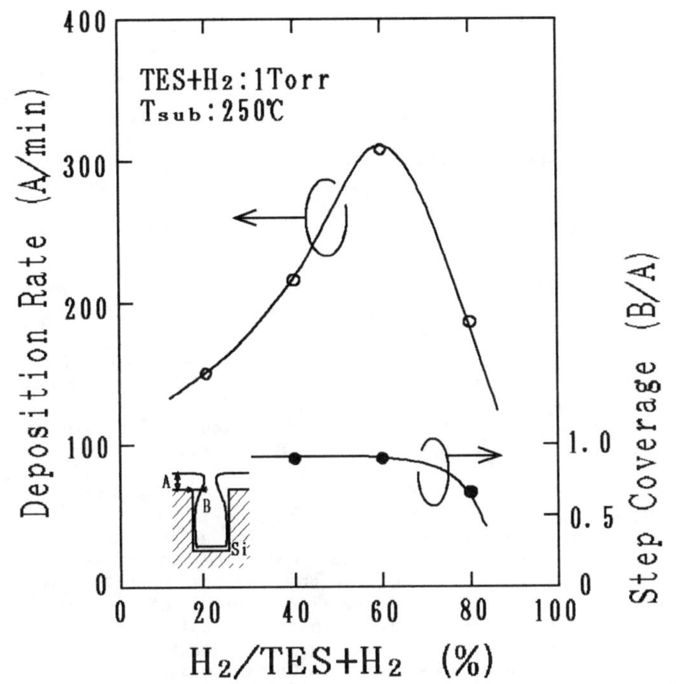

Fig.7 Deposition rate of films and step coverage vs H_2 concentration in TES.

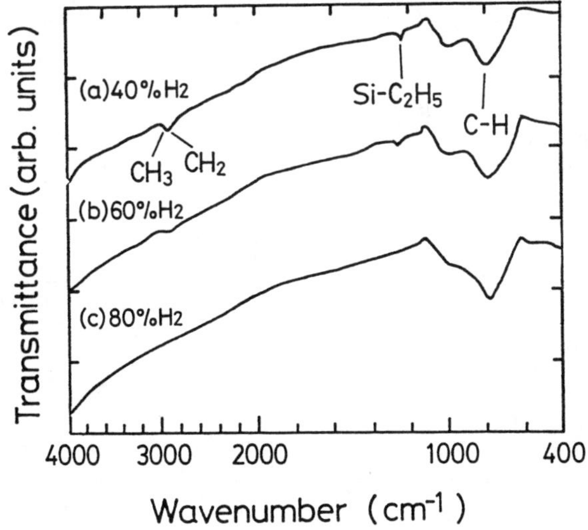

Fig.8 FTIR spectra of the TES/H reaction films as a function of H_2 concentration.

not observed at 80% H_2 concentration as shown in Fig.8. In addition, appreciable absorbance caused by C_xH_Y groups was observed in the film deposited at room temperature, which was deposited in the trench from the bottom to the upper surface. The changes of deposition features are considered to relate closely to the concentration of C_xH_Y groups involved in the film. That is, the inclusion of C_xH_Y species reduces the viscosity, flowing the film into a trench like water poured into a glass. Finally, the surface tension makes the film deposit conformaly on the trench wall. The organic species in the film which decrease the film viscosity, decrease with H_2 concentration. Thus, the film produced as a result of excess extraction reactions of the CH_2, CH_3 and C_2H_5 groups lacks the viscosity, reducing surface migration.

3.2. Digital CVD of Si oxide and nitride

Figure 9 illustrates the experimental apparatus of digital CVD and time sequence of the introduction of gases. A digital CVD system was carried out by repetitive process of a reaction of TES/H and subsequently oxidation or nitridation of this film with oxygen(O) or ammonium radicals which were generated by a remote microwave plasma of H_2, O_2 and NH_3 in an Al_2O_3 tube at a microwave power of 60W, respectively. TES/H_2 and O_2 or NH_3 gases were alternately blown on an Si wafer in the reactor. At first, the film, containing a high concentration of C_xH_y groups, was oxidized into a SiO_2 film. Oxidation characteristics of the film, whose initial layer thickness of 10Å was deposited by the TES/H reaction per one cycle, was investigated as a function of O_2 pulse widths. Both the film thickness per single cycle and ratio of integral absorption coefficient of Si-C_2H_5 to Si-O ($\int\alpha_{Si-C2H5}/\int\alpha_{Si-O}$) in the FTIR spectrum decreased with O_2 pulse width. This result demonstrates the removal of the C_xH_Y group through oxidation by oxygen radicals, leading to an improvement in SiO_2 film quality. Figure 10 shows a FTIR spectrum of the oxide film for initial thickness of 5Å, along with Si nitride which is described later. Although appreciable absorbance due to the Si-O stretching mode (1100cm^{-1}) is seen, a little Si-C_2H_5 (1250cm^{-1}) bond was seen still. The buffered HF (HF/H_2O = 1/50) etch rate was about 1000Å/min which was about 10 times higher than that of thermal SiO_2 with an etch rate of 110Å/min. Much more improvement is required for the film quality.

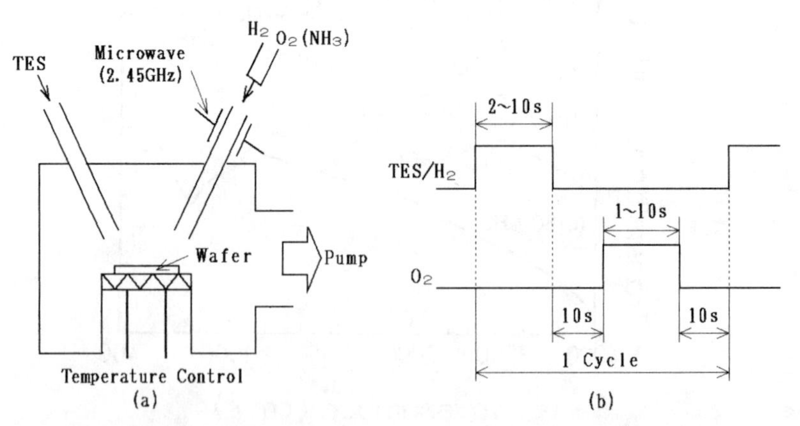

Fig.9 Schematic illustration of the experimental apparatus(a) and time sequence of the introduction of gases(b).

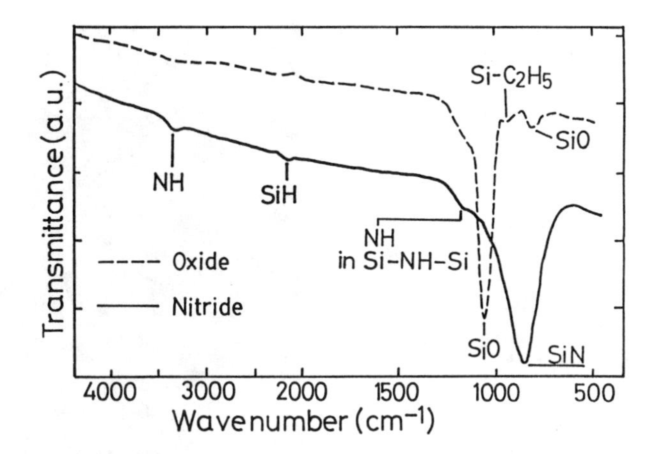

Fig.10 FTIR spectra of the silicon oxide and nitride film using the digital CVD.

Nextly, the digital CVD of Si nitride film was studied. First N_2 gas was used following inclusion of hydrogen into the film. However the TES/H reaction film was not nitrided. NH_3 was therefore used in this experiment. Deposition temperature was 300 °C. A FTIR spectrum for the Si nitride film is also shown in Fig.10. The nitride film shows typical characteristics of a plasma-enhanced CVD film, in which Si-N ($840cm^{-1}$), Si-H ($2150cm^{-1}$) and N-H (3330 and $1200cm^{-1}$) bonds are observed. The film quality depended on the initial film thickness and the NH_3 pulse width. The above film was obtained for initial thickness of 2Å, because the nitridation reaction was reduced rapidly for film thickness more than 2Å. N/Si atomic ratio of this nitride film was 0.6 defined from the XPS measurement, while a value of 0.9 is obtained for the typical pyrolytic Si_3N_4 films. However the present atomic ratio, lower than stoichiometric value of 1.3, might result from an inclusion of oxygen in the nitride film measured by XPS which was liberated from Al_2O_3 tube by NH_3 plasma.

In the digital CVD, organic Si layers with initial thicknesses of 5 Å for the Si oxide and 2 Å for the Si nitride are oxidized or nitrided. Therefore chemical bonding among the layers is assumed to be weak, because dangling bonds on each film are completely terminated by oxygen or nitrogen atoms. Any bonding is needed in order to construct three dimensional network structure. Thus we are now studying a process which first terminates the oxide surface with Si-O-H bonds and subsequently binds a upper layer with this oxide layer by removing H bonds.

3.3. Multilayer Stacked Film of Si Oxide/Nitride

The multilayer stacked film of Si oxide and nitride was tried by alternate oxidation and nitridation steps of the TES/H reaction film at the deposition temperature of 300 °C. Figure 11 shows a SEM micrograph of a cross-sectional view of the refilled ten layers of the stacked Si oxide and nitride with each 100Å thickness into the Si trench with aspect ratio of 6 (0.5μm width and 3μm depth). The Si nitride was slightly etched by a downstream method employing a $CF_4+O_2+N_2$ mixture so that we could observe the cross-section clearly. The electrical characteristic of the stacked film was also investigated. Figure 12 shows the breakdown characteristic of two type films of silicon oxide with 6.6Å and nitride with 11.6Å (film a), and silicon oxide with 13.3Å and nitride with 5.8Å (film b). The layers and total film thickness of both films were 10 layers and about 100Å respectively. The breakdown characteristic of the film a is much better than that of film b. It is considered that the stress in the Si nitride layer is relaxed effectively by thicker silicon oxide layer.

3.4. Diagnostic of Surface Reaction using in-situ FTIR-ATR

To make clear which is dominant process in the gas phase or surface reaction for the TES/H system, a) in-situ gas phase FTIR and b) ATR (Attenuated Total Reflection) of the TES/H reaction were measured. An in-situ FTIR-ATR system consisted of a radical inlet, an ATR prism and an evacuation system was set in an ATR accessory mounted on the standard FTIR (Perkin Elmer 1640). IR rays were incident normally on the 45 degree cut edge plane of the prism with 20x50 mm size. Since a Si prism does not provide any transmission lower than 1500cm^{-1}, due to Si self-absorption, a Ge substrate (n type, >60Ω·cm, (100)) was used as the prism. An amorphous Si film was prepared by an EB evaporation of Si on the one side of the Ge prisms.

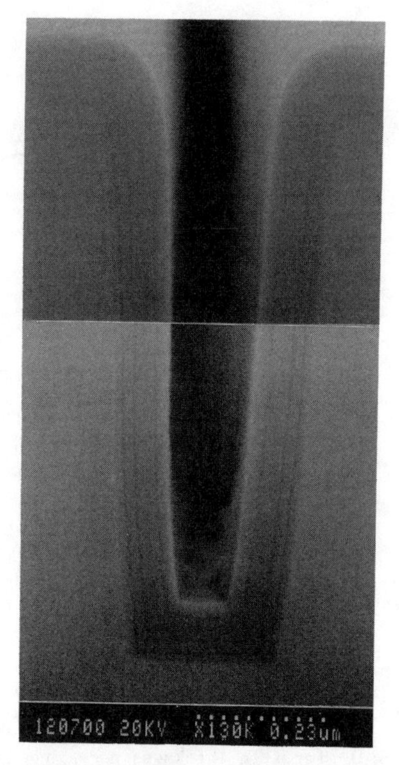

Fig.11 Cross-sectional SEM micrograph of the multilayer stacked silicon nitride and oxide film deposited in deep trench.

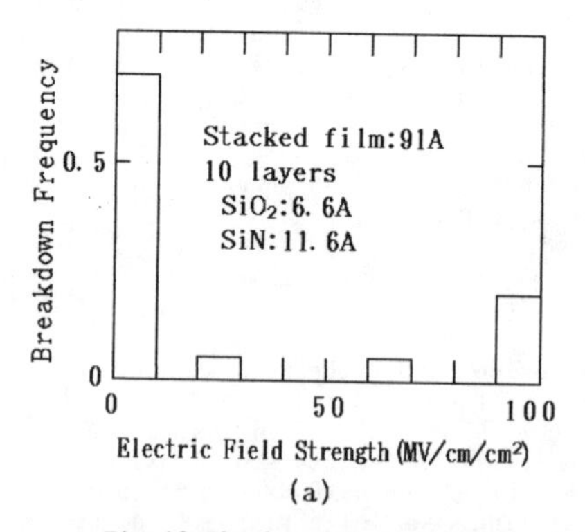

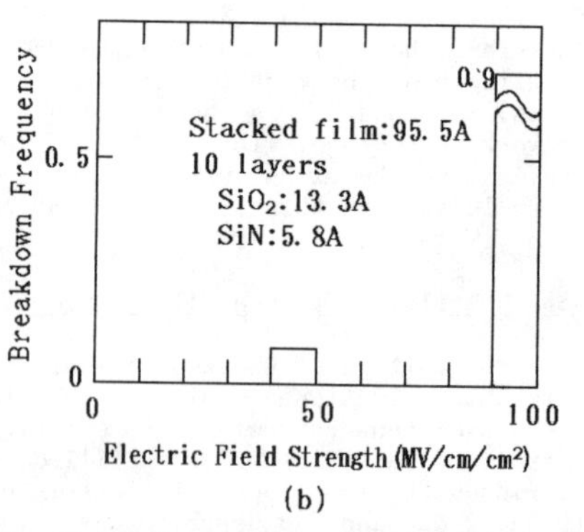

Fig.12 The breakdown characteristic of two type films of silicon oxide with 6.6Å and nitride with 11.6Å, (a), and silicon oxide with 13.3Å and nitride with 5.8Å, (b). The layers and total film thickness of both films were 10 layers and about 100Å.

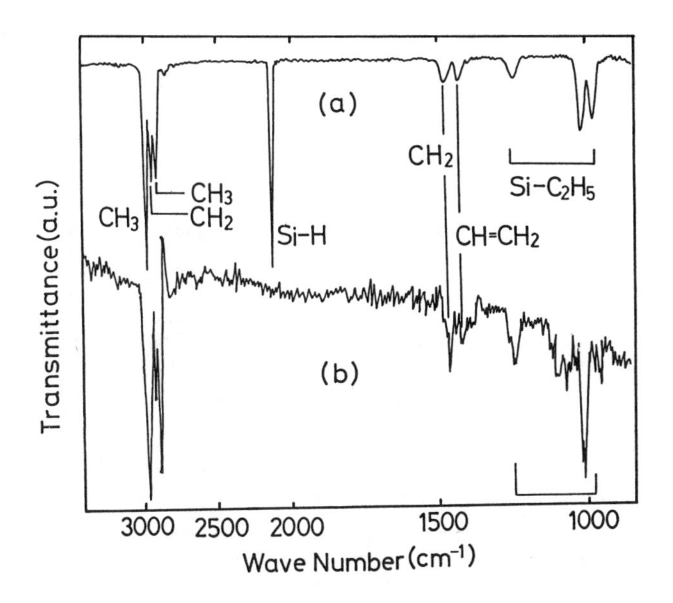

Fig.13 FTIR spectra of the TES/H reaction at gas phase (a) and on the H/F terminated Si surface (b).

The gas phase spectrum also was obtained by introducing TES and H_2 or H radicals into a reactor with a set of Si wafer windows which was arranged in an optical path of IR light. Almost same spectrum as one shown in Fig.13(a) was obtained in spite of change in a variety of parameters such as pressure and gas flow rates in both TES/H_2 and TES/H reactions. This suggests that H atoms do not react with TES in the gas phase. Fig.13(b) shows a different spectrum of the H/F terminated Si before and after TES/H reaction at R.T.. Hence Si-H peak, which results from H termination, is not observed because of compensation of both Si-H peaks. However Si-H bond of adsorbed TES is not also observed. This implys that H radicals react with methyl or ethyl groups of the adsorbed TES on the H/F terminated Si surface, releasing CH_4 , C_2H_6 and H_2 .

4. Conclusion

Digital etching and CVD and relevant surface reactions were studied. The novel method not only offers precisely controlled processes but enables us to elucidate reaction mechanisms, because fundamental reaction steps can be studied separately. In the etching, it is most likely that fluorine on cooled Si surfaces is in a physiosorbed state and the F/Si reaction is promoted by ion bombardment. The control of F coverage results in the atomic layer etching of Si. In the Cl/Si case, self-limiting reaction was achieved at the digital etch rate of 0.4 Å/cycle. Rapid increase in the etching was found for 40 V of substrate bias voltage. However origin of these values are not understood now. In the CVD, it was found that extraction of hydrogen from TES with H led to conformal deposition of a Si-C_xH_y film. Hence, the digital CVD which is done by repeating one cycle consisting of this film deposition and its subsequent oxidation or nitridation was developed for growing the Si oxide and nitride films. Thus obtained multilayer stacked oxide/nitride film was refilled in the deep trench with high resistance for breakdown. In-situ FTIR-ATR diagnostic of the TES/H reaction made it clear that it was dominated by the surface reaction.

References

[1] Y. Horiike, T. Tanaka, M. Nakano, S. Iseda, H. Sakaue, A. Nagata, H. Shindo, S. Miyazaki and M. Hirose, J. Vac. Sci. Technol., 8, 1844 (1990).
[2] H. Sakaue, S. Iseda, K. Asami, J. Yamamoto, M. Hirose, and Y. Horiike, Jpn. J. Appl. Phys., 29, 2648 (1990).
[3] J.W. Coburn and H.F. Winters, J. Vac. Sci. Technol., 16, 391 (1979).
[4] H.F. Winters: J. Vac. Sci, Technol., A3 (1985) 700.
[5] R.A. Haring, A. Haring, F.W. Saris, and A.E. de Vries, Appl. Phys. Lett., 41, 174 (1982).
[6] J.A. Mucha, V.M. Donnelly, D.L. Flamm and L.M. Webb, J. Phys. Chem., 85, 3529 (1981).
[7] Y. Nishimoto, N. Tokumatsu, T. Fukuyama and K. Maeda, Extended Abstracts of 19th Conf. on Solid State Devices and Materials, Tokyo, 1987 (Business Center for Academic Societies Japan, Tokyo, 1987) p.447.
[8] H. Sakaue, M. Nakano, T. Ichihara and Y. Horiike: Jpn. J. Appl. Phys., 30, L124 (1991).
[9] T. Ichihara, H. Sakaue, T. Okada and Y. Horiike, Proc. of Symp. on Dry Process, Tokyo, 1990 (The Inst. of Electrical Engineers of Japan, Tokyo, 1990), p.35.
[10] K. Tsujimoto, S. Tachi, S. Arai, H. Kawakami and S. Okudaira, Proc. of Symp. on Dry Process, Tokyo, 1988 (The Inst. of Electrical Engineers of Japan, Tokyo, 1988), p.42.
[11] F.R.Mcfeely, B.D.Silverman, J.A.Yarmoff and U.O.Karlsson, Proc. Int. Symposium on Plasma Chemistry, Vol.2, Tokyo, 1987 (1987) p. 927.
[12] Y. Horiike, T. Hashimoto, K. Asami, J. Yamamoto, Y. Todokoro, H. Sakaue, S. Shingubara and H. Shindo, Proc. of Microelectonic Engineering 13, 1991 (1991), p. 417.
[13] K. C. Pandey, T. Sakurai and H. D. Hagstrum, Phys. Rev., B16, 3648 (1977).
[14] L. J. Bellamy, The Infra-red Spectra of Complex Molecules, 3rd ed. (Chapman and Hall Ltd., London, 1975), p. 13, ibid, p. 374.

ATOMIC LEVEL CONTROL IN CRYSTAL GROWTH UTILIZING RECONSTRUCTION OF THE SURFACE SUPERSTRUCTURE

TAKASHI FUYUKI, TATSUO YOSHINOBU AND HIROYUKI MATSUNAMI
Kyoto University, Dept. of Electrical Engineering, Yoshidahonmachi, Sakyo, Kyoto 606-01, Japan

ABSTRACT

A novel mechanism of atomic level control in crystal growth utilizing reconstruction of surface superstructures is proposed. We have found a distinguished feature of surface reconstruction during crystal growth of 3C-SiC using an alternate gas molecular beam supply of Si_2H_6 and C_2H_2. When Si_2H_6 is supplied, Si atoms generated by thermal decomposition adsorb on the surface constructing superstructures. A fixed number of Si atoms forming surface superstructures can react with C_2H_2 yielding single crystalline 3C-SiC growth in the subsequent period, which realizes atomic level control in epitaxy of 3C-SiC. The reconstruction sequence is analyzed based on RHEED observations, and the obtained crystal quality is discussed.

INTRODUCTION

In conventional ALE (atomic layer epitaxy) techniques, crystal growth is controlled by so called "self-stopping mechanisms" by which sticking of extra atoms more than one monolayer is obstructed. There are many studies on the ALE of II-VI and III-V compound semiconductors using CVD and MBE techniques, but very few investigations upon the ALE of group IV semiconductors such as Si and SiC are reported up to now.

We have found a distinguished feature of surface reconstruction during crystal growth of 3C-SiC using an alternate gas molecular beam supply of Si_2H_6 and C_2H_2 [1]. When Si_2H_6 is supplied, generated Si atoms by thermal decomposition adsorb on the surface forming surface superstructures. When the C_2H_2 is supplied, the adsorbed Si atoms react with C_2H_2 molecules forming single crystalline 3C-SiC. The crystal growth is regulated by a fixed number of Si atoms constructing meta-stable surface superstructures.

In our previous study [2], 3C-SiC was homoepitaxially grown using an alternate supply of Si_2H_6 and C_2H_2 at a substrate temperature as low as 1000 ℃ , and the possibility of atomic level control in the crystal growth was proposed. In this report, surface reconstruction sequence is discussed in more detail based on RHEED (reflection high-energy electron diffraction) observations, and the crystal quality is also mentioned.

APPARATUS

Experiments were carried out in a UHV (ultra high vacuum) chamber designed for gas source molecular beam epitaxy (Fig.1). Cubic SiC(001) grown on Si(001) using a CVD method was used as a substrate. Si substrates employed for CVD growth have a slight inclination toward [110], with the resulting epitaxial layers being single domain (i.e. antiphase-domain free). Source gases are pure disilane (Si_2H_6) and pure acetylene (C_2H_2), and introduced into the chamber through individual nozzles each located 10cm distant from the substrate. The flow rates of these gases are controlled precisely by mass-flow controllers.

Prior to an experiment, the chamber is evacuated to a lower 10^{-10} Torr range by a sputter ion pump (SIP), which is detached from the system just

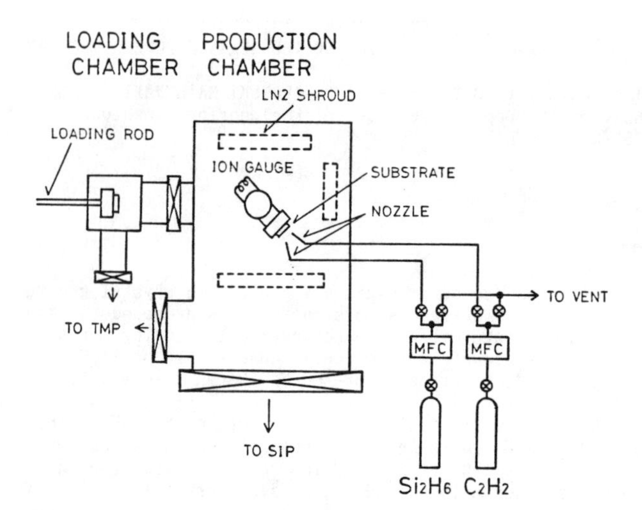

Fig.1. Gas source MBE apparatus.

before the introduction of source gases. Then, a turbo molecular pump (TMP) is employed for evacuation during the gas supply, and a liquid-nitrogen shroud is also employed to maintain a high vacuum.

The substrate was first heated up to 1100℃ in a high vacuum (below $3x10^{-8}$ Torr) for oxide removal, and then the substrate temperature was lowered to 1000℃, at which temperature the experiments were carried out.

SURFACE SUPERSTRUCTURE

The initial surface structure just after the oxide removal was identified to be c(2X2) by RHEED observations with <100> incidence. The reconstruction of the surface structures was monitored with the electron beam incident from the <110> direction. In our notation, [110] is parallel to the off-direction of the substrate, whereas [1$\bar{1}$0] is perpendicular to it. When the surface was exposed to a Si_2H_6 beam, extra streaks appeared in the [1$\bar{1}$0] azimuthal diffraction pattern, indicating the existence of two-fold periodicity along the [110] direction. The periodicity along the [1$\bar{1}$0] direction remained normal. This superstructure (2x1) is composed of Si atoms generated by thermal decomposition of Si_2H_6 molecules on the substrate surface.

Further supply of Si_2H_6 leads to (5x2) and (3x2) superstructures, successively as shown in Fig.2. These structures exhibit five-fold and three-fold periodicity along the [110] direction, respectively, and two-fold along the [1$\bar{1}$0] direction. These structures are expected to contain more Si atoms than the (2x1) structure, since they are formed after longer supply of Si_2H_6. No further transition from the (3x2) structure was observed for any continuous supply of Si_2H_6.

When the (3x2) surface was exposed to an C_2H_2 beam, those Si atoms adsorbed on the surface reacted with C_2H_2 molecules resulting surface structure change to the initial state in the reverse order.

By this surface reconstruction sequence, the crystal growth of 3C-SiC was confirmed[2] after some thousand numbers of recyclic procedures. The growth rate (thickness per cycle) is determined by the amount of Si atoms adsorbed on the surface, and so, the analysis of the surface superstructure

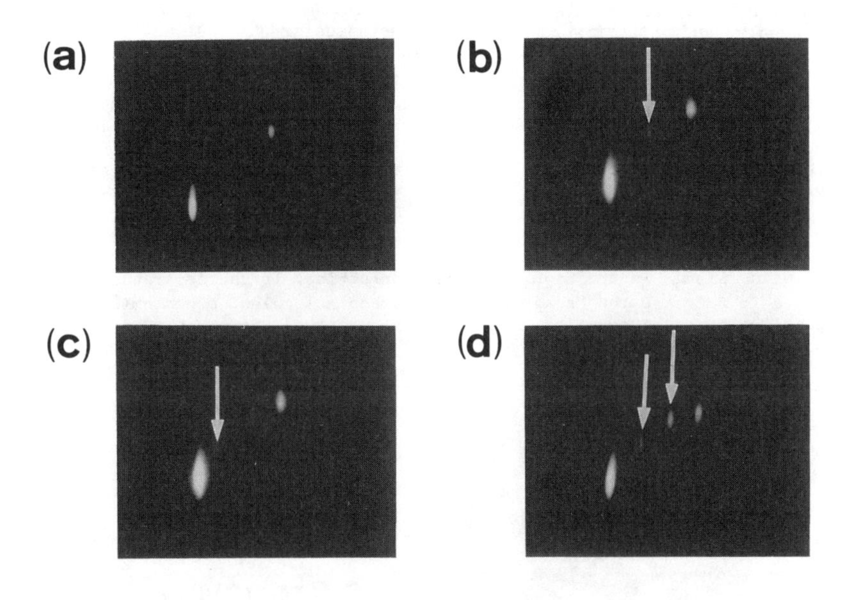

Fig.2. RHEED patterns with [110] incidence. (a) c(2x2) superstructure, (b) (2x1) superstructure, (c) (5x2) superstructure, (d) (3x2) superstructure.

is indispensable to atomic level control of the crystal growth. The change of the superstructure-related streaks were dynamically analyzed using a CCD camera and a pattern processing system, and discussed in detail elsewhere [3]. The durations t_1, t_2 and t_3 for the transitions c(2x2) → (2x1), c(2x2) → (5x2) and c(2x2) → (3x2), respectively, depended on the Si_2H_6 flow rate, but the ratios t_2/t_1 and t_3/t_1 are almost constant i.e. $t_2/t_1 \fallingdotseq 1.15$, and $t_3/t_1 \fallingdotseq 1.36$. The experimental values of their ratios are considered to be the ratios of the number of Si atoms in each superstructure, i.e. N(5x2)/N(2x1)≒1.15, and N(3x2)/N(2x1)≒1.36. These relations are in favor of the simple dimer model [4] for the (2x1) superstructure(1.0 overlayer of Si atom) and the additional dimer model [4] for the (5x2) and (3x2) superstructures (1.2 and 1.33 overlayer of Si, respectively). It should be noted that monolayer growth of 3C-SiC is expected when the (2x1) surface reacts with C_2H_2.

CRYSTAL GROWTH

Epitaxial growth of 3C-SiC was carried out by repeating the alternate supply of Si_2H_6 and C_2H_2 many times. This technique reduced the growth temperature to 1000℃, which is 350℃ lower compared with conventional CVD methods [5,6].

Growth parameters, in addition to the substrate temperature, are Si_2H_6 and C_2H_2 flux intensities, their durations and intervals. Experiments were carried out under various conditions, i.e. substrate temperature from

1000℃ (or 980℃) to 1050℃, flux intensities in the 10^{-6}-10^{-5} Torr range. The supply and cut-off sequence was conducted using an automatic valve operation, and one cycle is 20 sec - 2 min long.

RHEED patterns of the grown layers with about 500 nm thicknesses are shown in Fig.3. Growth rate (number of monolayer per 1 cycle) was deduced dividing the thickness by the number of repeated cycles. For the sample grown at a rate of 2.3 ML/cycle, clear streaks are observed in the RHEED pattern, whereas stacking faults and twins are observed for the sample grown at a larger growth rate of 3.5 ML/cycle. The surface morphology, on the other hand, exhibited no significant difference throughout the growth rate range of our experiments. Monolayer atomic epitaxy has not yet been realized since unintentionally desorbed hydrocarbons from the wall might react with Si_2H_6 in the Si_2H_6 introduction period. If the apparatus is improved to decrease the background hydrocarbon molecules, monoatomic layer epitaxy can be expected which leads to high-quality 3C-SiC.

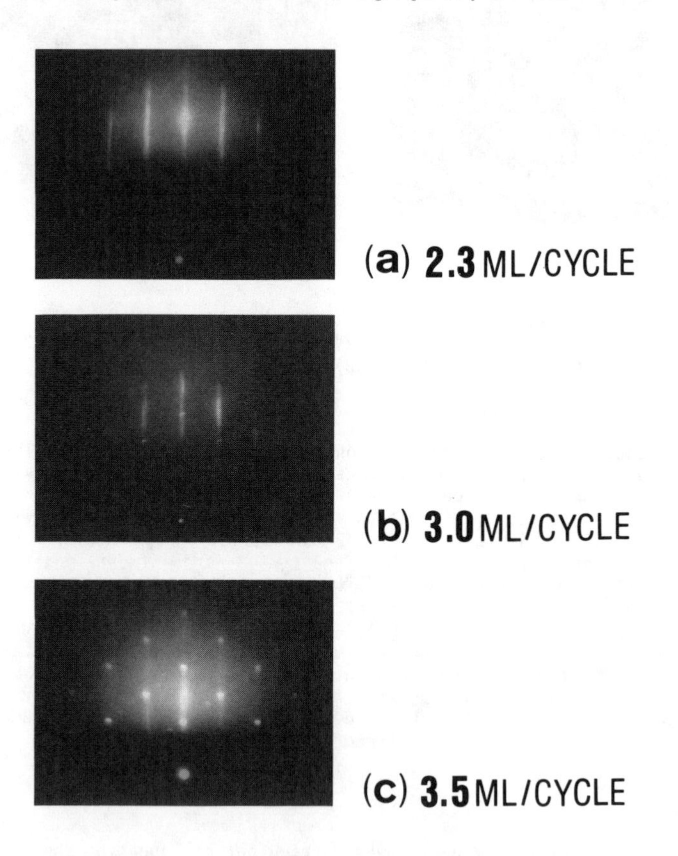

Fig.3. RHEED patterns of epilayers. (a) R_g=2.3ML/cycle, (b) R_g=3.0ML/cycle, (c) R_g=3.5ML/cycle.

CONCLUSION

A novel mechanism of atomic level control in crystal growth utilizing reconstruction of surface superstructures is proposed. Cyclic reconstruction of surface superstructures was found for the first time during crystal growth of 3C-SiC using an alternate gas molecular beam supply of Si_2H_6 and C_2H_2. When Si_2H_6 is supplied, Si atoms generated by thermal decomposition adsorb on the surface constructing superstructures. A fixed number of Si atoms can react with C_2H_2 forming single crystalline 3C-SiC in the subsequent period, which realizes atomic level control in epitaxy of 3C-SiC. The reconstruction sequence was analyzed based on RHEED observations. By repeating the alternate supply of Si_2H_6 and C_2H_2, epitaxial growth of single crystalline 3C-SiC was achieved at a substrate temperature of 1000℃. The growth rate was 2.3-3.5 ML/cycle.

This work was partly supported by a Grant-in-Aid for Scientific Research of the Ministry of Education, Science and Culture, Japan.

REFERENCES

[1] T. Fuyuki, M. Nakayama, T. Yoshinobu and H. Matsunami, J. Crystal Growth, 95, 461(1989).
[2] T. Yoshinobu, M. Nakayama, H. Shiomi, T. Fuyuki, and H. Matsunami, J. Crystal Growth 99, 520(1990).
[3] T. Yoshinobu, I. Izumikawa, H. Mitsui, T. Fuyuki, and H. Matsunami: to be published.
[4] S. Hara, PhD thesis, Waseda University, 1990.
[5] H. Matsunami, S. Nishino and H. Ono, IEEE Trans. Electron Devices ED-28, 1235(1981).
[6] S. Nishino, J. A. Powell and H. A. Will, Appl. Phys. Letters 42, 460(1983).

DECOMPOSITION OF ALKYLSILANES ON SILICON SURFACES USING TRANSMISSION FTIR SPECTROSCOPY

A.C. DILLON, M.B. ROBINSON, M.Y. HAN AND S.M. GEORGE
Department of Chemistry, Stanford University
Stanford, California 94305

ABSTRACT

Fourier transform infrared (FTIR) transmission spectroscopy was used to monitor the decomposition of alkylsilanes such as diethylsilane (DES) [$(CH_3CH_2)_2SiH_2$], di-t-butylsilane (DTBS) [$((CH_3)_3C)_2SiH_2$] and ethylsilane (ES) [$CH_3CH_2SiH_3$] on high-surface-area porous silicon samples. The FTIR spectra revealed that the alkylsilanes dissociatively adsorb on porous silicon at 300 K to form SiH and Si-alkyl species. As the silicon surface was progressively annealed, the Si-alkyl species decomposed and produced gas phase ethylene (DES,ES) or isobutylene (DTBS). The decomposition of the alkyl group was accompanied by the growth of additional SiH surface species. These reaction products were consistent with a β-hydride elimination reaction. Above 700 K, the SiH surface species decreased concurrently with the desorption of H_2 from the porous silicon surface. The uptake of surface species was also monitored at various adsorption temperatures to determine the optimal exposure temperatures for carbon-free silicon deposition. Carbon contamination was not detected at adsorption temperatures below 640 K prior to H_2 desorption. Because the alkylsilane adsorption process is self-limiting at temperatures below 640 K, alkylsilanes may be useful molecular precursors for the atomic layer epitaxy (ALE) of silicon.

INTRODUCTION

The need for greater control over epitaxial growth processes has motivated the development of atomic layer epitaxy (ALE) techniques. In the simplest form of ALE, the adsorption process is self-regulating and one monolayer of a compound is deposited per operational cycle (1). Silicon ALE has been performed with alternating cycles of SiH_2Cl_2 and H_2 at temperatures ranging from 1090 - 1100 K and 1160 - 1180 K for growth on Si(100) and Si(111), respectively (2,3). Unfortunately, the purity of the deposited layer was difficult to control because of autodoping during the high temperature growth (3,4). Moreover, SiH_2Cl_2 is difficult to handle due to its high flammability and toxicity (5). The simple silanes are also possible candidates for silicon ALE at temperatures below the H_2 desorption temperature at 800 K. However, the silanes are plagued by experimental hazards due to high flammabilities and toxicities (6,7).

Alkylsilanes may offer many advantages to silanes and chlorosilanes because they are less toxic and less flammable (8,9). In addition, many alkylsilanes are liquids with high vapor pressures which would facilitate their use for epitaxial growth. Unfortunately, very little is known about alkylsilane chemistry on silicon surfaces. In this study, the decomposition of diethylsilane (DES), ethylsilane (ES) and di-t-butylsilane (DTBS) on porous silicon surfaces was examined using transmission FTIR vibrational spectroscopy. The use of high-surface-area, porous silicon to characterize chemical reactions on silicon surfaces has been illustrated in several previous studies (13-15).

The FTIR spectra were employed to identify the surface species after a saturation alkylsilane exposure on porous silicon at 300 K. Annealing studies were performed to measure the thermal stability of surface reaction species and establish the decomposition mechanism. In addition, the growth of surface species at various adsorption temperatures was examined as a function of alkysilane exposure time. The results of these FTIR studies suggest the optimal alkylsilane adsorption temperatures and the thermal cycling required for silicon deposition without measurable carbon contamination (20-22).

EXPERIMENTAL

The electrochemical techniques used to prepare porous silicon have been previously described (13). These studies employed single-crystal Si(100) wafers with a thickness of

400 μm. These samples were *p*-type boron-doped with a resistivity of ρ= 0.1 - 0.3 Ωcm. The 2 μm thick porous silicon layer provided a surface area enhancement of x400 compared with a single-crystal sample. Prior to mounting the porous silicon sample in the UHV chamber, the sample was rinsed successively in ethanol, tetrachloroethylene and hydrofluoric acid. The hydrofluoric acid was used to remove any native oxide that may have formed on the porous silicon surfaces after anodization.

The samples were mounted at the bottom of a liquid-nitrogen-cooled cryostat on a differentially-pumped rotary feedthrough (23). The crystal mounting design and techniques for sample heating have been detailed previously (13). The UHV chamber for *in-situ* transmission FTIR studies have been described earlier (13). A Nicolet 740 FTIR spectrometer and an MCT-B infrared detector were employed in these studies.

The surface species produced during the alkylsilane adsorptions were studied as a function of alkylsilane exposure and silicon surface temperature. In these experiments, the silicon substrate was held at a constant temperature. FTIR spectra were recorded versus alkysilane exposure until a saturation coverage was obtained. The FTIR spectra were monitored at the adsorption temperature for the isothermal measurements below 500 K. Because of increased infrared absorption by free carriers at higher temperatures (24-26), FTIR spectra were recorded at 500 K for the isothermal measurements above 500 K.

In the thermal annealing studies, porous silicon samples were exposed to a particular alkysilane until a saturation coverage was obtained. Alkylsilane exposures at 1 x 10^{-5} Torr for 5 minutes with the porous silicon sample at 100 K, followed by 10 minutes with the porous silicon sample at 300 K were employed to achieve a saturation coverage. The silicon surface temperature was then increased using a heating rate of 7 K/sec. The sample was held at the annealing temperature for one minute. Subsequently, the sample was returned to 300 K before recording the FTIR spectra. This experimental sequence was performed repeatedly for annealing temperatures up to 860 K. A new porous silicon sample prepared under identical conditions was used for each adsorption study and each set of annealing experiments. As described earlier (13), the porous silicon sample was annealed in vacuum to remove the surface hydrogen and obtain a clean surface before alkylsilane exposure.

RESULTS

After a saturation DES exposure at 300 K, changes in the infrared absorption spectrum of porous silicon versus annealing temperature are shown in Fig. 1. Figure 1a shows the decrease in the infrared absorption of the C-H stretching modes at 2955-2879 cm^{-1} versus annealing from 300 to 760 K. Figure 1b shows the concurrent increase in the absorption of the Si-H stretch at 2088 cm^{-1} over the temperature range from 300 to 700 K.

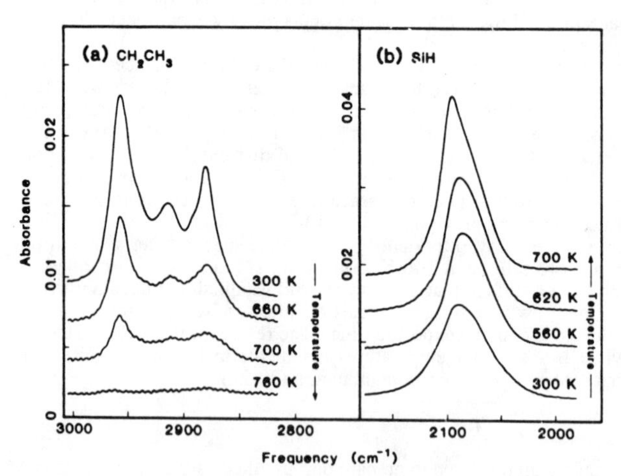

Figure 1 Infrared absorbance of a) the C-H stretching vibration and b) the Si-H stretching vibration versus annealing temperature after a saturation $(CH_3CH_2)_2SiH_2$ exposure at 300 K.

The integrated infrared absorbances versus annealing temperature after a saturation DES exposure at 300 K are displayed in Fig. 2. The C-H stretching modes corresponding to $SiCH_2CH_3$ surface species progressively decrease as the surface is annealed between 450 - 760 K. An increase by a factor of approximately 1.5 in the integrated absorbance for the Si-H stretching vibrations is also observed as the temperature is increased from 300 K to 720 K. At temperatures higher than 720 K, the integrated absorbance decreases for the Si-H stretching vibration. This decrease is consistent with the desorption of H_2 from the silicon surface (13,27,28).

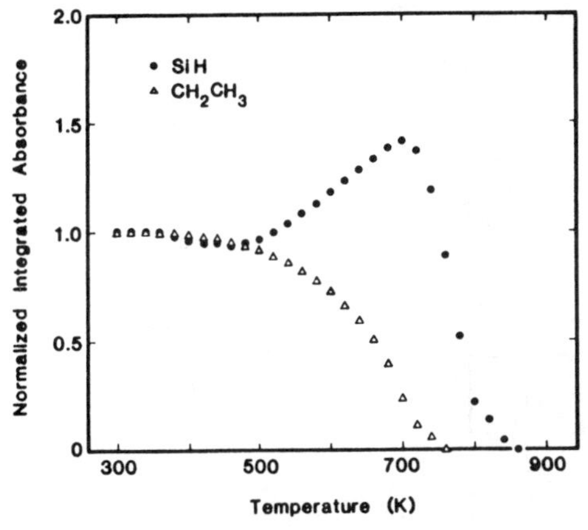

Figure 2 Integrated infrared absorbances of the Si-H and the C-H stretching vibrations as a function of annealing temperature after a saturation $(CH_3CH_2)_2SiH_2$ exposure at 300 K.

The integrated infrared absorbances versus annealing temperature after a saturation ethylsilane exposures at 300 K are displayed in Fig. 3. For ethylsilane adsorption on porous silicon at 300 K, the surface species are SiH, SiH_2 and $SiCH_2CH_3$. The decrease in the integrated absorbance for the $Si-H_2$ scissors mode in Fig. 3 indicates that the silicon dihydride surface species have disappeared by 540 K. The C-H stretching modes corresponding to $SiCH_2CH_3$ surface species have disappeared from the FTIR spectrum after annealing above 720 K. An increase by a factor of approximately 2.2 in the integrated absorbance of the Si-H deformation mode is observed as the temperature is increased from 300 K to 540 K. In addition, an increase in the integrated absorbances by a factor of approximately 1.25 for the Si-H deformation mode and approximately 1.2 for the Si-H stretching mode is observed as the temperature is increased further from 540 K to 660 K. The desorption of H_2 from the silicon surface above 680 K is also apparent in Fig. 4 (13,27,28).

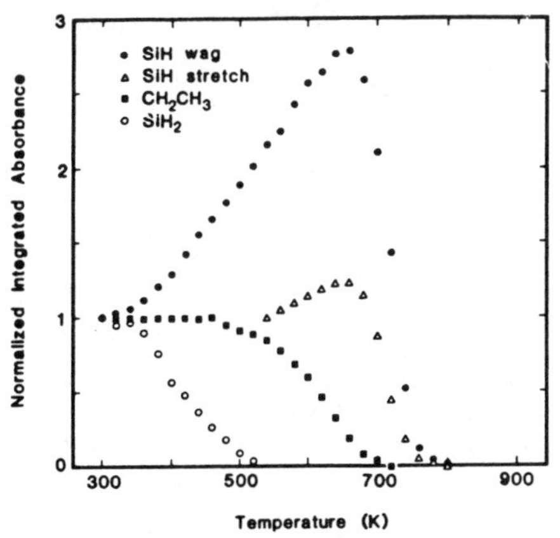

Figure 3 Integrated infrared absorbances of the Si-H wag, the Si-H stretch, the C-H stretches and the $Si-H_2$ scissors mode as a function of annealing temperature after a saturation $CH_3CH_2SiH_3$ exposure at 300 K.

The spectra of porous silicon after a saturation ES exposure at 580 K and an equivalent ES exposure at 800 K are shown in Fig. 4a and 4b, respectively. After ES exposure at 580 K, the porous silicon spectra exhibited pronounced infrared absorption features attributed to the surface hydrogen at 2088 and 627 cm^{-1}. At 800 K, carbon is incorporated into the silicon lattice and a saturation coverage cannot be defined. The porous silicon spectra exhibited a major infrared feature assigned to a Si-C vibrational mode at 750 cm^{-1}. Minor spectral features also appeared at 2119 and 620 cm^{-1}. Similar spectra were obtained after saturation exposures of DES at 640 K and 800 K.

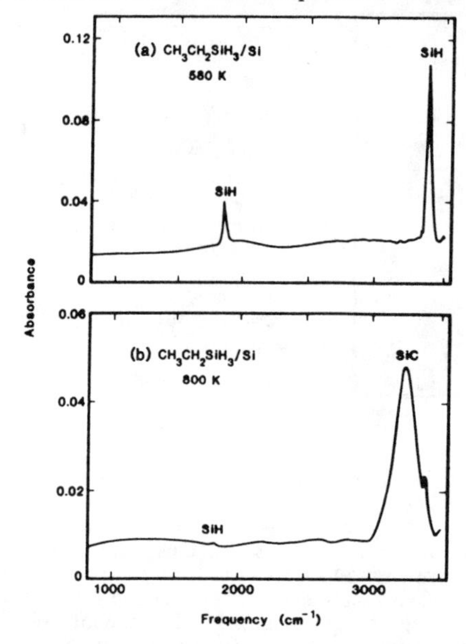

Figure 4 Infrared spectrum of porous silicon surfaces after a) a saturation $CH_3CH_2SiH_3$ exposure at 580 K and b) an equivalent exposure at 800 K.

The integrated infrared absorbances versus annealing temperature after a saturation di-t-butylsilane exposure at 300 K are similar to the integrated absorbances displayed in Fig. 2 for DES. For di-t-butylsilane adsorption, the surface species are SiH and $SiC(CH_3)_3$. The C-H stretching modes corresponding to $SiC(CH_3)_3$ surface species have disappeared from theelimination reaction: $SiCH_2CH_3$ --> SiH + CH_2=CH_2(g) (20). Because of a high background pressure for annealing temperatures between 500 - 760 K, an increase in the ethylene background pressure was not observed. However, in recent temperature programed desorption (TPD) studies of DES on Si(111) 7x7, an ethylene desorption signal was observed at 700 K with a heating rate of 9 K/s following DES adsorption at 200 K (29).

Since a saturation DESFig. 2 show that the decrease in the $SiCH_2CH_3$ species is accompanied by the concurrent growth of SiH species. In addition, Fig. 2 reveals that the maximum integrated absorbance for the Si-H stretching vibration occurs at 700 K and is slightly less than 1.5 times the initial Si-H integrated absorbance at 300 K.

Together, the increase in the integrated absorbance of the Si-H stretching vibration and the disappearance of the C-H stretching modes suggest the following β-hydride elimination reaction: $SiCH_2CH_3$ --> SiH + CH_2=CH_2(g) (20). Because of a high background pressure for annealing temperatures between 500 - 760 K, an increase in the ethylene background pressure was not observed. However, in recent temperature programed desorption (TPD) studies of DES on Si(111) 7x7, an ethylene desorption signal was observed at 700 K with a heating rate of 9 K/s following DES adsorption at 200 K (29).

Since a saturation DES exposure is expected to produce equivalent amounts of surface $SiCH_2CH_3$ and SiH species, the overall decomposition reaction is given by: SiH + $SiCH_2CH_3$ ---> 2SiH + CH_2=CH_2 (g). At the maximum absorbance in Fig. 2, the SiH species does not increase by a factor of 2 as expected. The lower increase may be attributed to some recombinative H_2 desorption prior to the decomposition of all the

$SiCH_2CH_3$ surface species (13,27,28).

In similarity to the DES results, thermal annealing studies of ethylsilane and di-t-butyl silane reveal that the decrease in the Si-alkyl species is accompanied by the concurrent growth of SiH species. These results are also consistent with a β-hydride elimination mechanism for the alkyl group decomposition (21,22). The normalized integrated absorbances for the thermal annealing study of ethylsilane shown in Fig. 3 also reveal the initial presence of SiH_2 surface species which have disappeared by 540 K. This surface reaction will be analyzed in detail elsewhere (21).

The thermal annealing results for ethylsilane shown in Fig. 3 show that the C-H stretching modes corresponding to the $SiCH_2CH_3$ surface species have disappeared by 720 K. The maximum integrated absorbance for the Si-H stretching vibration occurs at 660 K and is approximately 1.2 times the Si-H integrated absorbance at 540 K. Assuming that H_2 does not desorb during the decomposition of the SiH_2 species,a saturation ethylsilane exposure would be expected to produce an increase of 1.3 times the Si-H integrated absorbance at 540 K (21).

The thermal annealing results for di-t-butylsilane also demonstrate that the C-H stretching modes corresponding to the $SiC(CH_3)_3$ surface species have disappeared by 660 K. The maximum integrated absorbance for the Si-H stretching vibration occurs at 660 K and is approximately 2 times the initial Si-H integrated absorbance at 300 K. Given the overall reaction SiH + Si-t-butyl --> 2SiH + isobutylene (22), a saturation di-t-butylsilane exposure would be expected to display the observed increase of 2 times the initial Si-H integrated absorbance at 300 K. There is no evidence of recombinative H_2 desorption during Si-alkyl decomposition (13,27,28).

Temperature Dependent Adsorption Studies Suggest Silicon ALE Application

Alkylsilane adsorption studies at various isothermal temperatures provide an interesting comparison with the annealing studies that were performed after saturation alkylsilane exposures at 300 K. These experiments also measure the effects of adsorption temperature on the resultant surface species and their adsorption rates. These adsorption experiments are useful in determining the optimal temperature for silicon deposition without carbon contamination.

Figure 3a reveals that after saturation ES exposure at 580 K, the only surface species is SiH. The absence of absorption features between 2955 - 2879 cm^{-1} suggests that no surface $SiCH_2CH_3$ species are present. A dramatic change in the surface species occurs after a saturation ES exposure at 800 K. A large absorption is observed at 750 cm^{-1} as shown in Fig. 3b. This broad absorption feature is assigned to a Si-C vibrational mode. These results indicate that silicon epitaxial growth with ES at temperatures at or above 800 K is not desirable because of carbon incorporation. Very similar results were observed for DES at 640 and 800 K (20).

Upon completion of the absorption study conducted at 580 K, the porous silicon sample was annealed to 860 K for 4 minutes. The absorption features in Fig. 3a disappeared as expected following H_2 desorption from the surface (13,27,28). More importantly, there was no detectable Si-C absorption at 750 cm^{-1} which indicates that carbon was not present in the silicon lattice. Similar results were obtained in the DES adsorption study at 640 K (21). Isothermal adsorption studies of di-t-butylsilane should yield similar results for temperatures between 500-520 K (22). The lack of carbon incorporation is very encouraging for the application of alkylsilanes for silicon atomic layer epitaxy (ALE) (20). Based upon these results, the operational cycle for silicon ALE would be:

1) adsorption at 640 K (DES), 580 K (ES) and 500-520 K (DTBS) where only hydrogen and silicon are deposited
2) annealing to 800 K to desorb the hydrogen as H_2
3) cooling back down to the adsorption temperature to repeat the cycle

Recent studies on Si(111)7x7 indicate that each alkylsilane adsorption cycle should deposit approximately 0.25 ML of silicon on the silicon surface (29).

CONCLUSIONS

Alkylsilanes possess chemical properties which may prove useful in the atomic layer epitaxy of silicon. Fourier transform infrared (FTIR) transmission spectroscopy was used to monitor the decomposition of alkylsilanes such as diethylsilane (DES) $[(CH_3CH_2)_2SiH_2]$, di-t-butylsilane (DTBS) $[((CH_3)_3C)_2SiH_2]$ and ethylsilane (ES) $[CH_3CH_2SiH_3]$ on high-surface-area porous silicon samples. The FTIR spectra revealed that the alkylsilanes dissociatively adsorb on porous silicon at 300 K to form SiH and Si-alkyl species. As the silicon surface was progressively annealed, the Si-alkyl species decomposed via a β-hydride elimination reaction and produced gas phase ethylene (DES,ES) or isobutylene (DTBS). The decomposition of the alkyl group was accompanied by the growth of additional SiH surface species. Above 700 K, the SiH surface species decreased concurrently with the desorption of H_2 from the porous silicon surface. In order to determine an optimal ALE operational cycle, the uptake of surface species was monitored at various adsorption temperatures. These studies revealed that silicon ALE may be performed without detectable carbon contamination at adsorption temperatures below 640 K prior to H_2 desorption.

ACKNOWLEDGEMENTS

This work was supported by the U.S. Office of Naval Research under Contract No. N00014-90-1281. Some of the equipment utilized in this work was provided by the NSF-MRL through the Center for Material Research at Stanford University. We thank Dr. David Roberts of the J.C. Schumacher Co. both for supplying the diethylsilane and for numerous helpful discussions. SMG acknowledges the National Science Foundation for a Presidential Young Investigator Award and the A.P. Sloan Foundation for a Sloan Research Fellowship.

REFERENCES

[1] C.H.L. Goodman and M.V. Pessa, J. Appl. Phys. **60**, R65 (1986).
[2] J. Nishizawa, K. Aoki, S. Suzuki, and K. Kikuchi, J. Cryst Growth **99**, 502 (1990).
[3] J. Nishizawa, K. Aoki, S. Suzuki, and K. Kikuchi, J. Electrochem. Soc. **137**, 1898 (1990).
[4] S.M. Sze, *VLSI Technology* (McGraw-Hill, New York, 1988), 2nd ed., chapt. 2.
[5] Material Safety Data Sheet for dichlorosilane (OSHA, US Dept. of Labor).
[6] Material Safety Data Sheet for silane (OSHA, US Dept. of Labor).
[7] Material Safety Data Sheet for disilane (OSHA, US Dept. of Labor).
[8] Material Safety Data Sheet for diethylsilane (OSHA, US Dept. of Labor).
[9] Material Safety Data Sheet for ethylsilane (OSHA, US Dept. of Labor).
[13] P. Gupta, V.L. Colvin and S.M. George, Phys. Rev. B **37**, 8234 (1988).
[14] P. Gupta, A.C. Dillon, A.S. Bracker and S.M. George, *Surf. Sci. (in press)*.
[15] A.C. Dillon, P. Gupta, M.B. Robinson A.S. Bracker and S.M. George, *J. Vac. Sci Technol. A (in press)*.
[20] A.C. Dillon, M.B. Robinson, M.Y. Han and S.M. George, *(Submitted to J. Electrochem. Soc.)*
[21] A.C. Dillon, M.B. Robinson, M.Y. Han and S.M. George, *(in preparation)*
[22] A.C. Dillon, M.B. Robinson, M.Y. Han and S.M. George, *(in preparation)*
[23] S.M. George, J. Vac. Sci. Technol. A **4**, 2394 (1986).
[24] G.E. Jellison, *Semiconductors and Semimetals*: Pulsed Laser Processing of Semiconductors; edited by R.F. Wood, C.W. White and R.T. Young (Academic Press, New York), Chapter 3, Vol III.
[25] G.G. MacFarlane, T.P. McLean, J.E. Quarrington and V. Roberts, Phys. Rev. **111**, 1245 (1958)
[26] H.A. Weakliem and D. Redfield, J. Appl. Phys. **50**, 1491 (1979).
[27] G. Shulze and M. Henzler, Surf. Sci. **124**, 336 (1983).
[28] B.G. Koehler, C.H. Mak, D.A. Arthur, P.A. Coon and S.M. George, J. Chem. Phys. **89**, 1709 (1988).
[29] P.A. Coon, M.L. Wise and S.M. George *(in preparation)*

THE ETCHING OF SILICON BY OXYGEN OBSERVED BY in situ TEM

FRANCES M. ROSS AND J. MURRAY GIBSON*
AT&T Bell Laboratories, 600 Mountain Avenue, Murray Hill, NJ 07974
*Present address: University of Illinois Materials Research Laboratory, 104 S. Goodwin Ave., Urbana, IL 61801

ABSTRACT

We describe observations made in situ in a modified UHV transmission electron microscope of the process of etching of the Si (111) surface by oxygen. Etching occurs by the motion of individual bilayer steps across the surface and by analysing the step motion we discuss the etching mechanism in the context of macroscopic parameters.

INTRODUCTION

A clean silicon surface exposed to oxygen gas can either etch or oxidise depending on the temperature and oxygen pressure [1-3]. The etching is a result of the reaction of Si and O_2 to form the volatile oxide SiO. This reaction, as we demonstrate below, can create an atomically flat but active surface and so may be of use in device processing. We believe that an examination of the mechanism of this interesting reaction on a microscopic scale will complement the thermodynamic studies which have already been carried out [1-4], and we have therefore made observations of this reaction in situ in a transmission elecron microscope, using imaging techniques which are sensitive to the topography and atomic structure of the silicon surface.

Our equipment, described elsewhere [5], consists of a transmission electron microscope modified to attain UHV (10^{-9} Torr operating conditions) with a specimen heating stage and gas inlet. The silicon specimen is made from (111) oriented p-doped 10 Ωcm material chemically pre-thinned and then cleaned and further thinned in situ, either by heating under UHV to about 1200°C or by using the etching reaction itself. The resulting foil is thin, flat and clean enough to show the 7x7 superlattice spots clearly at temperatures below 860°C and the "forbidden" $^1/_3$ 422 diffraction spots at all temperatures [6, 7]. Images obtained using one of these forbidden reflections are sensitive to the positions of surface steps, even those only a monolayer in height, and the steps are clearly visible on both the top and bottom surfaces of specimens up to about 200nm thick (although the quality of the images is of course improved the thinner the specimen).

We firstly observe these steps on a clean surface and then introduce oxygen gas to a pressure of 10^{-7} to 10^{-5} Torr while maintaining the specimen at a fixed temperature in the range 600-1000°C. We find that the etching reaction then occurs by the removal of bilayer steps from the silicon surface. The reaction occurs fairly fast under these conditions and can be recorded either onto photographic plates or onto video (via an image intensifier), enabling accurate time resolved measurement of surface change. Finally, information about the state of the terraces between the steps can also be obtained by analysing any intensity changes of the spots in the 7x7 diffraction pattern, provided of course that the experiment is being carried out below the 7x7↔1x1 transition temperature (860°C).

We have observed etching by water vapour as well as oxygen, both above and below 860°C, but here we focus on etching by oxygen at the higher temperatures. We will firstly describe the interpretation of the step images we have obtained, and show the evidence from measured intensity levels that bilayer steps are removed during etching. We will then examine the mechanism of the etching process using parameters derived from observations of the step movement. Etching is hard to study microscopically using other methods since a high degree of surface sensitivity is needed (such as in STM or REM). However some other evidence is available, particularly from thermodynamic studies, and we will discuss our results in the context of other measurements.

IMAGING SURFACE STEPS DURING ETCHING

Figure 1 shows a typical surface-sensitive image obtained near the (111) normal using the $^1/_3$ 422 reflection. The positions of steps are shown very clearly by sharp changes in the intensity levels. In particular, the shape of steps and the way they interact with obstructions such as SiC particles can be seen.

The $^1/_3$ 422 reflection is forbidden in the bulk silicon structure. However if the specimen thickness in the beam direction is not an integral number of unit cells, the incomplete structure does not cancel all the amplitude in this reflection [8]. At every change in specimen thickness, caused by a step of even monolayer height ($^1/_6$ unit cell) on either the top or bottom surface of the specimen, this remaining amplitude can change strongly leading to the step sensitivity observed in this image. (The unit cell we refer to here is the hexagonal unit cell, described in figure 2.) Thus step movement can be followed accurately, and furthermore the intensity levels seen in images such as figure 1 can be analysed quantitatively by comparison with computer simulations to yield information about the nature of the steps seen, as we demonstrate below.

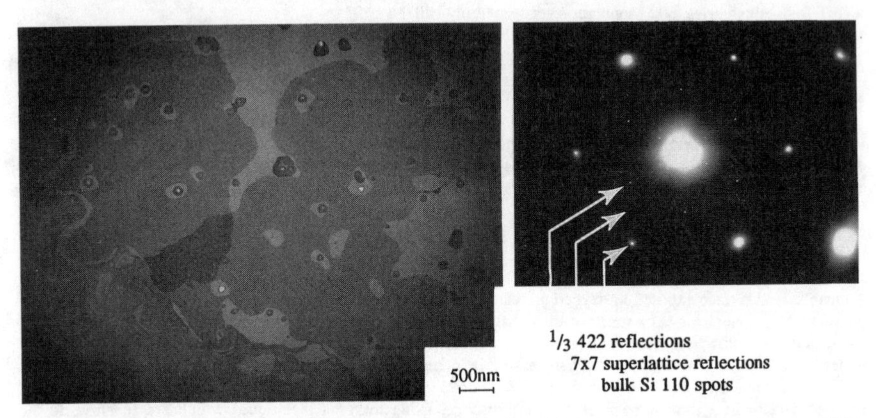

Figure 1. An image formed using the $^{1}/_{3}$ 422 reflection. The small particles are SiC formed during heating of the specimen. The diffraction pattern shows the 7x7 and $^{1}/_{3}$ 422 diffraction spots.

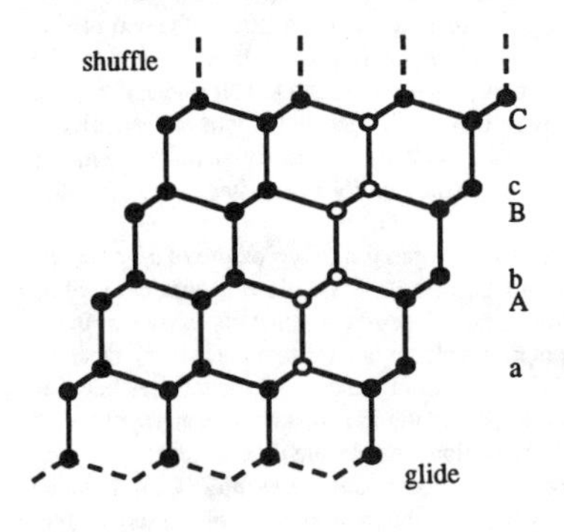

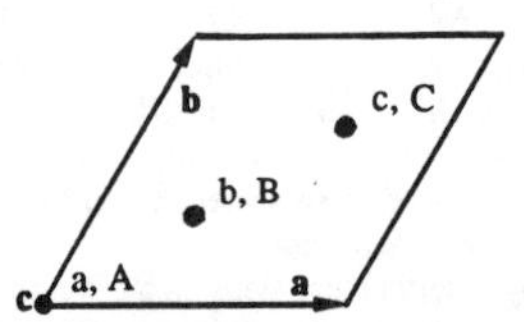

Figure 2. The hexagonal unit cell: $\mathbf{a}=^{1}/_{2}[1\bar{1}0]$ (0.38nm), $\mathbf{b}=^{1}/_{2}[10\bar{1}]$ (0.38nm), and $\mathbf{c}=[111]$ (0.94nm). This unit cell contains six atoms in three pairs (a-C), with atoms of the same pair directly above each other at heights of 0 and $^{3}/_{12}$, $^{1}/_{3}$ and $^{7}/_{12}$ and $^{2}/_{3}$ and $^{11}/_{12}$. The two possible surface terminations (shuffle and glide) achieved by cutting either between or within pairs are also shown (dangling bonds are shown as dashed lines).

RESULTS

a) Step movement during etching

Figure 3a shows a tracing of a step moving across the Si (111) surface during the etching process, taken from a video record of the experiment. It can be seen that the step nucleated at one point near the edge of a flat area and moved rapidly across, flowing round obstructing SiC particles and stopping when it reached the opposite wall. During this experiment, which lasted 50 minutes, we observed a sequence of 390 steps moving across the thin area in this way until the silicon foil eventually etched right through. At lower temperatures, particularly below 860°C, the steps seen are more angular, preferring the lower energy 110 direction rather than flowing isotropically in the way shown here. Very low etching rates at lower temperatures can result in irregular etching and roughening of the step edges particularly if the specimen surface is not very clean.

Initially the region shown in figure 3a was a small area at the bottom of a gently sloping valley, but as the etching proceeded the valley floor was flattened and enlarged with the walls becoming very steep (figure 3b). The apparent size of the SiC particles increased suggesting that they were left on the top of unetched cones of silicon of semi-angle about 60°. This morphology of etching suggests that the movement of steps is slowed at steep regions, i.e. by the presence of other steps nearby, a phenomenon consistent with theories of crystal growth to be discussed below. This results in preferential etching of smoother regions to give a potentially useful surface topography consisting of atomically flat terraces several microns across (larger for an initially flatter and cleaner surface) separated by steep walls.

It thus appears that etching proceeds by a movement of steps which, once they nucleate at a certain point on a flat terrace, move across the terrace rather quickly until they are obstructed at a steep or dirty part of the surface. The etching reaction under these conditions is limited by the step nucleation rate rather than by the speed at which the steps move once formed. The vast majority of the steps we observed in this sequence nucleated on one surface of the specimen at the point indicated in figure 3, though an increasing proportion (up to 10%) nucleated at other points along the steep walls as the foil became thinner. We can speculate that these sites are preferred for nucleation because of strain effects caused by the abrupt change in specimen thickness. In other experiments we clearly observe steps on both surfaces passing across each other (in projection) without interacting.

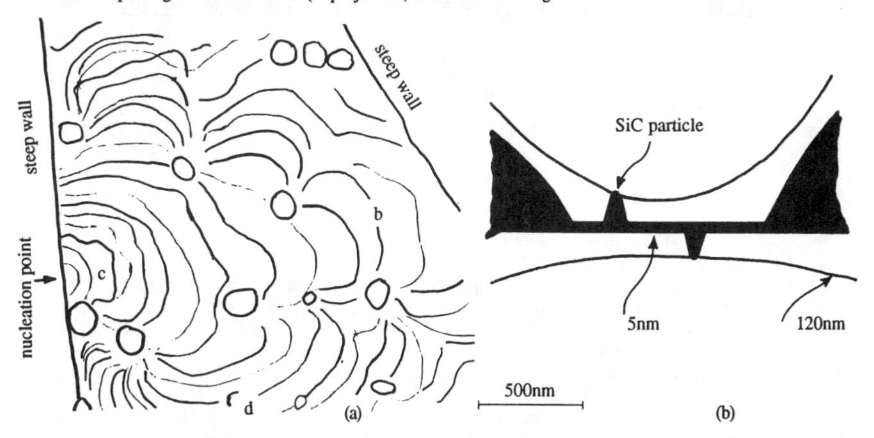

Figure 3. a) Tracing from successive frames ($^1/_{30}$ second apart) from a video recording of a step moving across the Si surface, and (b) a schematic diagram of the shape of the specimen before and during etching with vertical dimensions exaggerated. The etching took place at a specimen temperature of 875±50°C and an O_2 pressure of 2.8×10^{-6} Torr.

b) How high are the steps?

It will be recalled from the behaviour of the $^1/_3$ 422 reflection [7, 8] that intensity levels should repeat themselves after 6n monolayers (0.93n nm) have been removed from the specimen. For specimens which were well oriented with respect to the (111) zone we observed repeating contrast levels after every 3 layers had been etched away. This offers direct proof that the units removed during the etching process are two (or four) monolayers high, i.e. 0.31nm (0.63nm). Furthermore, static images of rows of parallel steps on a clean Si surface showed the same 3-fold repeat.

An exact 3-fold repeat is in fact only expected for an untilted specimen, and tilting the crystal modulates the expected intensity levels with an envelope function whose period depends on the tilt, leading to complicated intensity sequences with long repeat periods. For a specimen which was tilted off-axis the complex series of intensity levels predicted by a computer model was in fact seen experimentally. Figure 4 shows experimental and calculated intensity levels, demonstrating semi-quantitative agreement only if we assume that double steps are removed. (Non-linearity of the video recording makes it hard to compare intensities quantitatively.)

The removal of double rather than single steps makes sense on physical grounds. Depending on whether the crystal is cut between or within the pairs of planes described in figure 2, the exposed surface will have either one or three dangling bonds per atom. Above 860°C we label these two 1x1 surfaces the "shuffle" and "glide" terminations respectively, and removal of planes in pairs will preserve whichever termination was present initially. The two terminations are expected to show different contrast levels in the $^1/_3$ 422 reflection (figure 4), with the shuffle generally showing lower intensities, and using this we can in fact demonstrate that the *shuffle* is the favoured termination by comparing qualitatively the changes in overall contrast level seen on going from the 1x1 to the (known) 7x7 structure, firstly by cooling through 860°C and secondly by oxidising the clean 7x7 surface: details are given in [7] (although shuffle and glide are transposed there!). The good match in the pattern of levels in figure 4 confirms this choice of the shuffle. Although the shuffle is the termination we might expect on physical grounds, it is interesting that a direct experimental verification of this fact is rather hard using other methods. Below 860°C diffraction data shows that the 7x7 termination is preserved during etching: the surface reconstructs as each double layer is removed.

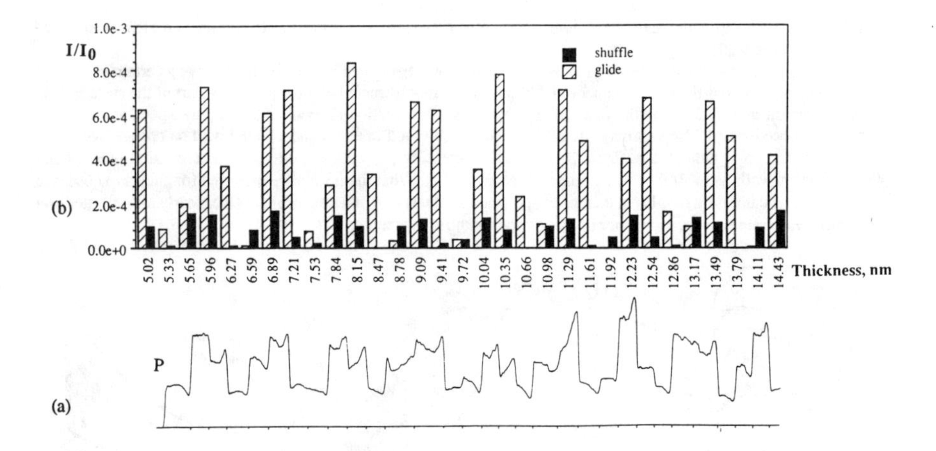

Figure 4. a) Intensity levels of successive terraces, recorded by measuring the brightness of the CRT with a photodiode while playing back the video described in figure 3. A change in intensity is seen whenever a step sweeps across the field of view (<0.5cm wide). This trace is from the last seconds of the experiment, at 875±50°C and 2.8×10^{-5} Torr of O_2, and the foil became very thin and broke at point P. The intensity scale is non-linear and the transient seen at each step may be due to the response of the image intensifier.
b) Computer simulations of the intensity of the $^1/_3$ 422 reflection. The specimen thickness increases in bilayers preserving either the shuffle or the glide termination on both surfaces. The simulation was done using a multislice algorithm including HOLZ effects [7] and the correct specimen tilt of 3.1° from the exact (111) axis. It can be seen that the experimental intensity levels match up with those predicted for the shuffle simulation and not well with the glide: note however that the exact thickness at P was not an experimentally determined parameter.

c) Step nucleation and movement

Assuming that we start with a stepped surface, etching may proceed according to two possible scenarios. Oxygen molecules impinge on the terraces, diffuse a certain distance and react with the Si to form Si-O bonds either uniformly over the *terrace* or preferentially at the *step edges.* In the first case the etching still proceeds by step movement, simply because evaporation of SiO is more likely to occur from a step edge than from the middle of a terrace. In the second case the terraces play no role except in collecting oxygen over a strip one diffusion length wide along each step. This latter scenario is closer to the spirit of crystal growth and evaporation models [9]. If no steps are present initially then step nucleation must occur directly from the terrace by evaporation of SiO, a process which is likely to be rate limiting, as we observed in these experiments.

It is clear that we must consider step nucleation and step movement separately, as different energy terms will dominate these processes and they will yield different information about the mechanism of etching. We thus show measurements of the step nucleation rate and the speed the steps move once nucleated in figures 5a and b. The nucleation rate (or the number of steps passing any point per second) appears to be proportional to the oxygen pressure, while the step speed does not bear a simple relationship to p_{O_2}.

First consider the nucleation of steps. The linear relationship between nucleation rate and p_{O_2} suggests that step nucleation requires a certain coverage of oxygen on the Si surface. It is important to note that this need not necessarily be a complete bilayer, but if it is, we can calculate from the oxygen flux a sticking coefficient of 11±2% for oxygen molecules impinging on the surface. (In this calculation one O_2 molecule must react at each surface site since *double* layers are removed.) Similar calculations have been done previously [1-3] using macroscopic measurements of the etching rate, obtaining results of the same magnitude at this temperature and range of pressures.

It thus appears that step nucleation occurs once the surface is "activated" by a certain coverage of oxygen. We can not at present say what the chemical state of the oxygen is at the time of nucleation, but analysis of diffraction patterns taken during etching below 860°C is currently in progress and we expect that this will yield information about where on the 7x7 structure the oxygen is first bonded [10]. (This method has successfully been used to determine the position of oxygen during adsorption at room temperature [11]). The adsorption of O_2 on Si (111) has also been examined by several spectroscopic techniques [12-14], although again only at lower temperatures. Extrapolating these results to high temperatures suggests that the reaction pathway in our experiments might proceed in two steps: firstly the formation of a molecular precursor consisting of a peroxy bridge formed by the reaction of an O_2 molecule with the dangling bonds of two Si surface atoms, and secondly a state in which either one or both oxygen atoms are in bridging positions between Si atoms in the top and second layers. The peroxy bridge precursor, with one oxygen atom per surface site, has a short lifetime particularly at higher temperatures [14], decaying rapidly to the atomic bridging configuration. Note that at the second stage more oxygen can react, increasing the oxygen coverage to the required two O atoms per surface Si site. From this latter configuration one can visualise how part of the terrace, supplied with oxygen already in bridging positions, can evaporate as SiO to nucleate a step.

The kinetics of step movement can now provide some further clues to the behaviour of oxygen on the terraces and step edges. Given the fully reacted surface postulated above, the first scenario described, in which SiO evaporates preferentially at steps once they have nucleated, appears to be the natural next stage of the process. One Si-Si bond and one Si-O bond must be broken as each SiO molecule evaporates from the step. The second scenario described above, that involving diffusion across the surface to the step, thus appears less important. Unfortunately for this simple picture, however, we expect in the first case that the step speed is independent of the oxygen pressure, since all the oxygen required is already present bonded into the surface. This was not seen experimentally, although one could argue that the speed is tending towards a limit at higher oxygen pressure. Instead, the slowing of the steps at lower pressures shows clearly that some other mechanism must be important in this régime. Not all the oxygen required for step movement can be present on the terrace. We suggest that some fraction of the oxygen must instead be supplied by diffusion from parts of the terrace away from the step. The step will act as a sink for diffusing oxygen since the oxygen is removed as the step propagates, so a concentration gradient will be set up around the step in the same way as is seen for adatom concentration during crystal growth according to the theory of Burton, Cabrera and Frank [9]. Under these circumstances we expect the step speed to be proportional to the oxygen pressure.

We suggest that this diffusion process limits the step speed at low oxygen pressures, but at higher pressures the rate limiting speed is that determined by the evaporation kinetics. The combination of these two processes can produce a graph of the form shown in figure 5b. Note that this double reaction can only explain the data if the nucleation process generates steps at coverages of less than two monolayers; coverages between 1 and 2 monolayers are likely as soon as the peroxy bridge precursor reacts to form atomic bridging oxygen.

We now consider briefly the information which can be derived from observations of the shapes of the steps during movement. Although the BCF theory mentioned above [9] is not applicable in its simplest form, it does suggest how several microscopic parameters can be estimated from examining step movement. Firstly we note that steps should move more slowly when curved because of changes in the chemical potential term, and can not in fact propagate below a certain critical radius which depends on the edge energy and the supersaturation. We observe slowing as the steps squeeze between pinning points in figure 3a, but do not find the expected straight line relationship between velocity and radius of curvature: it is hard to estimate step radii given the irregularity of the profiles. This type of analysis on a cleaner, more regular surface would be valuable as it is hard to calculate step energies even on a clean silicon surface due to the extensive reconstructions which can reduce step energies dramatically [15]; STM studies of step dynamics are however proving valuable in determining surface terms [16].

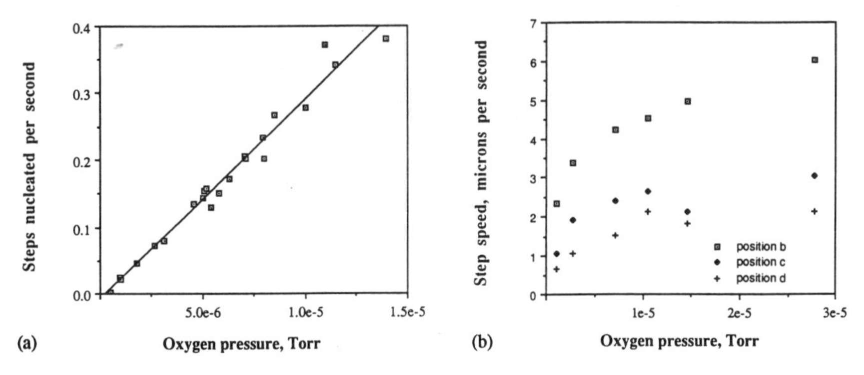

Figure 5 a) The nucleation rate of steps as a function of p_{O2}, and (b) the speed the steps move, measured at three different positions from data such as that shown in figure 3a.

A second prediction is that steps will slow down if closer to one another than a typical diffusion distance because of the interference between their diffusion fields. This phenomenon does not explain the data of figure 5b because the steps are much farther apart than typical diffusion distances, and in fact *decrease* in spacing as p_{O2}

increases (from 40μm apart at $2x10^{-6}$ Torr O_2 to 10μm at $1.5x10^{-5}$ Torr). However we do see a manifestation of this phenomenon at the steep sidewalls surrounding the flat area. It can thus be seen that step movement, when viewed within the framework of the BCF theory, can indicate how parameters such as diffusion length and step energy can be determined. However both the complexity of the reaction sequence and the relatively high subsaturations which are used here mean that this theory must be applied with great care.

DISCUSSION AND CONCLUSIONS

The data we have presented here demonstrates that the etching of silicon by oxygen proceeds by the movement of double layer steps, 0.32nm in height, and that above 860°C the exposed 1x1 surface has the shuffle termination. The results suggest a model of the etching process in which oxygen impinges on the surface, eventually reacting with it and moving into bridging positions. Steps can then nucleate once the oxygen coverage has reached a critical value which we believe to be between 1 and 2 monolayers. Steps then move by evaporation of SiO from step edges with at least some fraction of the oxygen supplied by diffusion across the surface. We feel that it has been important to separate the processes of nucleation and movement of the steps. Analysis of the movement of steps is potentially very valuable because it can yield data both on surface diffusion lengths and step energies. For example the more angular steps we see at lower temperatures suggest that the anisotropy expected in the energies of different types of step plays a more important role there.

Silicon etching has also been observed both by reflection electron microscopy [17, 18] and by STM [19]. REM images show nucleation and growth of "hollows" on terraces. In our experiments we usually see nucleation at certain favoured sites, where strain energy perhaps encourages nucleation, but it is likely that on an initially very uniform surface nucleation may occur simultaneously at many places on the terrace (provided they have reached the critical oxygen coverage) particularly for the slow step speeds expected at the lower pressures used in these studies. The STM results, like ours, show silicon preferentially removed from step edges. The ultimate surface morphology which etching produces can be useful, but depends on a balance of surface diffusion, step behaviour and line energy terms as well as the original state of the surface (for example etching has been observed to lead to roughening on the (100) surface [20]). Further experiments under different conditions will enable us to determine the useful range of this interesting reaction.

We would like to acknowledge the valuable help of D. Bahnck and M. L. McDonald.

REFERENCES

1. J. Lander and J. Morrison, J. Appl. Phys. 33, 2089 (1962)
2. C. Gelain, A. Cassuto and P. Le Goff, Oxid. Metals 3, 139 (1971)
3. F. Smith and G. Ghidini, J. Electrochem. Soc. 129, 1300 (1982)
4. C. Wagner, J. Appl. Phys. 29, 1295 (1958)
5. M. L. McDonald, J. M. Gibson and F. C. Unterwald, Rev. Sci. Instrum. 60, 700 (1989)
6. F. M. Ross and J. M. Gibson, in *Advances in Surface and Thin Film Diffraction*, edited by P. I. Cohen, D. J. Eaglesham and T. C. Huang, Mat. Res. Soc. Proc. 208, (1991)
7. F. M. Ross and J. M. Gibson, in *Proceedings of MSM 7*, edited by A. G. Cullis and N. J. Long, Inst. Phys. Conf. Ser. (1991), in press
8. D. Cherns, Phil. Mag. 30, 549 (1974)
9. W. K. Burton, N. Cabrera and F. C. Frank, Phil. Trans. Roy. Soc. 243, 299 (1951)
10. F. M. Ross and J. M. Gibson, in preparation
11. J. M. Gibson, Surf. Sci. 239, L531 (1990)
12. H. Ibach, H. D. Bruchmann and H. Wagner, Appl. Phys. A. 29 (1982) 113
13. P. Morgen, U. Höfer, W. Wurth and E. Umbach, Phys. Rev. B. 39, 3720 (1989)
14. U. Höfer, P. Morgen, W. Wurth and E. Umbach, Phys. Rev. B. 40, 1130 (1989)
15. J. E. Griffith and G. P. Kochanski, Crit. Rev. in Solid State and Mater. Sci. 16 (1990) 255, and references therein.
16. M. B. Webb, F. K. Men, B. S. Swartzentruber, R. Kariotis and M. G. Lagally, Surf. Sci. 242, 23 (1991)
17. N. Shimizu, Y. Tanishiro, K Takayanagi and K Yagi, Surf. Sci. 191, 28 (1987)
18. H. Kahata and K. Yagi, Surf. Sci. 220, 131 (1989)
19. U. Memmert and R. J. Behm, Adv. Solid State Phys., submitted
20. M. Offenbach, M. Liehr and G. W. Rubloff, J. Vac. Sci. and Tech., in press.

LAYER-BY-LAYER OXIDATION OF SILICON

T. YASAKA, M. TAKAKURA, S. MIYAZAKI AND M. HIROSE

Department of Electrical Engineering, Hiroshima University
Higashi-Hiroshima 724, Japan

ABSTRACT

Growth kinetics of native oxide on HF-treated Si surfaces terminated with Si-H bonds has been studied by angle-resolved x-ray photoelectron spectroscopy. The oxide growth rate in pure water for an n^+ Si(100) surface is significantly high compared to that of p^+, and the n or p type Si oxidation rate is in between. This is explained by the formation of O_2^- ions through electron transfer from Si to adsorbed O_2 molecules and the resulting enhancement of the oxidation rate. The oxide growth on Si(100) is faster than (110) and (111) as interpreted in terms of the steric hindrance for molecular oxygen adsorption on the hydrogen terminated silicon 1x1 surface structures.

INTRODUCTION

Complete removal of native oxide and adsorbed molecules such as hydrocarbon on Si surfaces is needed for advanced ULSI fabrication processes in addition to the elimination of fine particles and heavy metal contaminations on the wafer. The clean silicon surface is chemically active, so that the hydrogen termination of Si surfaces by HF treatment is one of realistic solutions for passivating the surface [1]. Si-H bond energy (70.4 kcal/mol) is significantly large compared to that of Si-Si (42.2 kcal/mol). Consequently, HF-treated Si is known to be stable against oxidant and the oxidation reaction starts from the backbond of surface Si-H bond [2].
About 10 at. % surface Si-F bonds are also existing on HF-treated Si surfaces [3]. Considering that the electronegativity of fluorine is the highest and Si-F bond energy (129.3 kcal/mol) is very large, Si-F bonds must change the chemical nature of Si surfaces [3]. This paper describes native oxide formation kinetics of the HF treated Si surfaces in pure water and a possible oxidation mechanism is discussed.

EXPERIMENTAL

P-type cz Si(100) and (111) wafers ($1.4 \times 10^{15} \sim 5.5 \times 10^{18}$ cm^{-3} acceptor) and n-type cz Si(100), (111) and (110) wafers ($8.0 \times 10^{14} \sim 5.0 \times 10^{18}$ cm^{-3} donor) were used as substrates. They were cleaned in an organic solution, boiled in H_2O:H_2O_2:HCl = 86:11:3 for 10 min and dipped in H_2O:H_2O_2:NH_4OH = 7:3:3 (p-type) or 4:1:1 (n-type) for 2min. The wafer was finally dipped in a 4.5 % HF solution. The HF-treated wafer was rinsed with pure water for 1 min (n-type) or 10 min (p-type) to minimize the surface Si-F bond concentration. The pure water rinse was carried out by using a 300 cm^3 teflon beaker at a flow rate of 900 cm^3/min at a temperature of 20 °C. The resistivity of pure water was held above 16 MΩ·cm. The wafers were stored in pure water at 19 °C with a DOC of about 9 ppm as estimated from the equilibrium dissolved oxygen concentration at this temperature. The wafer was stored in pure water for 10 to 8000 min and the chemical bonding features of the Si surface was measured by x-ray photoelectron spectra (XPS) of Si_{2p} core levels.The angle between the photoelectron detector axis and the direction normal to the Si surface was kept at 75° for the surface sensitive measurements.

RESULTS AND DISCUSSION

The fluorine coverage on Si surface exhibits no significant change in the HF concentration range 0.5 to 50 % as shown in Fig.1. If fluorine atoms randomly stick on the surface, the fluorine coverage must increase with increasing the HF concentration. The result of Fig. 1 suggests that fluorine atoms selectively remain at specific sites such as atomic steps or microfacets from which silicon might start to be slowly etched by HF. The native oxide thickness slowly increases with the time stored in pure water as shown in Fig. 2 regardless of the conduction types and impurity concentrations. The average oxide thickness is calculated from the integrated photoelectron intensity of the chemically shifted Si_{2p} spectrum and that of the metallic Si signal by taking into account the escape depths of the Si_{2p} photoelectrons from SiO_2 (25 A) and Si (23 A) as well as the spectrometer function [3]. The oxidation curve exhibits the plateaus where the oxide thickness apparently saturates at about 2, 4, 6 and 8 A. These saturated oxide thicknesses coincide well with a structural model of the SiO_2/Si interface proposed by Herman et al [4] (Fig. 3).

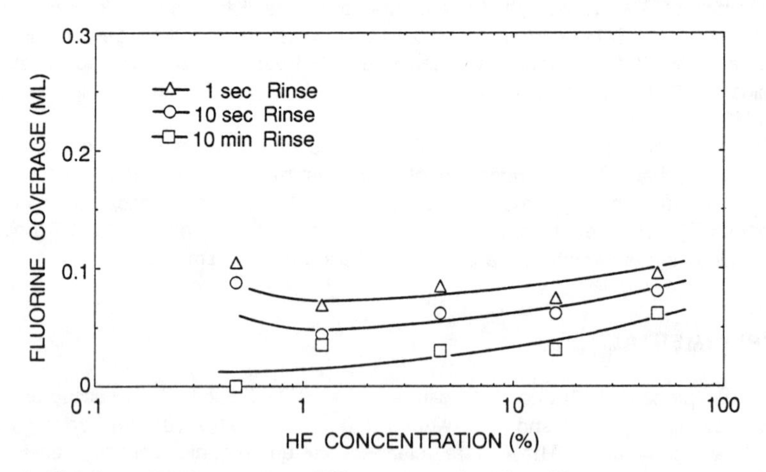

Fig. 1 Fluorine coverage versus HF concentration for different pure water rinse times.

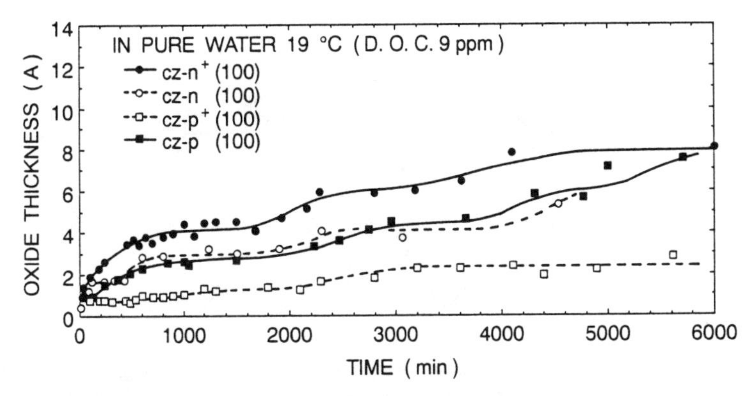

Fig. 2 Native oxide growth on Si(100) after 4.5% HF treatment as a function of oxidation time in pure water.

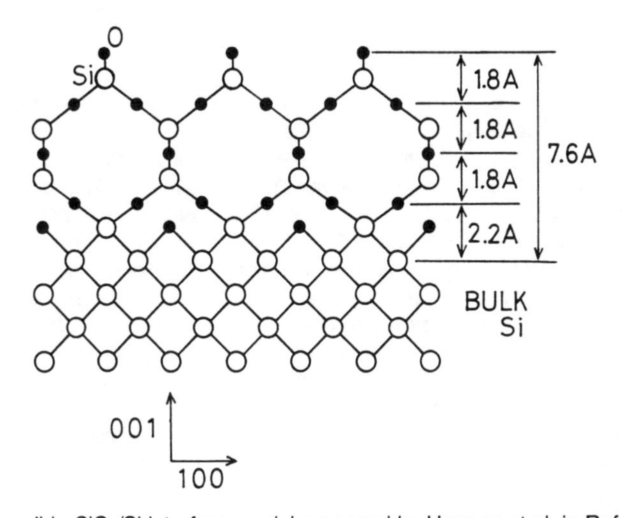

Fig. 3 A possible SiO_2/Si interface model proposed by Herman et al. in Ref. 4.

This implies that the layer-by-layer oxidation occurs on hydrogen terminated Si surface and that the oxidation reaction proceeds parallel to the surface. This idea is quite consistent with the fact that the oxidation of hydrogen terminated Si surface takes place from the backbond of surface Si-H bond. From these results it is likely that the oxidation is initiated at chemically reactive step edges and propagates along the surface as schematically shown in Fig. 4. The oxide thickness saturation observed in Fig. 2 could be explained by the transition time from the state (a) to (b) in Fig.4 because the oxidation of the second monolayer needs the incubation time for oxygen penetration from the step edges A and B to the sites E and F.

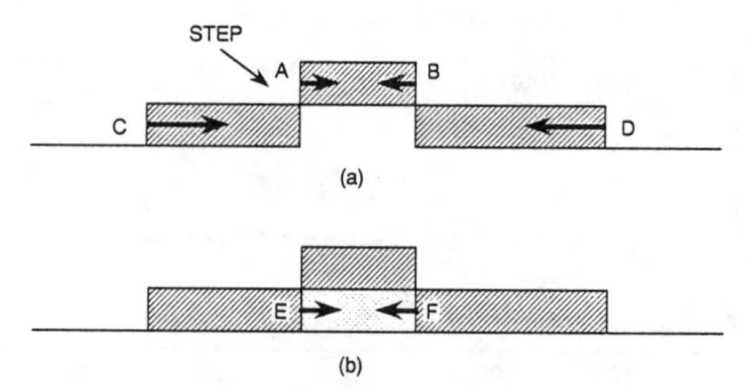

Fig. 4 The lateral oxidation of the first monolayer is thought to occur from the step edges A, B, C and D (a). A significant reduction of the lateral oxidation rate occurs at the beginning of the second monolayer oxidation (b). A plateau in an oxidation curve appears when the oxidation proceeds from the first monolayer to the second monolayer edges E and F.

In Fig. 2 the layer-by-layer oxidation rate of n^+ Si is significantly faster than that of p^+, while those of n and p-type Si are in between. The Si oxidation rate is known to be enhanced by forming O_2^- ions through free electron transfer from Si to adsorbed O_2 molecules. In case that the empty electron state of adsorbed O_2 is located near the Fermi level of p-type Si substrate (E_v + 0.2 eV), the electron transfer from Si easily occurs for n^+,n and p type Si in this order. The formation of O_2^- ions also induces the surface electric field which assists the oxidation. Calculated Debye length for a donor concentration of 5.0 x 10^{18} cm^{-3} is 18 A which provides the surface electric field as high as ~ 10^5 V/cm for the band bending of a few kT. This value is enough to significantly enhance the oxidation rate for n^+ Si [5].
The oxidation of (111) surface in pure water is also studied as shown in Fig. 5. The oxidation rate is slower than that of (100). Note that p^+ Si is oxidized up to about 1 A and remains unchanged while n and n^+ exhibit the plateaus corresponding to thicknesses of about 1, 2.2 and 3.5 A. The structural model to explain this result is not yet completely settled.
In order to reveal the influence of the crystalline orientation on the oxidation rates, n^+ Si(100), (110) and (111) surfaces are compared as shown in Fig.6. The (100) oxidation rate is faster than (111) and (110). This is consistent with the orientation dependence of thermal oxidation rate at a low oxygen partial pressure ($\leq 10^{-1}$ atom) [6]. However, room temperature oxidation of hydrogen terminated Si surfaces should be better understood on the basis of the topological aspects of the surface atom arrangement. Namely, the slower oxidation rate for the (111) orientation could be explained by two reasons: One is that the Si(111) surface has less reactive sites such as atomic steps or microfacets as evidenced by a previous work [1]. The other is that oxygen molecule with a size of 2.71 A easily penetrates through the (100) surface because the 1x1 surface has an atom void size of 2 r = 3.08 A as illustrated in Fig. 7. This implies that oxygen molecule can attack the back bonds of surface Si-

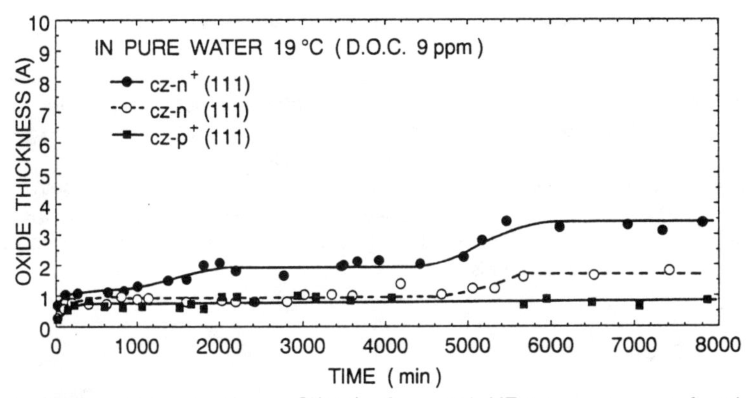

Fig. 5 Native oxide growth on Si(111) after 4.5% HF treatment as a function of stored time in pure water.

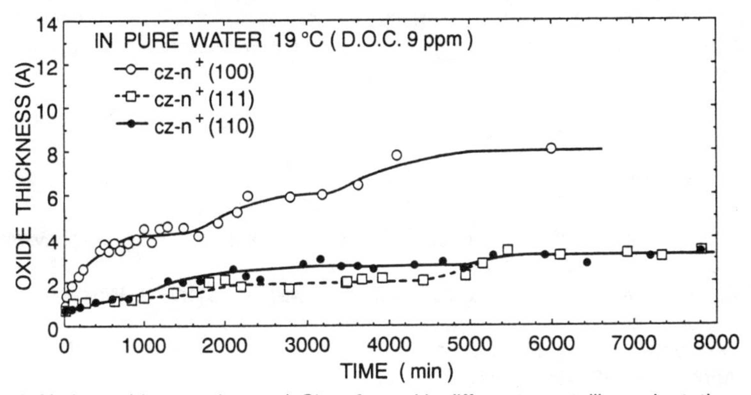

Fig. 6 Native oxide growth on n^+-Si wafers with different crystalline orientations after 4.5% HF treatment as a function of stored time in pure water.

(1 0 0)	(1 1 0)	(1 1 1)
r=1.54Å	r=1.37Å	r=1.04Å

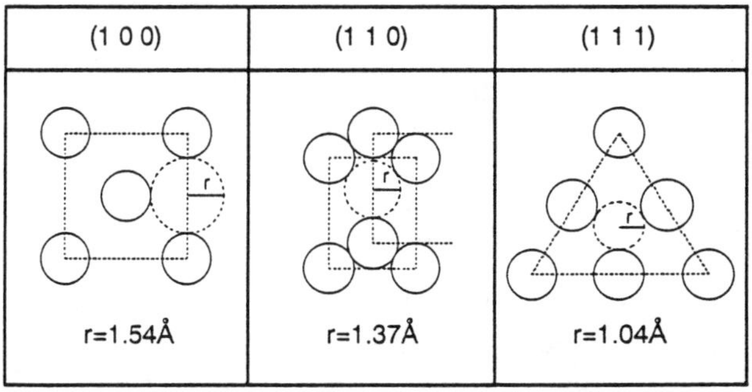

Fig. 7 Schematic diagram of the atom void radius r for crystalline Si surfaces.

H bonds without any steric hindrance by the top surface atoms. In contrast to this the surface atom void sizes of (110) and (111) are equal to or smaller than that of O_2 molecule. Hence, the surface steric hindrance for the penetration of molecular oxygen into the second layer of (110) and (111) surfaces is not negligible and lowers the charge transfer capability from Si to adsorbed O_2.

For examining the possibility of the oxide deposition originating from dissolved Si in pure water, an n^+ Si wafer was stored in pure water together with many bare Si wafers. The measured native oxide thickness is compared to the case of only one test n^+ wafer in pure water. The oxide thickness obtained for the both cases exhibits no significant difference.

CONCLUSIONS

We have shown that the layer-by-layer growth of native oxide on HF-treated Si surface in pure water depends on the Fermi level position and the crystallographic orientation. This is interpreted in terms of free electron transfer from Si to oxygen molecule and the steric hindrance of surface atoms for adsorbed oxygen. The lateral oxidation from the step edges is thought to be a key step of the observed layer-by-layer oxidation.

REFERENCES

1) G. S. Higashi, Y. J. Chabal, G. W. Trucks and K. Raghavachari: Appl. Phys. Lett., 56 (1990) 656.
2) T. Takahagi, A. Ishitani, H. Kuroda, Y. Nagasawa, H. Ito and S. Wakao: J. Appl. Phys. 68 (1990) 1.
3) T. Sunada, T. Yasaka, M. Takakura, T. Sugiyama, S. Miyazaki and M. Hirose: Jpn. J. Appl. Phys., 29 (1990) L2408.
4) F. Herman, J. P. Batra and V. Kasowski, The Physics of SiO2 and Its Interface: ed. by S. Pantelides (Pergamon Press, NY 1978) p.333.
5) P. J. Jorgensen: J. Chem. Phys., 37, 874 (1962).
6) S. I. Raider and L. E. Forget., J. Electrochem. Soc., 127 (1980) 1783.

PART V

II-VI Semiconductor Studies

ADVANCES IN SELF-LIMITING GROWTH OF WIDE BANDGAP II-VI SEMICONDUCTORS

M.Konagai, Y.Takemura, H.Nakanishi and K.Takahashi
Tokyo Institute of Technology, Department of Electrical and Electronic Engineering, 2-12-1 Ohokayama, Meguro-ku, Tokyo 152, JAPAN

ABSTRACT

Recent advances in understanding the ALE (atomic layer epitaxy) growth of ZnSe, ZnS and ZnTe are reviewed. The ideal ALE growth is obtained in the substrate temperature range of 250–350°C for ZnSe. In the ALE growth of ZnSe and ZnTe, a unique self-limiting mechanism is observed, in which the deposition rate saturates at 0.5 monolayer per cycle. Furthermore, applications of ALE of II–VI compounds to the growth of strained layer superlattices are also reviewed.

ATOMIC LAYER EPITAXY OF WIDE BANDGAP II–VI COMPOUNDS

In a II–VI compound, both group-II and group-VI elements have extremely high vapor pressures compared with that of the compound. Therefore, even if an excess molecular beam is incident on substrate, the ALE condition can be satisfied by thermal desorption of the adsorbates under proper growth conditions. Thus, II–VI compounds are potentially more suitable for the ALE than III–V compounds.

The first report on the ALE of a single-crystal II–VI compound was given by Pessa et al.[1]. Thereafter Yao[2] and our group[3] reported the ALE growth of ZnSe on GaAs using elemental sources. Our current approach is to use the MBE (molecular beam epitaxy) or the MOMBE (metalorganic molecular beam epitaxy) technique for the ALE growth of wide bandgap II–VI compounds. Table I summarizes the ALE growth of single-crystal wide-bandgap II–VI compounds. In addition to elemental sources, gas sources are widely used as reactants. MOVPE (metalorganic vapor phase epitaxy) technique is also used for the ALE growth of ZnSe. In this paper, the detailed ALE growth conditions for II–VI compounds are reviewed.

MBE-ALE

The successful ALE growth of ZnSe and ZnTe has been reported by several groups using the MBE technique. A conventional MBE system with solid sources is used for MBE-ALE.

Table I ALE growth of single-crystal wide-bandgap II-VI compounds

materials	sources	methods	Ref.
ZnSe	Zn+Se	MBE	[2]
	Zn+Se	MBE	[3]
	DEZn+DESe	MOMBE	[4]
	DMZn+H_2Se	MOMBE	[7]
	DEZn+H_2Se	MOVPE	[9]
	Zn+H_2Se	MOVPE	[10]
ZnS	DMZn+H_2S	MOMBE	[8]
ZnTe	Zn+Te	MBE	[2]
	Zn+Te	MBE	[3]

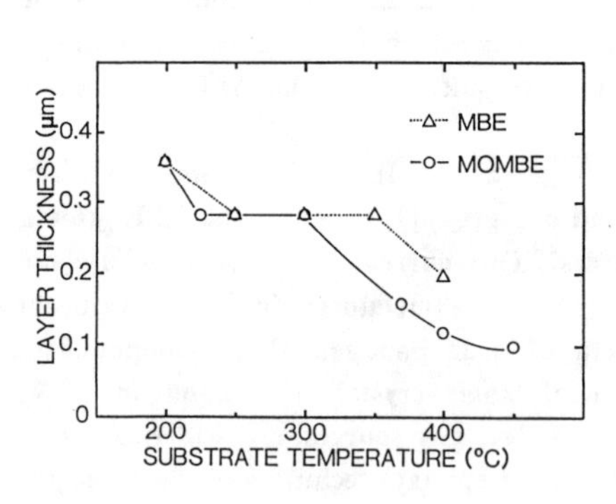

Fig.1 ZnSe thickness as a function of substrate temperature

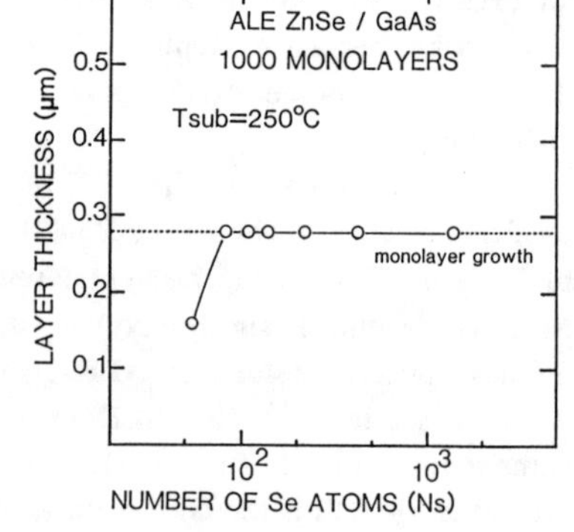

Fig.2 ZnSe thickness as a function of the number of supplied Se atoms

(1) ALE-ZnSe

ZnSe layers were grown on (100)GaAs substrates. The shutters of Zn and chalcogens were alternately opened and closed, typically, with an interval of 1 sec. Figure 1 shows the ZnSe thickness as a function of substrate temperature[3]. The growth rate of one monolayer/cycle is obtained in the substrate temperature range of 250–350°C for the Zn supply duration of 4–7 sec.

We also investigated the growth rate per cycle as a function of the number of Se atoms impinging on substrate by varying the Se beam intensity. Figure 2 shows the thickness of the ALE ZnSe grown with various Se beam intensities[4]. It is interesting to note that the growth rate per cycle is independent of the number of the supplied Se atoms in the range of $83N_s$–$1380N_s$, where N_s is the surface density of Zn or Se sites.

We achieved drastic improvements in the surface flatness and the optical properties of the ZnSe films, in contrast to the ZnSe films by the conventional MBE. Optical properties of the ALE-ZnSe films were studied by PL (photoluminescence) at 4.2K. The PL spectrum of the sample grown at 250°C are dominated by the donor-bound exciton emission as shown in Fig.3. However, for substrate temperatures above 300°C, the deep-level emission is dominant.

Yao et al. also reported the ALE-ZnSe films which exhibit PL spectra containing dominant excitonic emission lines[5].

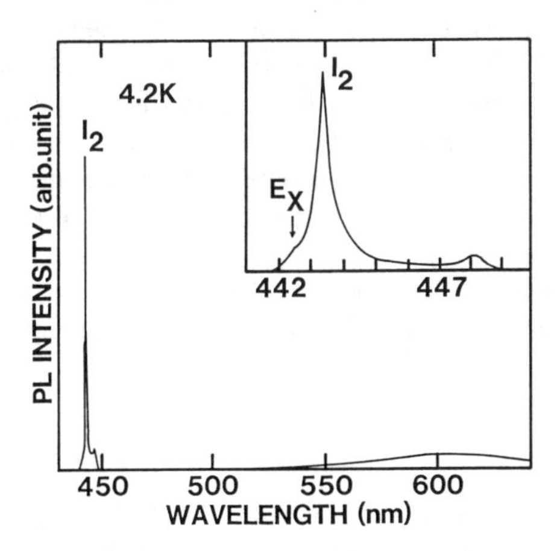

Fig.3 PL spectrum of ZnSe grown by MBE-ALE

(2) ALE-ZnTe

The ALE-ZnTe films were deposited on InP and GaAs substrates. The ideal ALE growth is obtained in the substrate temperature range of 240-280°C[3]. The processing window for the ALE growth is narrower for ZnTe than for ZnSe, because the vapor pressure of ZnTe is about two orders of magnitude higher than that of ZnSe. Surface morphology and RHEED patterns of the ZnTe samples grown with various Te beam intensities were studied. The surfaces are very specular for the samples grown under the ideal ALE conditions. However, the surface morphology is rapidly degraded with hillocks, and the RHEED patterns show rings when the deposition rate exceeds one monolayer/cycle. The number of hillocks varies with the Te beam intensity. The hillocks may be caused by Te precipitates.

Strictly speaking, the growth condition to obtain good crystallinity is not the same as the ideal ALE growth condition for ZnTe. If we adjust the Te beam intensity to achieve the ALE growth, surface morphology and crystallinity are degraded with the hillocks. In order to remove hillocks, the deposition with a low Te beam intensity is required (see section FRACTIONAL ALE).

(3) SURFACE PROCESSES

Yao et al. have made extensive studies of surface processes in the ALE growth of ZnTe and ZnSe using RHEED[2]. It is observed that three dimensional growth mode dominates at the initial stage of the heteroepitaxy(II-VI on GaAs), while quasi-two-dimensional growth mode dominates after deposition of more than 1000 monolayers. However, the growth of ZnSe on ZnTe and vice versa is dominated by the two-dimensional growth mode.

Yao also measured the sublimation time of Se and Zn using the RHEED intensity variation[6]. It was found that the sublimation time of Se decreases exponentially with increasing temperature. The activation energy for the Se sublimation is estimated to be about 0.6 eV. The adsorption time of Zn is almost constant at lower temperatures, while it seems to decrease with temperature at higher temperatures due to the Se sublimation from the surface.

MOMBE-ALE

In MBE, if we use two different kinds of high vapor-pressure source materials, especially S and Se, we suffer from elemental cross contamination. In contrast, MOMBE is known as an excellent epitaxial growth technique to deposit materials with high vapor-pressure elements. Thus, ZnSe and ZnS were also successfully grown by ALE using MOMBE.

Our group used diethylzinc (DEZn) and diethylselenium (DESe) as source gases for Zn and Se, respectively[4]. The pyrolysis of DEZn and DESe was carried out in cracking cells at the outlets of the tubes. The substrate temperature dependence of the ZnSe thickness grown by MOMBE-ALE is also shown in Fig.1. These results are very similar to those obtained for the MBE-ALE ZnSe, because in either case, both Zn and Se exist on the surface. However, the intensity of the excitonic emission from the ZnSe grown by MOMBE-ALE is very weak, and the deep-level luminescence is dominant for all the studied substrate temperatures. We speculate that thermal decomposition of the metalorganic sources at the outlets of the tubes is critical to obtain high-quality ALE-ZnSe. In the experiment, the cracking temperature of DESe is 700°C, which is not high enough for decomposition of DESe.

Yoshikawa et al. and Y.Wu et al. used metalorganic Zn and hydrides for the MOMBE-ALE of ZnSe and ZnS, respectively. Yoshikawa has grown ZnSe layers in both the MBE-like and the CVD-like ALE by MOMBE using DMZn and H_2Se[7]. It is found that the MBE-like or the CVD-like ALE can be achieved depending on whether the source materials are cracked before reaching substrate or not. In the MBE-like ALE, since the source materials are supplied on the substrate surface in the form of constituent elements, they can migrate over a fairly long distance on the surface. Thus, PL properties of the ZnSe grown by the MBE-like ALE is superior to those by the CVD-like ALE.

Y.Wu et al. demonstrated the growth of single-crystal ZnS onto GaAs substrates for the first time using MOMBE-ALE[8]. The self-limiting process is obtained in the substrate temperature range of 250-310°C. The ZnS layers grown by ALE show good surface morphology and exhibit strong near-bandedge photoluminescence. Although the ideal ALE growth was achieved by several research groups, surface reactions in MOMBE-ALE are not yet well understood.

MOVPE-ALE

The ALE growth is also obtained by MOVPE. Shibata et al. reported the ALE growth of ZnSe films on GaAs substrates using alternating supplies of DEZn and H_2Se[9]. The reactor pressure was 30 Torr during growth. Figure 4 shows the growth temperature dependence of the ZnSe thickness per unit cycle. For temperatures between 350 and 400°C, the growth rate per cycle is constant at one monolayer. This temperature range is about 100°C higher than those for MBE-ALE and MOMBE-ALE. Although there exists a very strong reaction between group-II alkyl molecules and group-VI hydride molecules, the substrate temperature above 350°C is required to decompose the adsorbed reactant molecules. They also reported the ALE-ZnSe with high purity and crystallinity.

Koukitu et al. very recently reported the ALE growth of ZnSe at atmospheric pressure using a Zn metal as group II source and H_2Se as a group VI source[10]. The ALE growth of ZnSe is obtained in the substrate temperature range of 350-500°C.

FRACTIONAL ALE

ALE is commonly considered as a growth method producing "just" one monolayer per cycle. On a surface, however, the density of atoms might be different from the density of atoms in a bulk crystal[11]. Thus we may observe different saturation at the number of monolayers less than one due to surface reconstructions. We call this type of ALE "Fractional ALE". Table II summarizes the published Fractional ALE in II-VI compound semiconductors. By changing the temperature and the beam intensity, one might be able to select various surface reconstructions. The surface reconstruction strongly depend on the growth parameters.

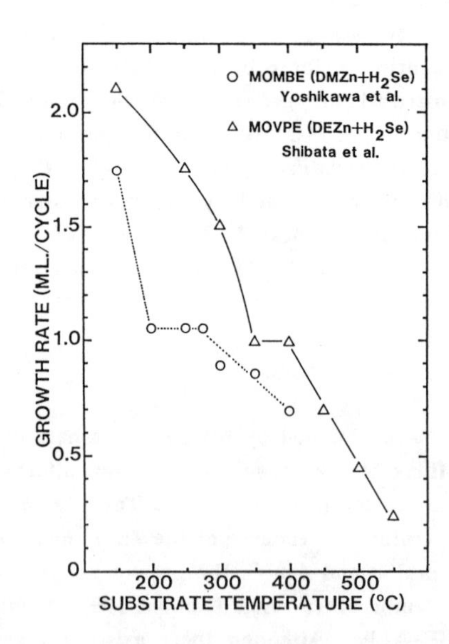

Fig.4 ZnSe thickness as a function of substrate temperature

The first clear observation of the Fractional ALE was reported by Faschinger et al. for CdTe deposited on (100) GaAs substrates using hot-wall epitaxy[12]. In the substrate temperature range between 260 and 290°C, they obtained one monolayer per cycle. However, at higher temperatures, the rate decreases rapidly to 0.5 monolayers per cycle, and stays at this value until it decreases exponentially at temperatures near 400°C.

Kobayashi also reported the growth rate of 0.5 monolayer per cycle in the migration enhanced epitaxial (MEE) growth of ZnSe[13]. A conventional MBE system was used in their experiments. At the substrate temperature of 250°C, the self-limiting growth of 0.5 monolayer per cycle is obtained independent of the Zn or Se beam intensity in the range between $1.0x10^{-7}$ and $1.8x10^{-7}$ Torr. We already reported the ideal ALE growth of ZnSe at the substrate temperature of 250°C. The only difference in the deposition parameters for obtaining the fractional (0.5 monolayer/cycle) ALE and the ideal ALE is in the beam intensity.

Recently we obtained the self-limiting growth of 0.5 monolayer per cycle for ZnTe using MBE-ALE[14]. Figure 5 shows the growth rate per cycle as a function of the number of total Zn atoms impinging on substrate during each supply. The Te beam intensity was kept constant at $2.0N_s$. As shown in the figure, the growth rate increases with increasing the number of supplied Zn atoms, and the growth rate is constant at 0.5 monolayer per cycle for the number of supplied Zn atoms between $6.6N_s$ and $44N_s$.

To explain the Fractional ALE, a missing Zn array structure of the c(2x2) reconstructed structure was proposed along the [100] and $[\bar{1}00]$ directions by Kobayashi[13]. The c(2x2) reconstruction pattern is caused by the half-coverage of the Zn sites, which results in the self-limiting growth with 0.5 monolayer per cycle. The c(2x2) RHEED pattern was also observed after the Zn supply in the ALE growth of ZnTe[2]. This may be explained in the same way as for the MEE growth of ZnSe.

Table II Fractional ALE in II-VI compounds

materials	ML/cycle	methods	Ref.
CdTe	0.5	Hot-wall	[12]
ZnSe	0.5	MBE	[13]
ZnTe	0.5	MBE	[14]

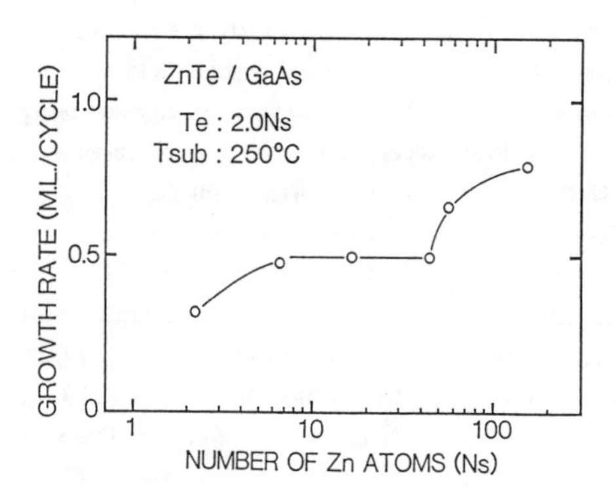

Fig.5 ZnTe thickness as a function of the number of supplied Te atoms

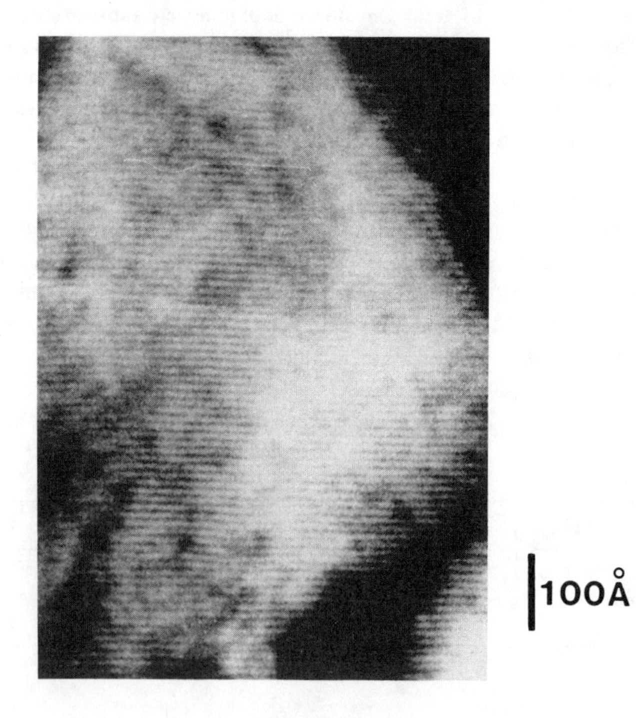

Fig.6 Dark-field image of $(ZnSe)_2$–$(ZnTe)_2$ SLS

SUPERLATTICES PREPARED BY ALE

A present interest in ALE of II–VI compounds is primarily for epitaxial growth of layered structures, such as superlattices. II–VI compound semiconductors, especially strained-layer superlattices(SLS's) consisting of ZnS, ZnSe and ZnTe are promising for light emitting devices with the visible wavelength. We have been investigating the ZnSe–ZnTe SLS's for several years. The use of ALE for preparing superlattices is motivated by several benefits, which include the followings; (1) growth temperature lower than those for the conventional epitaxial growth techniques, (2) precise control of the thickness with monolayer accuracy due to the saturated surface reactions.

$(ZnSe)_m$–$(ZnTe)_n$ (m,n= 1–4) short-period SLS's were grown on InP substrates by MBE–ALE[15]. The substrate temperature was 250°C, which is 70°C lower than that for the conventional MBE. Transmission electron microscopy (TEM) was performed to investigate the grown SLS structures. Figure 6 shows the dark-field image of the $(ZnSe)_2$–$(ZnTe)_2$ SLS for the (200) diffraction beam. High-contrasted stripes, which consist of a dark stripe for ZnSe and a bright stripe for ZnTe, are clearly observed. The TEM micrograph shows presence of a fine superlattice structure with an abrupt interface with at most one-monolayer steps. It is also confirmed by the X-ray diffraction satellites that the actual numbers of ZnSe and ZnTe monolayers were equal to the intended numbers. Raman scattering measurements showed that each layer is strained in the free standing with substrate.

SUMMARY

The ALE growth of wide bandgap II–VI compound semiconductors using MBE, MOMBE and MOVPE has been reviewed. The growth temperature of the MBE–ALE ZnSe films is about 70°C lower than the optimum growth temperature of the conventional MBE. Drastic improvements were achieved in surface morphology and optical properties using the ALE techniques. The ALE growth of ZnS and ZnSe using MOMBE has been also discussed. However, the surface reactions in MOMBE–ALE are not yet well understood. Furthermore, the Fractional ALE has been reviewed. Up until now, the Fractional ALE was observed for ZnSe, ZnTe and CdTe. Although the surface of ZnTe films prepared by the ideal ALE degrades with hillocks, the ZnTe films grown by the Fractional ALE show smooth surfaces without hillocks.

ACKNOWLEDGEMENTS

This work was supported in part by a Grand-in-Aid for Scientific Research on Priority Area, New Functionality Materials-Design, Preparation and Control, the Ministry of Education, Science and Culture, No.02204011.

REFERENCES

[1] M.Pessa, P.Huttunen and M.A.Herman, J.Appl.Phys.,54(1983)6047
[2] T.Yao and T.Takeda, Appl.Phys.Lett.,48(1986)160
[3] S.Dosho, Y.Takemura, M.Konagai and K.Takahashi, J.Appl.Phys.,66(1989)2597
[4] M.Konagai, Y.Takemura, R.Kimura, N.Teraguchi, H.Nakanishi and K.Takahashi, Acta Polytechnica Scandinavia, Chemical Technology and Metallurgy Series No.195(1990)81
[5] T.Yao, T.Takeda and R.Watanuki, Appl.Phys.Lett.,48(1986)1615
[6] Z.Zhu, M.Hagino, K.Uesugi, S.Kamiyama, M.Fujimoto and T.Yao, J.Crystal Growth, 99(1990)441
[7] A.Yoshikawa, T.Okamoto, H.Yasuda, S.Yamaga and H.Kasai, J.Crystal Growth,101(1990)86
[8] Y.Wu, T.Toyoda, Y.Kamakami, S.Fujita and S.Fujita, Jpn.J.Appl.Phys.,29(1990)L727
[9] N.Shibata and A.Katsui, J.Crystal Growth, 101(1990)91
[10] A.Koukitu, A.Saegusa, M.Kitho, H.Ikeda and H.Seki, Jpn.J.Appl.Phys.,29(1990)L2165
[11] T.Suntola, Acta Polytechnica Scandinavia, Chemical Technology and Metallurgy Series No.195 (1990)93
[12] W.Faschinger and H.Sitter, J.Crystal Growth,99(1990)566
[13] N.Kobayashi and Y.Horikoshi, Jpn.J.Appl.Phys.,29(1990)L236
[14] Y.Takemura, H.Nakanishi, M.Konagai and K.Takahashi, Jpn.J.Appl.Phys.,30(1990)L246
[15] Y.Takemura, S.Dosho, M.Konagai and K.Takahashi J.Crystal Growth,101(1990)81

ATOMIC LAYER CONTROLLED SUBSTITUTIONAL DOPING WITH LITHIUM IN ZnSe

Ziqiang ZHU*, Mitsuo KAWASHIMA* AND Takafumi YAO**
*Sumitomo Metal Mining Company, Ltd, Electronics Materials Laboratory, Suehiro-cho, Ohme-shi, Tokyo 198, Japan
**Hiroshima University, Department of Electrical Engineering, Higashi-Hiroshima 724, Japan

ABSTRACT

The detailed observation of dynamical behaviors of reflection high energy electron diffraction (RHEED) patterns during the adsorption processes of Li, Se and Zn is carried out. It is found that the RHEED intensity variation reflects the Li surface coverage during Li adsorption process on a Se-covered surface. This fact enables one to control quantitatively the doping of Li "in situ". A new method for atomic-layer controlled substitutional doping of ZnSe layers with lithium is proposed based on the RHEED investigations. The method allows the incorporation of Li dopants on Zn-sites of ZnSe by monitoring the RHEED patterns and intensities, and is expected to suppress the compensation by Li interstitials. Photoluminescence spectrum shows the growth of high quality p-type layers.

Introduction

One of the most important aims of the research on MBE growth of II-VI compounds would be achievement of good bipolar conductivity for fabrication of efficient blue light emitting p-n junctions[1,2]. In ZnSe, compensation and self-compensation effects related to residual impurities and to deviation from stoichiometry have made it very difficult to obtain p-type materials. Recent developments in the growth of ZnSe epitaxial films using molecular beam epitaxy (MBE)[3,4] have shown promise for the production of low resistivity p-type materials. The use of Li as a p-type dopant have recently yielded significant success and have achieved the highest hole concentrations ($8x10^{16}$-$9x10^{17}cm^{-3}$)[5,6]. But, another compensation arises from the amphoteric role of impurity of Li[7]. The impurity of Li substituting on the Zn site acts as a shallow acceptor, while the Li interstitials behave as a shallow donor. This compensation has limited the ability to obtain highly doped p-type material with sufficiently low resistivity. It is necessary to develop an atomic-scale controlled doping technique to minimize the compensation effect related to the interstitial Li.

Reflection high-energy electron diffraction (RHEED) has proven to be an important technique for studying MBE and ALE (atomic layer epitaxy) growth processes and related surface phenomena. Detailed monitoring of the intensity of the RHEED features gives important information relevant to surface processes during MBE and ALE growth of ZnSe[8-10]. The temporal behavior of specular beam intensity of RHEED during ALE reflects the adsorption processes of impinging molecular beams[10-11]. The adsorption times during Zn and Se adsorption processes for a monatomic layer have been estimated from the temporal dependence of RHEED intensity[9-10].

Very recently, we investigate the detailed dynamical behaviors of RHEED pattern during Li, Se and Zn adsorption processes, respectively. In this paper, a new method for atomic layer controlled substitutional doping of ZnSe layers with lithium is proposed based on the RHEED investigations. The method allows the incorporation of Li dopants on Zn-sites of ZnSe by the "in situ" observation of RHEED patterns and intensities. The doping is achieved by impinging Li beam on a Se-covered surface and Se beam on a Li-covered surface to form stable Li-Se bonds during ALE or MBE growth of ZnSe.

Experimental

The experiment was carried out using a conventional MBE growth apparatus equipped with a 20 keV RHEED system and a facility for monitoring the RHEED intensity. (001)GaAs was used as the substrate material. After being degreased, a GaAs wafer was chemically etched and then mounted on a molybdenum block with In. A ZnSe buffer layer of about 1 μm was deposited on GaAs substrate using ALE technique at the initial heteroepitaxial growth and MBE technique during successive homoepitaxial growth. The flatness of the top layer of ZnSe allows observation of the RHEED intensity variation during ALE growth or the RHEED intensity oscillation during MBE growth. The intensity of the specular spot of the RHEED pattern was recorded with an optical fiber system coupled with a photomultiplier. The incidence angle of the electron beam was about 1.2^{o}, which yields an off-Brag diffraction condition. Li-doping was achieved by using elemental Li-sources in a standard Knudsen cell operated at temperatures between 135 and 170°C.

Results and Discussion

The detailed investigation on dynamical behaviors of RHEED patterns was carried out during the adsorption processes of Li, Zn and Se, respectively. Fig.1 shows the temporal response of the specular beam intensity during the adsorption processes of Li and Zn on Se-covered surfaces and Se on Li and Zn-covered surfaces (a), Li on Zn (b) and Zn on Li (c), where the azimuth of the electron beam is along the [110] direction. The Zn, Se and Li-cell temperatures are kept at 280, 175 and 140°C, respectively. The substrate temperature is at 320°C. When the surface is covered with Se and the Se beam is turned off, the next impinging Li beam increases the specular beam intensity, which eventually saturates at a certain value during the impingement of the Li beam (Fig.1a). When the Li beam is stopped and the Se beam is turned on again, the intensity decreases and saturates at the same intensity level as that of the previous Se-covered surface. When the Se beam is turned off after the formation of the Se-covered surface and the Zn beam is turned on, the intensity increases again and saturates at the same intensity level as the previous Li-covered surface. Simultaneously with the change in intensity, the surface reconstruction pattern changes completely from (2x1) to (7x1) or a mixture of (7x1) and c(2x2) as the intensity changes from the level of the Se-covered surface to the Li (Zn)-covered surface and vice versa, as shown in Fig.2. These facts suggest

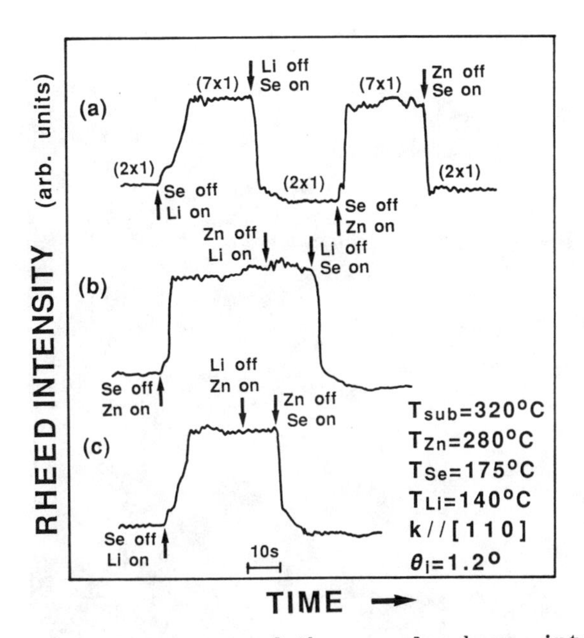

Fig.1 Temporal response of the specular beam intensity during the adsorption processes of Li and Zn on Se-covered surfaces and Se on Li and Zn-covered surfaces (a), Li on Zn (b) and Zn on Li (c), with k//[110].

that there exists a stable Li-layer on a Se-layer or between two Se-layers and that a stable Li-Se bond is formed. Furthermore, these Li atoms are believed to be located on the Zn-sites from the fact that the reconstruction pattern for the Li-covered surface is the same with that for Zn-covered surface. Therefore, by means of "in situ" observation of RHEED pattern and intensity, the atomic-scale controlled doping is possible, which is expected to allow the incorporation of Li dopants on Zn-sites of ZnSe.

On the other hand, when the Li beam impinges on the Zn-covered surface, both the RHEED pattern and intensity do not change (Fig.1b) and vice versa (Fig.1c). These facts indicate that the Li atoms hardly adsorb on Zn-covered surface to form Li-Zn bonds.

Fig.2 shows the RHEED patterns of the Se-covered surface with (2x1) reconstruction (a), Zn-covered surface with (7x1) (b) and a mixture of (7x1) and c(2x2) (d), and Li-covered surface with (7x1) (c) and a mixture of (7x1) and c(2x2) (e) during Li-doped ALE process. The doping is achieved by impinging the Li beam on the Se-covered surface and the Se beam on the Li-covered surface during ALE growth of ZnSe. The RHEED for the Se-covered surface displays (2x1) reconstruction, indicating Se-stabilized surface, which has been observed in undoped MBE or ALE growth processes[1,10]. The Zn-covered surface exhibits (7x1) reconstruction or a mixture of (7x1) and c(2x2), indicating Zn-stabilized surface. These patterns have not been observed during either undoped or Li-doped epitaxial

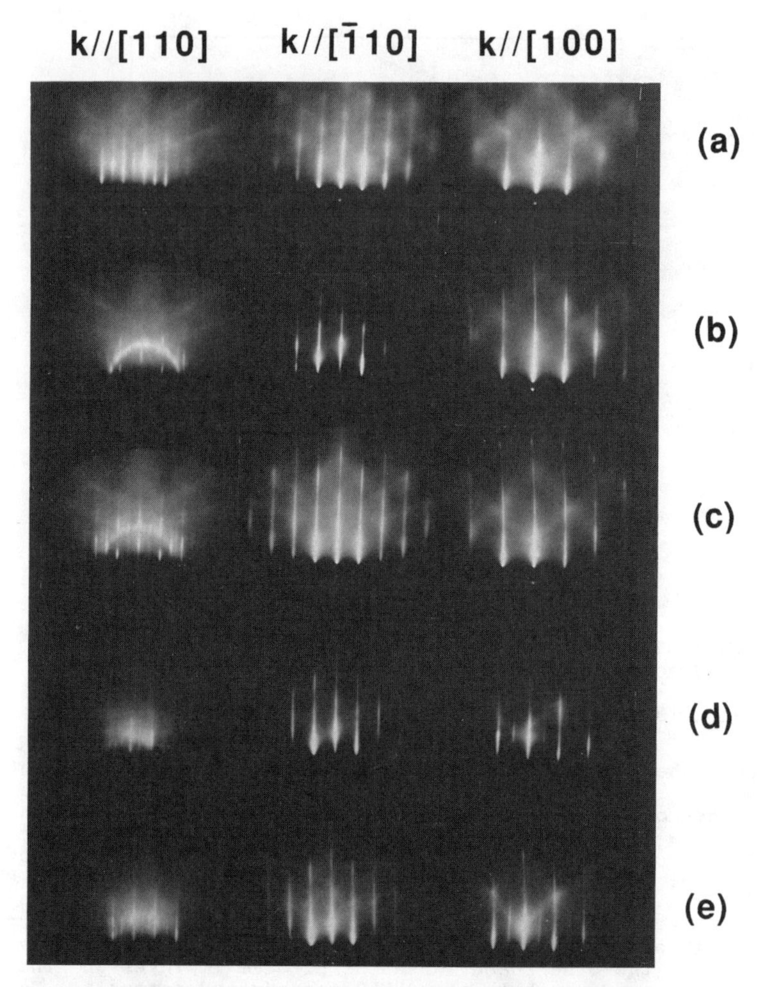

Fig.2 RHEED patterns of the Se-covered surface with (2x1) reconstruction (a), Zn-covered surface with (7x1) (b) and a mixture of (7x1) and c(2x2) (d), and Li-covered surface with (7x1) (c) and a mixture of (7x1) and c(2x2) (e) during Li-doped ALE process.

growth by MBE. The Zn-stabilized surface shows c(2x2) reconstruction during undoped epitaxial growth[1,10], and c(2x2) for lower Li-concentration doping and (2x1) for higher Li-concentration doping in conventional MBE processes[3,12]. The difference in surface reconstruction of Zn-stabilized surface between Li-doping ALE and MBE processes may result from that the incorporated Li atoms distribute randomly in toplayers during MBE process, while most of Li-atoms are confined to the Li-layer between the two Se-layers underlying the Zn-stabilized surface during ALE process. It should be noted that the Zn-covered surface always shows the same reconstruction (either (7x1) or a mixture of (7x1) and c(2x2)) with the Li-covered one throughout ALE growth. Curved streaky patterns in the [100] azimuth are observed, indicating the existence of one-dimensional disorder[13].

We have indicated that the temporal behavior of specular beam intensity of RHEED during ALE of ZnSe reflects the surface coverage of impinging molecular beams[11], and that the adsorption time constant for impinging atoms can be estimated from the temporal dependence of RHEED intensity[9-10]. The time constant for formation of a Li-stabilized layer can be measured and hence the "in situ" control of Li concentration becomes possible. Fig.3 shows the measured temporal dependence of RHEED intensity variation of the specular spot for various Li-cell temperatures during the adsorption process of Li onto the Se-covered surface, where the substrate temperature is kept at 320°C and the Li cell temperature is varied from 145 to 170°C. The electron beam azimuth is set to be [110]. When the Se beam is turned off after the formation of Se-covered surface and the Li beam is turned on, the specular beam intensity gradually increases and eventually saturates at a specific level related to the Li-covered surface. The surface reconstruction simultaneously changes from (2x1) to (7x1). As the Li-cell temperature increases, the intensity saturates rapidly, implying that the time for the formation of Li-covered surface becomes shorter. Thus, the time constant for the change of these two intensity levels should be correlated with the Li

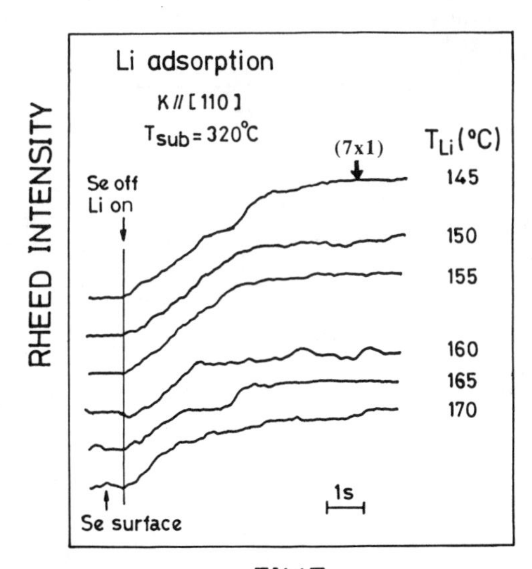

Fig.3 Temporal dependence of RHEED intensity variation of the specular spot for various Li-cell temperatures with k//[110] during the adsorption process of Li onto the Se-covered surface. The Li-stabilized surfaces show (7x1) reconstruction.

adsorption time. Using these intensity variations of the specular beam during the deposition of Li atoms, "in situ" measurements of the Li adsorption time can be made[8,9,14]. These facts confirm that the RHEED intensity variation corresponds to the variation of surface coverage as was confirmed in the cases of ALE and MBE of Zn chalcogenides[8-9]. The RHEED intensity variation technique[9,10] can be also applied to doping during epitaxial growth. Therefore, the atomic-layer controlled doping with lithium of ZnSe is possible.

The photoluminescence (PL) measurement has been used to characterize the layers grown by the atomic-layer controlled doping technique. The sample is fabricated by 1) depositing a ZnSe buffer layer of about 1 μm on GaAs substrate, and then 2) impinging the Se, Li, Se, and Zn beams alternately at 320°C. During one period, the Li, Se and Zn beams are impinged for times required for complete formation of Li, Se and Zn-stabilized surfaces, respectively, which is controlled by monitoring RHEED intensity. The total number of the periods are 100 and the layer thickness is estimated to be <600Å. Fig.4 shows a PL spectrum of the sample at 4.2K measured at an excitation power of less than 20mW/cm^2 of the 3250Å line from a He-Cd laser. The PL spectrum shows a dominant acceptor-bound excitonic emission (I_1) and no detectable donor-bound ones (I_2) or deep level emission, indicating the growth of high quality p-type layers.

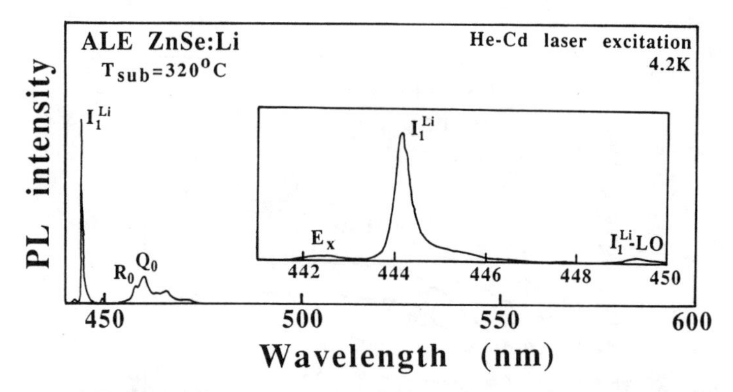

Fig.4 PL spectrum of Li-doped ALE-ZnSe grown at 320°C.

Summary

We have made detailed investigation on dynamical behaviors of RHEED patterns during Li, Se and Zn adsorption processes. It is found that through the observation of the RHEED intensity variation during impingement of Li onto the Se-covered surface, the Li adsorption process as well as the Li surface coverage on Se-covered surface can be elucidated. Therefore, it is possible to control quantitatively the Li doping "in situ". A method for the atomic-layer controlled substitutional doping of ZnSe layers with lithium is proposed. The method allows the incorporation of Li dopants on the Zn-sites of ZnSe by monitoring the RHEED patterns and intensities, which is expected to suppress the formation of Li interstitials. The PL spectrum shows the growth of high quality p-type layers.

References

[1] T. Yao, The Technology and Physics of Molecular Beam Epitaxy, (Plenum, New York, 1985), Chap.10.
[2] R.N. Bhargava, J. Cryst. Growth 86, 873(1988).
[3] M.A. Haase, H. Cheng, J.M. DePuydt, and J.E. Potts, J. Appl. Phys. 67, 448(1990).
[4] K. Ohkawa, T. Mitsuyu, and O. Yamazaki, J. Crystal Growth 86, 329(1988).
[5] T. Yasuda, I. Mitsuishi, and H. Kukimoto, Appl. Phys. Letters 52, 57(1988).
[6] H. Cheng, J.M. DePuydt, J.E. Potts, and M.A. Haase, J. Cryst. Growth 95, 512(1989).
[7] C.H. Henry, K. Nassau, and J.W. Shiever, Phys. Rev. B4, 2453(1971).
[8] T. Yao, Z. Zhu, K. Uesugi, S. Kamiyama, and M. Fujimoto, J. Vac. Sci. Technol A8, 997(1990).
[9] Z. Zhu, M. Hagino, K. Uesugi, S. Kamiyama, M. Fujimoto, and T. Yao, J. Cryst. Growth 99, 441(1990).
[10] T. Yao and T. Takeda, J. Cryst. Growth 81, 43(1987).
[11] Z. Zhu, M. Hagino, K. Uesugi, S. Kamiyama, M. Fujimoto, and T. Yao, Jap. J. Appl. Phys. 28, 1659(1989).
[12] Z. Zhu, et. al. (unpublished).
[13] P.J. Dobson, J.H. Neave, and B.A. Joyce, Surf. Sci. 119, L339(1982).
[14] Z.Zhu, M. Kawashima, and T.Yao, (submitted to J. Appl. Phys.)

P-TYPE CONVERSION OF NITROGEN DOPED ZnSe FILMS GROWN BY MOCVD

Babar A. Khan, Nikhil Taskar, Donald Dorman and Khalid Shahzad
Philips Laboratories, Briarcliff Manor, NY

ABSTRACT

We have obtained p-type zinc selenide films by nitrogen doping combined with a post growth anneal. These films were grown on <100> gallium arsenide substrates using a low-pressure MOCVD process. Ammonia gas was used as the source for nitrogen. The as-grown films were annealed and then studied by photoluminescence (PL) and capacitance-voltage (CV) techniques. The PL data is dominated by the acceptor-bound exciton peak and donor-acceptor pair spectrum associated with the nitrogen acceptor. The C-V data shows that the films are p-type, with the highest measured net acceptor concentration of $3\times10^{16}/cm^3$.

1. Introduction

Zinc selenide is one of the most promising materials for blue laser applications. The eventual aim of making a semiconductor diode laser which emits in the blue region of the spectrum could be realised with zinc selenide. Such a laser would increase the storage density of optical disks by a factor of four. However, until recently, such a goal was considered unrealistic because of the difficulty in achieving p-type doping in zinc selenide. Even though it was known for some time that nitrogen and lithium behave as acceptors in bulk zinc selenide, electrically measurable levels of p-type conversion have only recently been achieved in zinc selenide epitaxial layers grown on gallium arsenide substrates. The development of epitaxial growth techniques such as Molecular Beam Epitaxy (MBE), Metal Organic Chemical Vapor Deposition (MOCVD) and related techniques such as Chemical Beam Epitaxy (CBE), has led to the recent breakthroughs in p-type doping of zinc selenide.

Nitrogen is potentially a more promising dopant because it is more stable. Recent results for p-type doping of zinc selenide epitaxial films in MBE and MOMBE system are very encouraging [2,3]. In Ref.2, an RF plasma source was used to create atomic nitrogen, before introducing the gas into the MBE growth chamber. In Ref.3, ammonia was introduced into a CBE machine after heating it up to 350 C. Net acceptor concentrations in the $10^{17}/cm^3$ range were achieved by these processes. On the other hand the highest reported net acceptor concentration achieved with the MOCVD process is on the order of $10^{14}/cm^3$ [4].

In this work we report on the p-type conversion of nitrogen doped films grown by MOCVD after subjecting them to a Rapid Thermal Anneal (RTA) process. Net acceptor concentrations in the low $10^{16}/cm^3$ have been achieved.

2. Experiment and Discussion

Zinc selenide epitaxial layers were grown on gallium arsenide substrates in a vertical MOCVD reactor using a photo-assisted process. A schematic of the reactor is shown in Fig.1 . The reactor is equipped with automated pressure balancing between the run and vent lines which is useful for process runs which involve switching between sources. Automatic pressure control for the OM sources provides good control over the flow of source material and prevents surges if

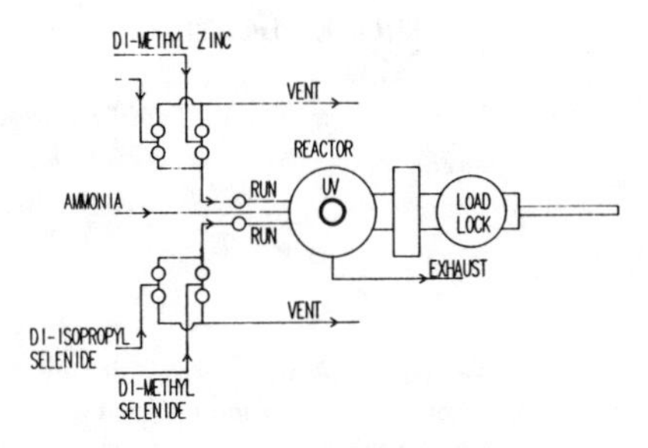

Fig.1: Schematic of MOCVD reactor. The reactor is equipped with an optical window, two seperate source lines, ammonia line and load lock.

the flow is changed suddenly. A turbo-pumped load lock keeps the system isolated and clean. An optical port on the reactor is used to expose the substrate to UV light from a mercury lamp. The reactor was constructed by EMCORE Corporation and its features such as high speed rotation, controls etc. are typical of EMCORE's basic design.

Epitaxial layers of zinc selenide are grown on gallium arsenide <100> substrates with a photo-assisted process (375 C) that uses di-methyl zinc and di-methyl selenium [5]. This photo-assisted growth process requires that a thin layer of zinc selenide be grown on top of the gallium arsenide so that the UV light can be absorbed in this layer and the photo-assisted process initiated. If the initial layer is not there then no visible growth is observed with the photo-assisted process (375 C), in agreement with earlier results by other groups [6]. In order to grow this initial layer (1000 A) at 375 C, we developed a novel process using di-isopropyl selenide and di-methyl zinc which react at this temperature. The final layers which were grown over this initial layer were doped with nitrogen using ammonia gas. The processing conditions for the initial and final layers are shown in Table 1. The deposited films were single crystal (FWHM 350 arc secs for .8 micron film...comparable to the best MBE grown layers) and had good surface morphology as indicated by the specular surfaces. The growth rate was $1\mu m/hr$. Films ranging in thickness from 1-4 microns were deposited. A TEM cross-section of the as-grown zinc selenide epitaxial layer is shown in Fig.2.

Photoluminescence measurements were carried out on the as-grown nitrogen doped films (see Fig.3...before anneal). The photoluminescence spectrum is dominated by the $A^o x$ peak associated with the exciton bound to the neutral acceptor (nitrogen). The peak associated with the donor bound exciton ($D^o x$) can also be observed, indicating that donors are also incorporated during the growth process. These donors might be associated with chlorine incorporation, although the source has not yet been isolated. The donor to acceptor (D-A) pair spectrum can also be observed. These features of the spectrum clearly indicate that some nitrogen atoms are being incorporated substitutionally on selenium sites.

Capacitance-Voltage (C-V) measurements were carried out to measure the net acceptor concentration in these films. This technique has been used previously to measure net acceptor concentration in lithium doped films and donor concentration in chlorine doped films [7,8]. The as-deposited films are fully depleted, indicating that the net activated dopant concentration is less than $10^{15}/cm^3$.

TABLE 1

PROCESS PARAMETERS										
Layer	Temp (C)	Press (Torr)	UV Int (mw/cm^3)	Time (min)	Thick (μm)	Flow (sccm)				
						DMZn	DIPSE	DMSe	NH_3	H_2
Buffer	375	300	-	10	0.1	3	3	-	-	7000
PhotoAssist	375	300	50	180	3	0.8	-	0.8	25	7000

Fig.2: TEM cross section of as-grown ZnSe epitaxial film. The density of dislocations at the zinc selenide gallium arsenide interface varies from $10^6/cm^2$ to $10^8/cm^2$.

In previous work, RTA has been used to activate implanted nitrogen in epitaxial zinc selenide layers [9]. The zinc selenide epitaxial layers cannot be annealed at high temperatures because of the loss of zinc atoms and the resulting degradation of the films [9]. In order to prevent this, capping layers of silicon dioxide and silicon nitride deposited by plasma enhanced CVD were used in the earlier work [9]. However, RTA annealing experiments carried out on nitrogen implanted zinc selenide layers with these capping layers did not achieve p-type conversion [9]. In this earlier RTA work on implanted zinc selenide layers, PL data led to the conclusion that the optimum RTA temperature for maximizing the acceptor bound exciton peak in nitrogen implanted epitaxial layers without creating other defects was 550 C [9].

Even though the RTA results on implanted samples were not that encouraging, we thought that it was possible that the effects of RTA on layers grown by MOCVD and doped in-situ might be different. These layers would have less lattice damage than the implanted layers for equivalent levels of nitrogen incorporation. Therefore, the nitrogen doped as-deposited MOCVD layers, were subjected to RTA using a lamp annealing system. In order to prevent degradation, sputtered Silicon Dioxide layers were used for the caps. The sputtered silicon dioxide layers were used instead of plasma deposited silicon dioxide layers to avoid the reactive environment

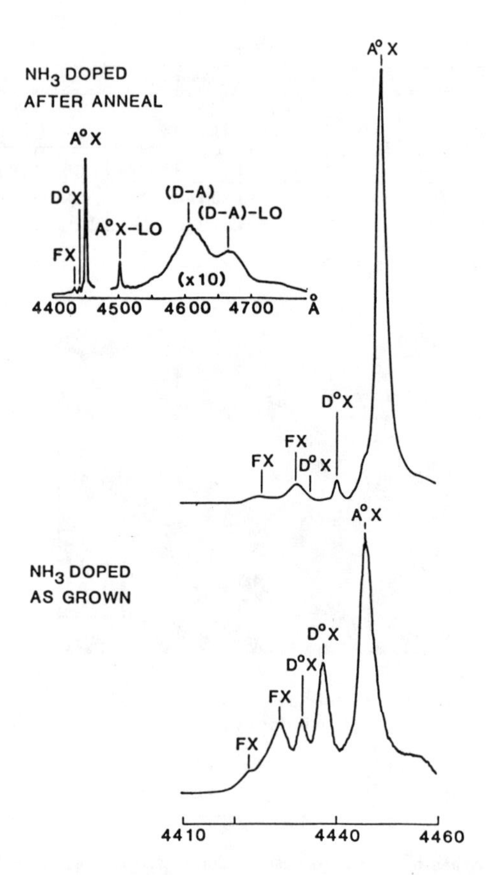

Fig.3: PL data for nitrogen doped ZnSe films, before and after RTA. The $A^{\circ}x$ peak intensity due to the nitrogen acceptor increases by a factor of 10 after annealing, in agreement with the increase in substitutional doping measured by C-V techniques.

of plasma deposition. The oxide thickness was 0.1μ. The films were then subjected to RTA with peak temperatures ranging from 600-950 C. C-V measurements were then carried out on the annealed layers. The films annealed below 650 C remained fully depleted and no significant increase in the net acceptor concentration was observed. However, the films annealed at higher temperatures were found to be sufficiently p-type so that the films were no longer depleted and C-V measurements could be carried out. We found that time and temperature both played a role in this conversion. If the films were annealed at a high temperature for a long time (960 C for 15 secs), they were found to be n-type (particularly if they were thin..less than 2μ), presumably due to gallium diffusion from the substrate. Therefore, an optimum process in time-temperature space would be such as to maximize the activation of nitrogen atoms and keep gallium diffusion from the substrate to a minimum. In order to minimize the time spent at the maximum temperature, the samples were exposed to triangular annealing profiles. The ramp-up and ramp-down rates above 500C are about 50C/sec. The peak temperatures for various anneals and measured net acceptor concentrations on identically grown samples are shown in Table 2. The triangular annealing profiles allowed the time at the highest temperature to be minimized, but they may not necessarily yield the best results. In fact, the highest net acceptor concentration ($3\times10^{16}/cm^3$) was measured on a sample which was annealed with a rectangular profile, where the peak temperature was held constant for some time (700C for 10 secs). However, we have not yet carried out a comprehensive study of the various annealing possibilities.

The PL data for the annealed samples also indicates the increased activation of the acceptors (see Fig.3...after anneal). It can be seen from the figure that the $A^o x$ peak intensity increases by a factor of 10, whereas the $D^o x$ does not change. Therefore, the incorporation of Nitrogen atoms as acceptors is substantially increased by this annealing procedure. Clearly, this method could be used to increase the net acceptor concentrations in those nitrogen doped zinc selenide films in which all the nitrogen atoms are not electrically active. The measured net acceptor concentration in the annealed films followed the same trend as the ratio of acceptor bound exciton to free exciton in the as-grown films. It is possible that the concentration of nitrogen atoms available for substitutional doping is higher in the films with the higher acceptor bound to free exciton ratio. In earlier work, it has been shown that the concentration of nitrogen atoms incorporated into the zinc selenide epitaxial layers can be much higher ($10^{19}/cm^3$) than the net acceptor concentration ($10^{17}/cm^3$) [3].

3. Conclusion

Nitrogen doped zinc selenide epitaxial layers when capped with sputtered silicon dioxide and subjected to RTA using a lamp annealing system were converted from semi-insulating to p-type. The PL data showed a marked increase in the acceptor bound exciton peak associated with the nitrogen acceptor, after the anneal. The net acceptor concentration, measured by C-V techniques was as high as $3\times10^{16}/cm^3$. Films annealed below 600 C (10 secs or less), were found to be fully depleted (semi-insulating). This technique could be used to increase the concentration of acceptors in nitrogen doped films in which some of the nitrogen atoms are not electrically active.

TABLE 2

Net Acceptor Concentration vs Peak Temperature				
Peak Temp. (C)	725	780	825	900
$N_A - N_D (cm^{-3})$	1×10^{15}	1.5×10^{16}	2×10^{16}	2×10^{16}

4. References

1. H.Cheng, J.M.DePuydt, J.E.Potts and T.L.Smith, Appl. Phys. Lett. 52, 147 (1988).
2. R.M.Park, M.B.Troffer, C.M.Rouleau, J.M.DePuydt and M.A.Haase, Appl. Phys. Lett. 57, 2127 (1990).

3. A.Taike, M.Migita and Y.Yamamoto, Appl. Phys. Lett. 56, 1989

4. A.Ohki, N.Shibata and S.Zembutsu, Jap. J. Appl. Phys., Vol.27, #5, May 1988, L909-L912.

5. S.Fujita, A.Tanabe, T.Sakamoto, M.Isemura and S.Fujita, Jap. J. Appl. Phys. 66, 1753 (1989).

6. A.Yoshikawa, T.Okamoto, T.Fujimoto, K.Onoue, S.Yamaga and K.Kasai, Jap. J. Appl. Phys. 29, L225 (1990).

7. T.Marshall and D.A.Cammack, J. Appl. Phys. , April 1991.

8. T.Marshall, S.Colak and D.Cammack, J. Appl. Phys. 66, 1753 (1989).

9. B.J.Skromme, N.G.Stoffel, A.S.Gozdz, M.C.Tamargo and S.M.Shibli, Mat. Res. Soc. Proc., Vol. 144, 391, 1989.

Growth of ZnSe and ZnSSe at Low Temperature with Aid of Atomic Hydrogen and Alternate Gas Supply

Jun GOTOH, Hajime SHIRAI, Jun-ichi HANNA and Isamu SHIMIZU

The Graduate School at Nagatsuta, Tokyo Institute of Technology
4259 Nagatsuta, Midori-ku, Yokohama 227, Japan

ABSTRUCT

High quality ZnSe films were successfully grown on GaAs(100) at low temperatures, 200 °C or lower by Hydrogen Radical-enhanced Chemical Vapor Deposition (HRCVD). Defects were makedly eliminated by the following factors: selection of source materials; avoidance of ion bombardment; and suppression of formation of adducts by alternate gas supply. Strained-layer superlattice (SLS) consisting of ZnSe as the well and $ZnS_{0.1}Se_{0.9}$ as the barrier was made by this technique. Emission line attributed to the free exciton was dominantly observed in the SLS.

INTRODUCTION

In recent years, techniques of crystal growth at low temperatures have become important in making materials with pseudomorphic structures, i.e., superlattices, modulation doping[1], or other artificial profilings in their potentials, in which there are many interfaces at the hetero-contacts. The following benefits are expected in making these pseudomorphic structures at low temperatures: formation of abrupt interfaces, inhibition of inter-diffusion of impurities and reduction in lattice mismatch caused by differences of their expansion coefficients.

The primary aim of this study is to develop a novel technique, which enables us to make these pseudomorphic materials under precise control over chemical reactions for the purpose of creating quantum functions in materials.

ZnSe-ZnSSe strained-layer superlattice(SLS) made in this study is an example of the pseudomorphic structures and has been widely investigated for its application to optoelectric devices such as the light emitting diode(LED) and injected laser diode (LD) in blue region.[2-6] In addition, the SLS is also used to suppress mis-fit dislocations caused at the hetero-interfaces by absorbing the stress.[7]

So far, the authors have reported the growth of ZnSe on GaAs(100) at about 200 °C from metalorganic compounds with the aid of atomic hydrogen by so-called Hydorogen Radical enhanced (HR-)CVD.[8-11] Recently, high quality ZnSe thin films were successfully grown at 150 °C or lower by this technique which was improved as follows: adopting new source molecule, i.e.,

Diethyldiselenide (DEDSe) was adopted; the substrate was floated electrically to avoid ion-bombardment; and supply of Zn-source gas was alternated with that of Se-source to eliminate formation of adducts.

In this paper, we report properties of ZnSe-ZnSSe SLS made by HRCVD at 230 °C after giving a short account of the HRCVD.

EXPERIMENTAL

Fig.1 shows a schematic diagram of HRCVD system. Hydrogen radical was made from H_2 in microwave plasma (2.45GHz) at power of 380 W. Diethylzinc (DEZn) and new materials, diethyldisulfur (DEDS) and dietyldiselenium (DEDSe), were used as the source materials in this study. All these molecules are able to be decomposed by collision with atomic hydrogen in a gas phase. Flow rates of DEZn, DEDS, and DEDSe were set at 1.8, 1.1, and 14 μmol/min, respectively. The source materials were introduced through a nozzle, and the supply of gases was controlled by computed pneumatic valves. In the alternate gas supply mode, each source material was alternately supplied to the reactor for 3 s at the intervals of 1 s, which resulted in the deposition of mono-layer of each component.

All the films were grown at 230 °C on a substrate on a susceptor floated electrically and connected with an ammeter, a volt meter and a power source, as shown in this figure. The reactor was cleaned by pumping-down with a turbo molecular pump before crystal growth, and its pressure was kept constant at 200 mtorr during deposition by using a set of a rotary pump and a mechanical booster pump. GaAs(100):Cr,O was cleaned with an organic solvent, etched with a mixture of NH_3,H_2O_2,H_2O in a ratio of 1:1:10, and treated with a solution of $(NH_4)_xS$[12]. The treated wafer was used as the substrate after evaporating an excessive sulfur by pumping down to $3x10^{-7}$ torr at 270 °C. During the preparation of SLS, the composition and the thickness of the barrier layer

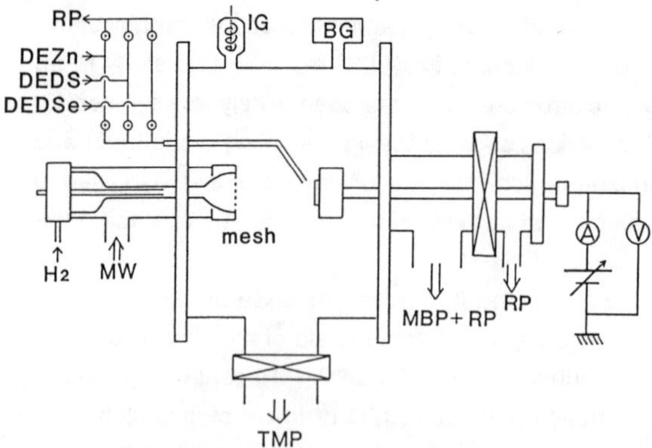

Fig.1 Schematic diagram of apparatus for HRCVD system. MW : microwave, RP : rotary pump, TMP : turbo molecular pump, MBP : mechanical booster pump, IG : ion gauge, BG : baratron gauge.

were kept constant, $ZnS_{0.1}Se_{0.9}$ and 140 A, respectively. SLS films 1600 A in thick consisting of the barrier ($ZnS_{0.1}Se_{0.9}$) with a constant thickness of 140 A (50 layers) and the well layer (ZnSe) with various thicknesses from 28 A (10 layers) to 140 A (50 layers) were grown directly on GaAs(100).

Measurements of photoluminescence (PL) at low temperatures, optical reflectance spectra and double crystal X-ray rocking curve (DCRC) were made to evaluate the qualities of films prepared. Monochromatic lights 365 and 325 nm in wavelength from an Ar-ion laser fitted with an interference filter and a He-Cd laser, respectively, were used as the light source for PL measurements at the low temperature of 20 K.

RESULTS AND DISCUSSION

(1) ZnSe prepared by HRCVD

Figure 2 shows PL spectra of ZnSe films prepared under various conditions; (a) continuous gas supply--grounded susceptor, (b) continuous gas supply--floated susceptor, (c) alternate gas supply--grounded susceptor and (d) alternate gas supply--floated susceptor. In (a) and (b), emission bands related to the deep levels at around 2.1 eV were dominant in the PL spectra, as seen in the spectra (a) and (b). On the other hand, sharp lines at the near-band-edge (NBE) were distinctively observed in the spectrum (c), and became dominant in the films prepared by a combination of the alternate gas supply and the floated substrate, resulting from a marked reduction in deep level emission, as seen in the spectrum (d). With regard to the growth temperature, we have confirmed that high quality of ZnSe was made at 150 °C by this technique.

(2) Strained Layer Superlattice

The appearance of the satellite peaks by the 2nd order in XRD observed for all SLS films prepared in this study gives support to the formation of accurate periodic structures with sharp interfaces. In the PL spectrum for a SLS consisting of the well layer 112 A thick, excitonic emission lines were dominantly observed and their intensity is markedly increased in the SLS by 30 -- 50 times compared with that of ZnSe films prepared under the same condition. In addition, the emission related to the deep levels were considerably weakened in the SLS. Accordingly, the intensity ratio (I d/Ie) of the deep level emission to the excitonic ones are reduced from 10^{-3} in the ZnSe to 10^{-4} --2×10^{-5} in the SLS. Figure 3 shows fine structures of the excitonic emission lines and the reflectivity spectrum for the SLS. The emission lines observed at 2.8333 eV and 2.8175 eV are able to be assigned to the light-hole exciton [FE(lh)] and the heavy-hole exciton [FE(hh)], respectively. On the other hand, the line at 2.8117 eV may be related to the donor-bound exciton (BE).

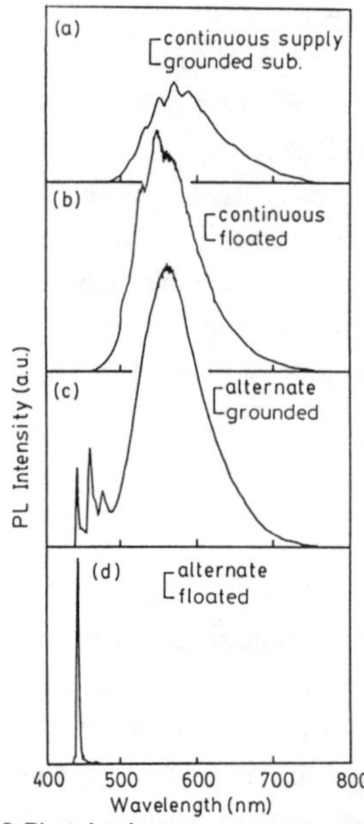

Fig.2 Photoluminescence spectra at 20 °K of ZnSe films grown under various conditions; (a) continuous gas supply--grounded susceptor, (b) continuous gas supply--floated susceptor, (c) alternate gas supply--grounded susceptor, and (d) alternate gas supply--floated susceptor.

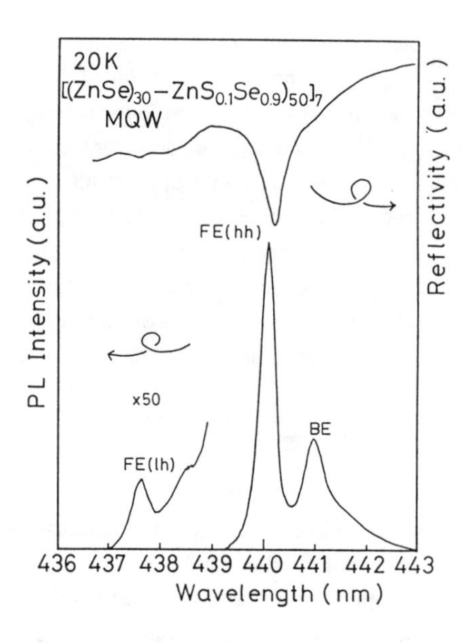

Fig.3 20 K photoluminescence and reflectivity spectrum at near band edge for SLS. FE(lh) : light-hole free exciton, FE(hh) : heavy-hole free exciton, BE : bound exciton.

The intensity ratio of the BE to the FE(hh) is markably decreased in the SLS compared with that of the ZnSe single layer.

Fig.4 illustrates energy shift of the FE(hh) for the SLS films plotted as a function of the thickness of the well layer. The solid curve in the figure shows the shift of the FE(hh) calculated on the Kronig-Penney model by varying the thickness of the well layers with the effective mass as 0.6 (heavy-hole) and the energy offset as 60 meV at the valence band.[2] With regard to the electrons at the conduction band, a constant energy of 2 meV is added to them as the quantization energy because of its small energy offset (--3 meV) at the conduction band.[2] In this calculation, we did not take into account the effects of the strain on the shift. A rapid fall of the shift energy with an increase in the thickness of well layer is observed similarly in both the calculated and the experimental ones. Together with the results of the XRD described above, this good agreement leads us to a conclusion that the superlattice structures having accurate periodicities with sharp interfaces were successfully made by this technique. There are, however, small inconsistencies in the energy between the experimental and the calculated ones (2--3 meV), which may be arisen from the strain effects in the lattices which we have neglected.[2,3]

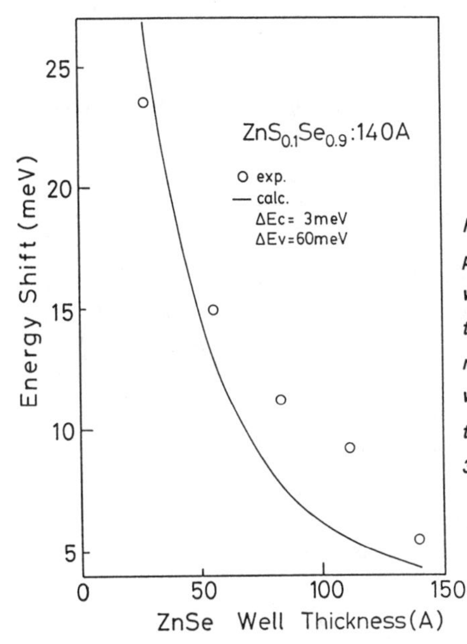

Fig.4 Energy shift of FE(hh) for the SLS films plotted as a function of the thickness of the ZnSe well thickness. The solid curve shows the shift of the FE(hh) calculated on the Kronig-Penney model by varying the thickness of the well layer with the effective mass as 0.6m_0 (heavy hole) and the energy offset as 60 meV at valence band and 3 meV at conduction band.

The line-widths of the SLS films were narrower than that of the ZnSe single layer and decreased smoothly with increasing the well thickness. The minimum width of 2.1 meV was obtained for the SLS with the well 140 thick with the aid of the quantum size effect. On the other hand, small fluctuations in the thicknesses of both the well and the barrier layers are deduced to be responsible for the slight broadening in the lines with increasing the numbers of interfaces.[13,14)]

In summary, we have confirmed that high quality SLS consisting of ZnSe and $ZnS_{0.1}Se_{0.9}$ was made by this new technique termed HRCVD. The specific features of this technique are :

(a) The source gases such as DEZn, DEDSe and DEDS are decomposed partly in a gas-phase by collision with atomic hydrogen, which results in the formation of chemically active precursors.

(b) High quality films, ZnSe or ZnS_xSe_{1-x}, are epitaxially grown on GaAs(100) substrate at low temperatures, around 150 °C, by adopting the alternate gas-supply and the biased substrate to avoid ion-bombardment.

(c) The chemical compositions of alloys such as ZnS_xSe_{1-x} are smoothly changed by simply alternating the gas supply, namely, by changing the mixing ratio or the periods of gas supply.

(d) No additional defects are formed at the hetero-interface since the strain stored at the interfaces is minimized by lowering the growth temperature.

(e) Clean atmosphere necessary for HRCVD is made mainly by a continuous flow of atomic hydrogen and not by additional expensive apparatus.

Consequently, this is a proper technique for making materials with finer pseudomorphic structures such as short-period superlattices or graded structures for the purpose of creating novel quantum functions.

REFERENCES

1) H.Kunzel, G.H.Dohler, PRuden and K.Poog, Appl.Phys.Lett. 41, 852 (1982).
2) K.Shahzad, D.J.Olego and C.G.Van de Wall, Phys. Rev. B38, 1417 (1988).
3) K.Mohammed, D.J.Olego, P.Newbury, D.A.Cammack,R.Dalby and H.Cornelissen, Apply.Phys.Lett. 50,1820 (1987).
4) K.Shahzad, D.J.Olego and D.A.Cammack, Apply.Phys.Lett. 52, 1416 (1988).
5) S.Fujita, Y.Matsuta and A.Sasaki, Appl.Phys.Lett. 47, 955 (1985).
6) Y.Wu, Y.Kawakami, Sz.Fujita and Sg.Fujita, Jpn.J.Appl.Phys. 30, L451 (1991).
7) I.J.Frits, S.T.Picraux, L.R.Dawson, T.J.Drummond, W.D.Laiding and N.G.Anderson, Appl.Phys.Lett. 46, 967 (1985).
8) S.Oda, R.Kawase, T.Sato, I.Shimizu and H.Kokado, Appl.Phys.Lett. 48, 33 (1986).
9) J.Gotoh, H.Shirai, J.Hanna and I.Shimizu, Proc. Jpn. Symp. Plasma Chem. 1, 61 (1988).
10) J.Gotoh, T.Kobayashi, H.Shirai, J.Hanna and I.Shimizu, Jpn.J.Appl.Phys.Lett. 29, L1767 (1990).
11) J.Gotoh, H.Shirai, J.Hanna and I.Shimizu, submitted to Jpn.J.Appl.Phys.
12) Y.Wu, T.Toyoda, Y.Kawakami, Sz.Fujita and Sg.Fujita, Jpn.J.Appl.Phys. 29, L144 (1990).
13) L.Goldstein, Y.Horikoshi, S.Tarucha and H.Okamoto, Jpn.J.Appl.Phys. 22, 1489 (1983).
14) J.Singh, K.K.Bajaj and S.Chaudhuri, Appl.Phys.Lett. 44, 805 (1984).

ACKNOWLEGEMENT

Authors would like to thank Professor H.Kukimoto for helpful discussion on the PL and reflectivity spectra. This work was supported in part by a Grant-in-Aid for Scientific Research on Priority Area, "New Functionality Materials-Design, Preparation and Control" from the Ministry of Education, Science and Culture (No 2205039).

CADMIUM SULPHIDE THIN FILMS GROWN BY ATOMIC LAYER EPITAXY

AIMO RAUTIAINEN, YRJÖ KOSKINEN, JARMO SKARP AND SVEN LINDFORS
Microchemistry Ltd., Box 45, SF-02151 Espoo, Finland

ABSTRACT

Polycrystalline cadmium sulphide (CdS) thin films were grown by Atomic Layer Epitaxy (ALE) using indium tin oxide and tin oxide coated glass substrates. Some of the experiments were made using elemental reactants, and others with inorganic compounds as reactants. Films were characterized using various techniques such as XRD, SEM and optical transmission spectroscopy. Growth rate of CdS films was observed to be 1/4 - 1/3 monolayer per cycle with elemental reactants. A full monolayer/cycle coverage was obtained when using $CdCl_2$ and H_2S as reactants. The crystalline structure of the CdS films was ß-cubic (111) when using elemental reactants. The mixed structure was observed when inorganic compounds were used as reactants. Only the hexagonal phase was observed, when substrate surface was pretreated before CdS deposition.

INTRODUCTION

Cadmium sulphide thin films have been prepared by several types of growth methods e.g. physical vapor deposition, chemical bath deposition, electrodeposition, close-spaced vapor transport, spray pyrolysis and sputtering techniques [1-7]. ALE-grown CdS thin films have also been mentioned in the literature [8]. A review of Atomic Layer Epitaxy is in ref. [9]. CdS films can be grown by ALE using 1) elemental cadmium and sulphur, or 2) inorganic compounds, such as cadmium chloride $CdCl_2$ and hydrogen sulphide H_2S, or 3) organometallic compounds as reactants.

Cadmium sulphide can be used as a window material in photovoltaic devices such as CdS/Cu_xS, CdS/CdTe and $CdS/CuInSe_2$ solar cells. In CdS/CdTe cells CdS forms the n-type side of the CdS/CdTe heterojunction. The lattice mismatch between CdS and CdTe is 9.7%. The purpose of this work was to develop a process for CdS for CdS/CdTe thin film solar cells [10].

EXPERIMENTAL

CdS thin films were grown in a lateral flow type ALE-reactor (Fig. 1) using Cd + S or $CdCl_2$ + H_2S as reactants. The temperatures used for sulphur, cadmium and cadmium chloride were 120, 320 and 470°C, respectively. For hydrogen sulphide the gas flow used was 5 sccm. The pressure in the reactor was 3-5 mbar. The transport and valving of the reactant gases was made with computer controlled inert gas flows. The carrier gas used was nitrogen with a total flow of 720 sccm.

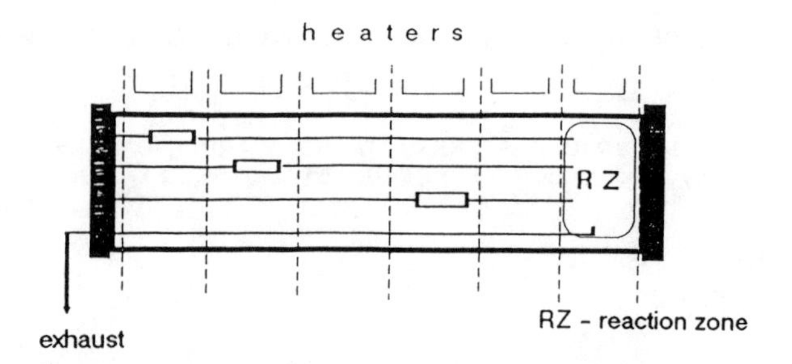

Fig. 1. Schematic picture of the ALE-reactor used.

Pulse and purge times used for elemental Cd and S were 200 and 400 ms, respectively. The temperature at the reaction zone was varied in the range 350-450°C. When $CdCl_2$ and H_2S were used as reactants, the reaction temperature was increased to 480°C due to the low vapor pressure of $CdCl_2$.

CdS thin films were grown on glass/TCO-substrates, where the transparent conducting oxide layer was indium tin oxide (ITO) or tin oxide (SnO_2). The sheet resistance of the TCO was approx. 10 Ω/square, and the glass thickness was 1.1 mm. The size of the substrates were 50 x 50 mm^2.

CdS film thicknesses were measured using a profilometer (Tencor). Scanning Electron Microscopy (SEM) pictures taken by JEOL JSM-T20 apparatus were used for qualitative analysis of the surface structure. Crystalline phase and orientation of the CdS films were analyzed by x-ray diffraction (XRD) measurements made by Siemens Diffrac 500. Optical transmission was measured using a Hitachi spectrophotometer with a wavelength range of 300-1100 nm.

RESULTS AND DISCUSSION

Cadmium sulphide thin films were first grown using elemental Cd and S as reactants. Figure 2 shows the CdS film thickness as a function of the number of cycles at 420°C [10]. The growth rate does not increase linearly while the number of cycles increase. Maximum CdS growth rate, 1/3 monolayer (ML) per cycle [β-cubic structure measured by XRD], was observed at the growth temperature of 380-390°C (Fig. 3). The decreasing growth rate below 380°C indicates that the dissociation of sulphur may be the rate-limiting factor.

Figure 4 shows the CdS film thickness as a function of the number of cycles when using $CdCl_2$ and H_2S as reactants. As seen in the graph, a full monolayer coverage per cycle is achieved. This result is different from the case of ALE-grown ZnS using $ZnCl_2$ and H_2S as reagents [11].

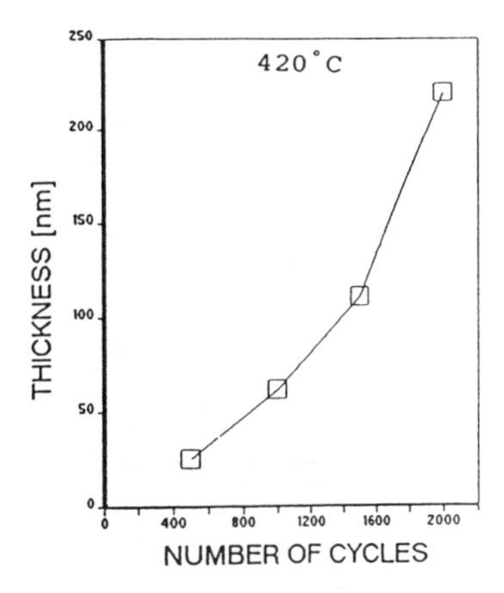

Fig. 2. CdS film thickness versus number of cycles used (from elements).

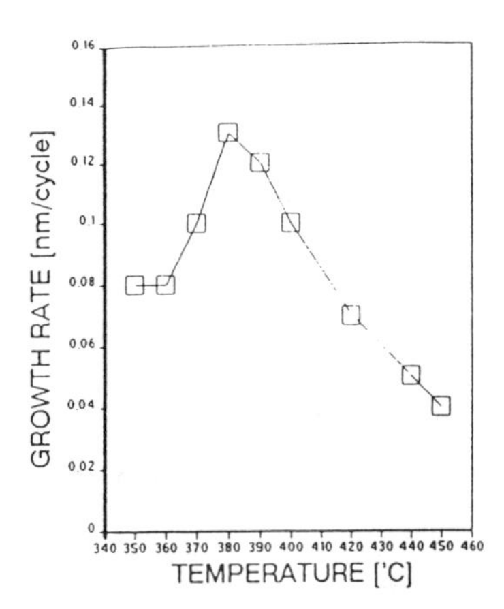

Fig. 3. CdS growth rate versus reaction temperature (from elements).

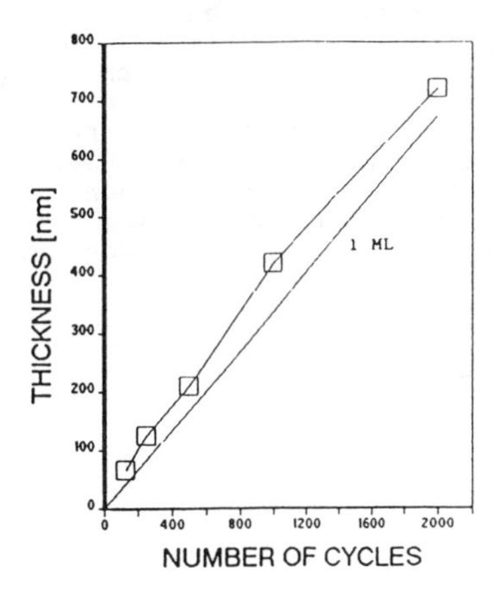

Fig. 4. CdS film thickness versus number of cycles used (from compounds).

X-ray diffraction studies reveal that CdS films grown from elements have ß-cubic crystal structure oriented primarily as the (111) crystal face. No hexagonal phase was observed. Figure 5 shows a typical XRD spectrum of the sample in which the CdS layer has been grown using $CdCl_2$ and H_2S as reactants. The highest peak (2θ = 26.5) corresponds to ß-cubic (111) and hexagonal (002) orientation [12]. No other peaks for ß-cubic structure are seen, but there are some minor peaks {(2θ;hkl): (47.8;103), (54.5;004) and (60.8;104)}, which belong to the hexagonal phase. The peak with two-theta value of 30.5 comes from the substrate material. The CdS thin film grown from compounds is a mixture of cubic and hexagonal form.

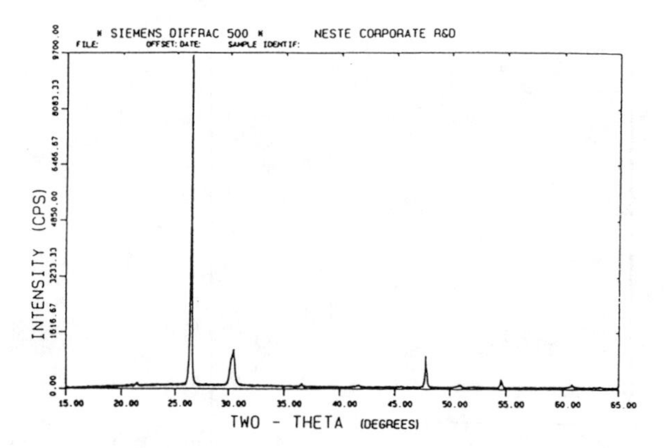

Fig. 5. X-ray diffraction spectrum of ALE-grown CdS thin film (compound reactants).

CdS layers were grown on Al_2O_3/glass substrates, too. The crystalline structure was observed to be the same as before. The effect of the starting layer was also studied. A very thin indium oxide layer was first grown onto the Al_2O_3 substrate. The indium source used was $InCl_3$ at 300°C, and the oxygen source was H_2O. After that the CdS layer was grown using elemental sources. Figure 6 shows an XRD spectrum of the sample grown. The spectral peaks have been compared to the theoretical lines for hexagonal CdS in the figure. The comparison shows a good match. The highest peaks (2θ: 24.8 and 28.1) correspond to hexagonal (100) and (101) orientations. No peaks for ß-cubic CdS and indium oxide are observed.

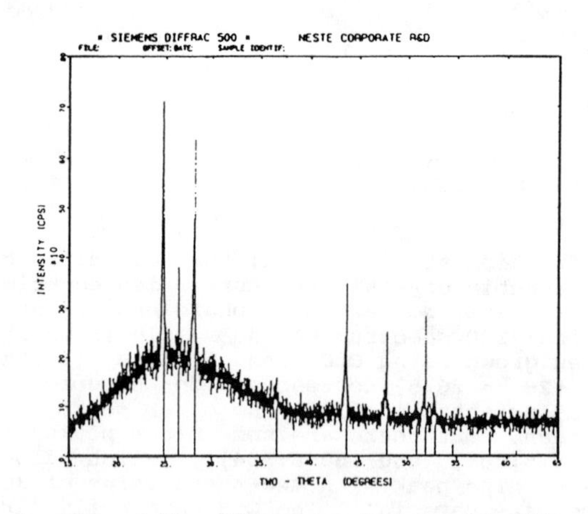

Fig. 6. XRD of the ALE-grown hexagonal CdS.

Spectral transmission studies show a sharp edge at optical band gap, which indicates good crystallinity. The surface morphology of the films is rather smooth according to the SEM-pictures, and the crystalline size is small.

CONCLUSION

Cadmium sulphide thin films have been successfully grown by Atomic Layer Epitaxy on TCO and Al_2O_3 substrates. CdS growth rate using $CdCl_2$ and H_2S as reactants is shown to be one monolayer/cycle. When elements are used, a growth rate of 1/4 - 1/3 ML/cycle is observed. The crystalline structure of the CdS films is ß-cubic (111) when using elements, and mixed ß-cubic + hexagonal when using inorganic compounds as reactants. On Al_2O_3/ALE-indium oxide substrates, only hexagonal CdS has been observed.

ACKNOWLEDGEMENTS

The authors thank Dr. T. Suntola for valuable discussions, and S. Kaipio and J. Marles for technical assistance.

REFERENCES

1. R. Hill and J.D. Meakin, Current Topics in Photovoltaics, Academic Press, London, 1985, p. 223.
2. L.P. Deshmukh, A.B. Palwe and V.S. Sawant, Solar Energy Mater., 20, 341 (1990).
3. G.F. Fulop and R.M. Taylor, Ann. Rev. Mater. Sci., 15, 197 (1985).
4. G. Perrier, P. Philippe and J.P. Dodelet, J. Mater. Res. 3, 1031 (1988).
5. M. Krunks, E. Mellikov and E. Sark, Thin Solid Films 145, 105 (1986).
6. M. Lo Sovio and M.E. Oliveri, Appl. Phys., A 50, 17 (1990).
7. F. El Akkad and M. Abdel Naby, Solar Energy Mater. 18, 151 (1989).
8. T. Suntola, Acta Polytech. Scand., Chem. Technol. Metall. Ser., No. 195, Helsinki, 1990, p. 93 (Proc. of 1st Intl. Symp. on Atomic Layer Epitaxy).
9. T. Suntola, Materials Science Reports 4, 261 (1989).
10. J. Skarp, Y. Koskinen, S. Lindfors, A. Rautiainen and T. Suntola, Proc. of 10th European Photovoltaic Solar Energy Conference, 8-12 April, 1991, Lisbon, Portugal (in press).
11. T.A. Pakkanen, V. Nevalainen, M. Lindblad and P. Makkonen, Surface Science 188, 456 (1987).
12. Joint Committee on Powder Diffraction Standards 1971, 6-0314 (CdS,greenockite) and 10-454 (CdS,hawleyite).

CADMIUM TELLURIDE THIN FILMS GROWN BY ATOMIC LAYER EPITAXY

ARLA KYTÖKIVI, YRJÖ KOSKINEN, AIMO RAUTIAINEN AND JARMO SKARP
Microchemistry Ltd., P.O.Box 45, SF-02151 Espoo, Finland

ABSTRACT

Polycrystalline CdTe films up to 2 μm thick were grown by Atomic Layer Epitaxy (ALE) at 350-450°C. The growth was carried out in a lateral flow reactor, using the elements as source materials and 25 cm² glass/ITO and glass/ITO/SnO_2 as substrates. A growth of CdS film by ALE preceded the growth of CdTe.

Profilometry, X-ray diffraction analysis and scanning electron microscopy were used to characterize the films.

The relatively high vapor pressure of CdTe determined the upper limit of the processing window, while the small vapor pressure of Te_2 set a practical lower limit of 390-400°C.

The CdTe films were smooth up to 100-500 nm depending on the temperature and the substrate surface. Development of surface roughness was detected as the growth process proceeded. At the same time there was an increase of the effective surface area of the film, observed as a significant increase in the macroscopic growth rate. The greater surface roughness was also evident in the reduced degree of (111) orientation.

The CdS/CdTe structure was investigated for its potential in solar cell applications.

INTRODUCTION

Polycrystalline thin film cadmium telluride is a promising material for low-cost photovoltaic cells because of its nearly ideal band-gap and high optical absorption coefficient. It is mainly used for heterojunction structures where the most common n-type semiconductors are CdS and $Cd_{1-x}Zn_xS$. CdS/CdTe cells have been produced by a variety of deposition techniques, including close space sublimation, screen printing-sintering, chemical vapor deposition, electrodeposition [1] and spraying [2]. In this paper Atomic Layer Epitaxy is added to the list of possible deposition techniques.

The growth of ZnS for electroluminescent thin film display devices has shown that ALE is an advantageous technique in commercial production [3]. In CdTe solar cells ALE offers the possibility to grow both CdS and CdTe in a single process and to prepare structures where CdS gradates into CdTe. Processing demands include also properties like uniformity over large areas and low pinhole density.

CdTe has earlier been grown by ALE on single crystals of CdTe [4], BaF_2 [5] and GaAs [6] using the elements as source materials under ultra high vacuum conditions. The ideal ALE growth rate, one monolayer per reaction cycle, is achieved both for the (111) CdTe on BaF_2 and for the (110) CdTe on GaAs substrates, at substrate temperatures of 260-285° and 260-290°C, respectively. Additional temperature regions with temperature-independent growth rates are found at higher substrate temperatures. For example, average coverages of 0.85 and 0.6 monolayers per cycle are found for (111) oriented CdTe on BaF_2 before the growth ceases at temperatures near 390°C [5].

In this paper we report the growth behavior of CdTe on polycrystalline substrates in a lateral flow reactor.

EXPERIMENTAL

CdTe films were prepared in a lateral flow reactor [3] using nitrogen for transporting the reactants, for valving the sources and for separating the dosing pulses (Fig. 1). The nitrogen pressure in the system was about 1000 Pa.

Films were grown on 5x5 cm^2 glass/indium tin oxide (ITO) or glass/ITO/SnO_2 substrates. A growth of CdS film by ALE preceded the growth of CdTe. The substrate temperature range studied was from 350°C to 460°C. Cadmium was vaporized at a source temperature of 340°C, and tellurium at 340°C to 400°C depending on the substrate temperature. The substrate temperature was always at least as high as the source temperature to prevent condensation of the elements. At lower substrate temperatures this meant that the pulse duration of Te was increased from 0.4 to 8 sec to ensure a sufficient dose of the element. For Cd the pulse and purge times were 0.2 and 0.4 sec, respectively. The purge times for Te varied between 0.4 and 4 sec.

Cubic, (111) oriented CdS was also grown using the elements as source materials. The growth was performed at the same temperature as the growth of CdTe. Source temperature for sulphur was about 90°C. Pulse and purge times were 0.2 and 0.4 sec. The number of reaction cycles was 1000.

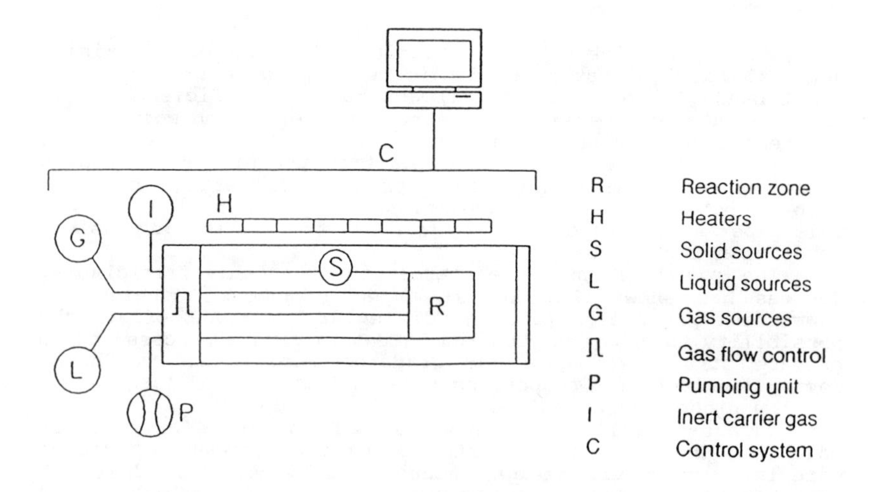

Fig. 1. The ALE reactor using inert gas as carrier.

The thickness of the films was measured with a Tencor Alpha-Step 2000 profilometer. The crystalline orientation of the films was measured with a Siemens Diffrac 500 diffractometer using Cu K_α radiation. Scanning electron micrographs were taken with Jeol JSM-T20 equipment.

RESULTS AND DISCUSSION

CdTe layers were successfully grown in the investigated temperature range between 350°C and 450°C. However, the practical processing window was from 390-400°C to 440°C in the reactor type used. The relatively high vapor pressure of CdTe determined the upper limit of the processing window. The small vapor pressure of Te_2 set in turn the lower limit.

To study further the growth behavior of CdTe on glass/ITO/SnO_2/CdS substrates, the thicknesses of the CdTe layers were measured versus the number of reaction cycles at a substrate temperature of 420°C (Fig. 2). The results showed that at the beginning of the growth the growth rate was about 0.5 monolayers per cycle. Uniform longitudinal thickness profiles of the layers indicated that saturation of the surface was achieved during each reaction cycle. Significant increase in the macroscopic growth rate was detected as the growth process proceeded. At the same time the surface of the films became rougher. We assume that the high macroscopic growth rates resulted, at least partly, from the increase in the effective surface area of the films due to development of the surface roughness. The increased surface roughness was also seen as a reduced degree of cubic (111) orientation.

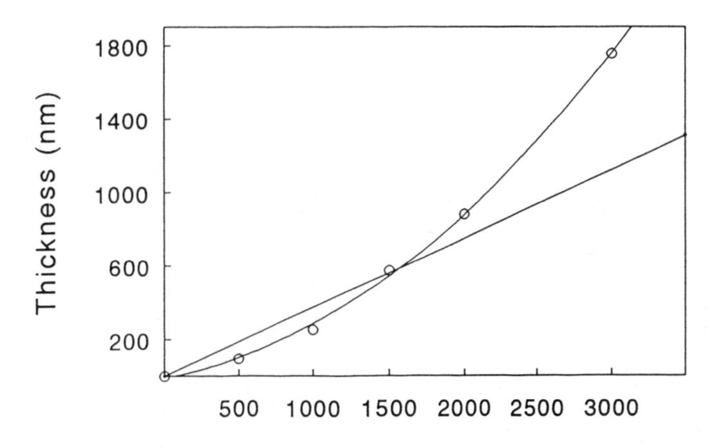

Fig. 2. The thickness of CdTe layers grown on glass/ITO/SnO2/CdS plates at a temperature of 420°C as a function of the reaction cycle number. The straight solid line shows the growth rate of one monolayer in the (111) direction per cycle.

Figure 3 shows the average macroscopic growth rate for CdTe in the temperature range 400°C to 450°C. The growth rate varied between 1.4 and 0.2 monolayers per cycle when the number of reaction cycles was 1000. Figure 3 further shows the strong temperature dependence of the growth. The effect of the increased surface roughness is already apparent in the growth rates. The growth behavior is also affected by the substrate surface under the CdS layer. For example, when glass /ITO/CdS substrates were used the growth rate was lower than for glass/ITO/SnO_2/CdS substrates in the same temperature range. The surfaces of 1-2 μm thick CdTe layers were rough on both substrates (Fig. 4).

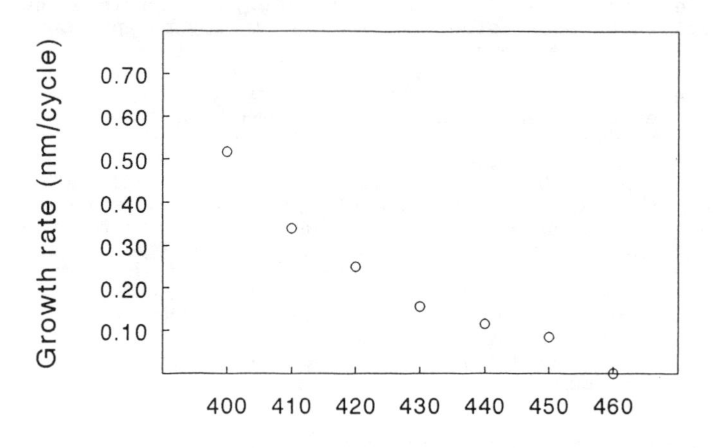

Fig. 3. Average growth rate of CdTe per reaction cycle as a function of temperature. Substrates were glass/ITO/SnO_2/CdS substrates.

Fig. 4. Scanning electron micrograph. 1,4 μm thick CdTe layer grown at 430°C on glass/ITO/CdS substrate. Magnification 10000x.

The photovoltaic characteristics of the CdS/CdTe solar cell samples were studied. In CdS/CdTe samples a heat treatment with $CdCl_2$ was made to improve further the crystallinity of CdTe. The highest efficiency measured was 14% (V_{oc} = 804 mV, I_{sc} = 23.8 mA/cm^2, FF = 0.73, active area = 12 mm^2, AM = 1.5 [7]), which is one of the highest ever reported for thin film solar cells.

CONCLUSIONS

The ALE growth behavior of CdTe on polycrystalline substrates under the experimental conditions used differed from the growth behavior reported on single crystals [4-6]. On the polycrystalline substrates a narrow, but still useable practical processing window was observed. Surface roughness and effective surface area of the films increased with increasing thickness.

Excellent conversion effiencies were measured on the CdS/CdTe solar cell samples made by ALE.

ACKNOWLEDGEMENTS

The authors thank Dr. T. Suntola for valuable discussions and S. Kaipio and J. Marles for technical assistance.

REFERENCES

1. Chu, T.L., in Current topics in photovoltaics, vol. 3, edited by Coutts, T.J. and Meakin, J.D., (Academic Press, London, 1988), pp. 264-300.
2. Banarjee, A., Saha, H., Guha, R., Indian J. Phys. 65A, 326 (1989).
3. Suntola, T. Materials Science Reports 4, 261 (1989).
4. Pessa, M., Huttunen, P., Herman, M.A., J. Appl. Phys. 54, 6047 (1983).
5. Faschinger, W, Sitter, H. Juza, P., J. Appl. Phys. 53, 2519 (1988).
6. Faschinger, W. and Sitter, H., J. Cryst. Growth 99, 566 (1990).
7. Skarp, J., Koskinen, Y., Lindfors, S., Rautiainen, A., Suntola, T., Proc., 10th European Photovoltaic Solar Energy Conf., 8-12 April, 1991, Lisbon, Portugal (in press).

ANALYSIS AND CALIBRATION OF THE FLOW CHARACTERISTICS OF A PRESSURE CONTROLLED VAPOR SOURCE FOR GAS SOURCE DOPING: CdTe:I

D. Rajavel, B.K. Wagner, K. Maruyama*, R.G. Benz,II, A. Conte and C.J. Summers
Physical Sciences Laboratory, Georgia Tech Research Institute, Atlanta, Georgia, 30322.
* Visiting Scientist, Fujitsu, Japan

ABSTRACT

A pressure controlled vapor source was developed for the gas source doping of (Hg,Cd)Te alloys. The dopant source has been subjected to extensive tests, and the flow characteristics determined. The dopant source was used to control the flow rates of ethyliodide for the n-type doping of CdTe. Highly conductive CdTe:I films were grown by molecular beam epitaxy.

INTRODUCTION

The range of extrinsic carrier concentrations used in semiconductor devices vary from 10^{14}-10^{19} cm^{-3}, and often must be achieved in the same device over very short distances. The need to realize both very low, and very high, levels of dopant concentration places stringent requirements on the doping system. The flow rate of the dopant must be stable and range over six orders of magnitude. Other requirements include a quick flux response to grow graded and abrupt dopant profiles. In conventional molecular beam epitaxy (MBE), the dopant and source materials are contained in crucibles heated by radiation from a heater assembly surrounding the crucible. The exponential dependence of the vapor pressure on temperature makes it possible to obtain a wide doping range. However, due to the large thermal mass of the source material the response time to change and to attain a steady flux is long. Pressure controlled vapor sources (PCVS) are becoming increasingly common in gas source MBE applications. Desired flow rates can be achieved by regulating the source material pressure upstream of an exit orifice. A pressure controlled vapor source which operates on the principle of choked flow was previously reported to control the flow of Hg vapor [1]. This Hg source has demonstrated excellent flux stability and response characteristics. In this paper we report the development of gas sources for the n- and p-type doping of (Hg,Cd)Te alloys.

A dopant PCVS has been developed and subjected to extensive tests, and results of the calibration and measurement of the flow rate and flow characteristics are presented. The PCVS was used for the gas source iodine doping of MBE grown CdTe, and highly conductive CdTe:I films were obtained. The dependence of electron concentration on the dopant flow rate and results of variable temperature Hall effect measurements are discussed.

THEORY OF OPERATION OF THE PRESSURE CONTROLLED VAPOR SOURCE

In the chemical beam epitaxy (CBE) system at Georgia Tech, the flow rates of group II and VI organometallics, which are controlled by MKS Model 1150 mass flow controllers, are approximately 0.25 sccm, for a growth rate of 0.5 μm/hr. Assuming 100% incorporation and activation of the dopants, the dopant flow rate required to achieve carrier concentrations between 10^{14}-10^{19} cm^{-3} ranges from 10^{-7} to 10^{-2} sccm. A flow rate of 10^{-7} sccm corresponds approximately to a beam equivalent pressure (BEP), as measured by an ionization gauge, of 1×10^{-10} torr. The typical operating base pressure in a CBE system used to grow Hg compounds is < 10^{-5} torr. Hence unlike the group II and VI source materials, the BEP of the dopant cannot be accurately measured by an ionization gauge. Therefore, the flow rate of the dopant has to be measured and monitored separately. The dopant system must also have the flexibility to accommodate any of the potential dopant source materials, the vapor pressures of which range from 0.01-100 torr at room temperature.

The flow of gas in a channel depends on the nature of the gas, the pressure and the geometry of the channel. Depending on the ratio between the pressures upstream (P_{up}) and downstream (P_{dn}) of an orifice, and the magnitudes of P_{up} and P_{dn}, flow across the orifice is termed molecular, transition or viscous. The Knudsen's number, Kn, generally is used to characterize the nature of the flow pattern of a gas in a channel [2]. Kn is a dimensionless number, given by the ratio of the mean free path of a gas to a characteristic dimension (example, the diameter of a pipe) of the channel. When Kn >1, the flow properties are governed by gas-wall collisions and the flow is termed molecular. The flow rate or throughput, Q, for molecular flow is expressed as

$$Q = C_1 (P_{up} - P_{dn}) \quad (1)$$

where C_1 is a constant. For Kn < 0.01, gas-gas collisions dominate and the flow is termed viscous, and the flow rate is given by

$$Q = C_2 \ (P_{up}^2 - P_{dn}^2) \quad (2)$$

where C_2 is a constant. In the region 1 > Kn > 0.01 the flow is neither molecular nor viscous and cannot be accurately modelled.

The principle of choked viscous flow has been reviewed by Santeler [3], and is of particular interest here. Under choked viscous flow conditions, the flow rate (Q) across a circular orifice is directly proportional to the upstream pressure (P_{up}) and independent of the downstream pressure (P_{dn}). Q is therefore given by

$$Q = C P_{up} \quad (3)$$

The proportionality constant C is the conductance of the orifice, and is a function of several variables, and is given by

$$C = C(\gamma, T, M, a) \quad (4)$$

where γ is ratio of specific heats, M is the molecular weight, T is the absolute temperature of the gas and the symbol a denotes the orifice radius. Thus for choked flow conditions, the desired flow rate can be achieved by regulating the pressure upstream of an orifice. The flow rate across an orifice can also be expressed as

$$Q = \frac{dn}{dt} \quad (5)$$

where n is the number of molecules flowing across the orifice and t denotes time. For low pressures (< 10 torr) on the upstream side of an orifice, the ideal gas law is a good approximation to calculate the number of molecules within a volume at a defined temperature and pressure, and is given by

$$P_{up} V = nRT \quad (6)$$

Here V is the upstream volume which is a constant, R is the gas constant and T is the absolute temperature. Rearranging terms in Eq.6 and substituting in Eq.5 yields,

$$Q = \frac{-V}{RT}\left(\frac{dp_{up}}{dt}\right) \tag{7}$$

The negative sign signifies flow out of the upstream volume, which results in a decrease in P_{up} with time. Substituting Eq. 3 in Eq. 7 gives

$$CP_{up} = \frac{-V}{RT}\left(\frac{dp_{up}}{dt}\right) \tag{8}$$

Rearranging,

$$\frac{dp_{up}}{p_{up}} = \frac{-CRT}{V}dt \tag{9}$$

Integrating, one obtains,

$$P_{up} = P_0 \exp\left(\frac{-CRTt}{V}\right) \tag{10}$$

where P_o is the upstream pressure at t=0. Thus for choked flow conditions, $\log(P_{up})$ varies linearly with time. Furthermore, by measuring the variation of P_{up} with time for choked flow, the conductance of an orifice can be calculated from Eq. 10. By similar reasoning the corresponding expression for molecular flow is

$$P_{up} = (P_0 - P_{dn}) \exp\left(\frac{-C_1RTt}{V}\right) + P_{dn} \tag{11}$$

For viscous flow, the expression is

$$\frac{P_{up} - P_{dn}}{P_{up} + P_{dn}} = \frac{P_0 - P_{dn}}{P_0 + P_{dn}} \exp\left(\frac{-2C_2RTP_{dn}t}{V}\right) \tag{12}$$

where the symbols have the usual meanings as defined previously. By comparing Eq. 1 and 3, and Eq. 10 and 11, note that when $P_{dn} << P_o$, the expression for molecular flow reduces to that for choked flow.

DESCRIPTION OF THE DOPANT-PRESSURE CONTROLLED VAPOR SOURCE

A schematic of the dopant PCVS is shown in Fig.1. A control valve admits dopant vapor into the upstream side of a variable orifice (Granville-Phillips Series 203 variable leak valve) and maintains the pressure at the desired value. A capacitance manometer (MKS 390HA) was used to monitor the upstream pressure, P_{up}, and to provide feedback to the control valve. The capacitance manometer was used to measure and control pressures between 10^{-4}-10^{1} torr with an accuracy better than $\pm 5\times10^{-5}$ torr. The conductance of the variable orifice can also be varied. Thus by controlling the upstream pressure and the conductance of the orifice it is possible to obtain a wide range of flow rates. Although the variable leak valve was supplied with a mechanical counter to indicate the position of the valve, its conductance as a function of the counter readout is not known. Thus it was necessary to calibrate the variable leak valve and the

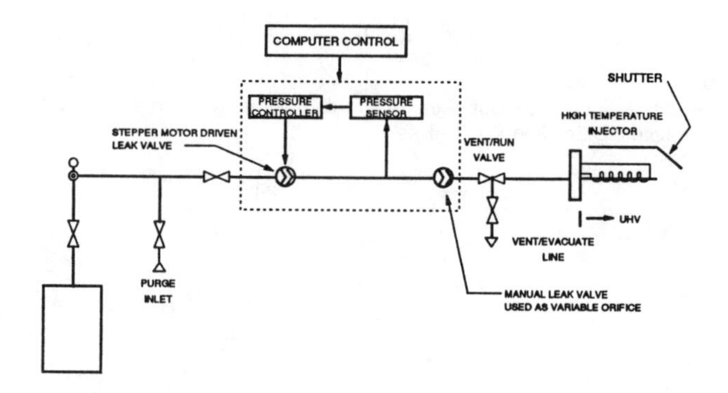

Figure 1. Schematic of the dopant-pressure controlled vapor source

system as a whole for flow rate and flow stability.

MEASUREMENT AND CALIBRATION OF FLOW CHARACTERISTICS

Verification of choked flow and calibration of flow rate

In order to measure and calibrate the flow characteristics of the PCVS, nitrogen gas was allowed to flow via the variable orifice into a vacuum chamber. The vacuum chamber was pumped by molecular drag and titanium sublimation pumps to maintain a base pressure of 10^{-5} torr, as measured by an ionization gauge. N_2 was used as the test gas in order to minimize problems associated with handling and scrubbing of toxic dopant sources. An absolute pressure regulator was used to maintain the N_2 pressure at the inlet at 100 torr. Before any measurements were made, the PCVS was thoroughly checked for leaks and outgassed by baking the system.

To measure the throughput, Q, for a given orifice opening, the upstream side of the variable orifice was filled to a desired pressure by admitting N_2 through the control valve and then the control valve was closed. The flow rate through the variable orifice was then determined by measuring P_{up} as a function of time. The capacitance manometer was interfaced to a personal computer for data acquisition. Fig. 2a and 2b show the variation of the upstream pressure with time for different orifice openings. The points represent experimentally measured data. For the sake of clarity only every tenth data point has been shown in the graph. Data collected at each orifice opening were fitted by an exponential function as predicted by the theory of choked flow (Eq. 10), and are shown as continuous lines in Fig.2a and 2b. The excellent agreement between experiment and theory confirms choked flow conditions of the PCVS when the downstream pressure was $< 10^{-4}$ torr during the experiment. On a semi-log plot, the slope of the straight line is -(CRT/V), as given by Eq. 10. The actual upstream volume, V, was determined by measuring the slopes of the $\log(P_{up})$ versus t plots obtained for two different upstream volumes. First the slope was determined at the actual volume V, and then the slope was once again measured for a larger upstream volume $(V + V_1)$, where V_1 is the volume of a mini-conflat nipple. The other variables C and T were the same for the two sets of data. From Fig.2a and 2b it can be seen that as the valve position was changed from counter position 17 to 60, the corresponding slopes of the straight lines varied by about four orders of magnitude. The values of the slopes and Eq.10 and 3 were then used to calculate the conductance of the valve and the throughput of the system. Conductance values and throughputs calculated in this manner for different orifice openings are shown in Table I. The estimated dopant concentration, assuming 100% incorporation of dopants is also listed.

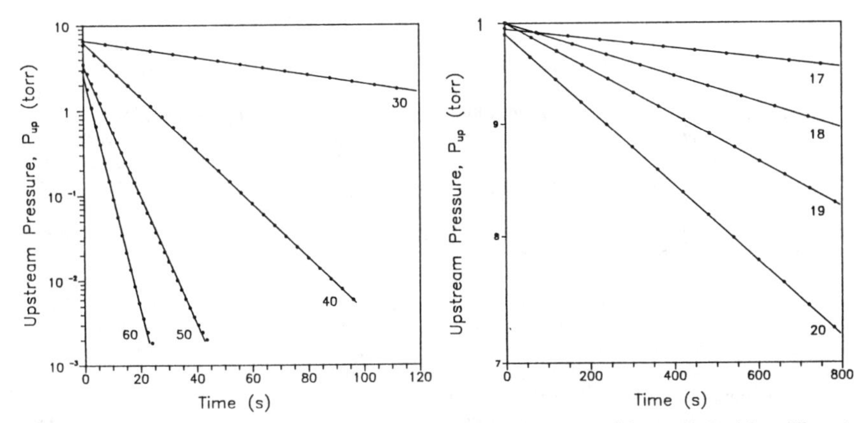

Figures 2a, 2b. Variation of P_{up} with time under choked flow conditions, plotted for different orifice openings. The points represent experimental data and the straight lines were calculated from the theory of choked flow. The counter readout of the variable orifice is listed near the respective lines.

Table I. Estimated values of orifice conductance, throughput and dopant concentration

Counter Readout	Conductance sccm torr^{-1}	Throughput[a] sccm	Dopant Concentration[b] cm^{-3}
60	2.2×10^{-1}	2.2×10^{-3}	2×10^{20}
50	1.4×10^{-1}	1.4×10^{-3}	1×10^{20}
40	5.9×10^{-2}	5.9×10^{-4}	4×10^{19}
30	9.8×10^{-3}	9.8×10^{-5}	7×10^{18}
20	2.9×10^{-4}	2.9×10^{-6}	2×10^{17}
19	2.0×10^{-4}	2.0×10^{-6}	1×10^{17}
18	1.1×10^{-4}	1.1×10^{-6}	7×10^{16}
17	4.1×10^{-5}	4.1×10^{-7}	3×10^{16}

a. Throughput estimated for $P_{up} = 0.01$ torr
b. Dopant concentration estimated for a growth rate of 0.5 μm/hr

Dependence of flow characteristics on downstream pressure

When the PCVS was connected directly to an injector, the downstream pressure was approximately equal to the pressure in the CBE system, which is $< 10^{-5}$ torr. However, when the same injector was shared by two or more sources, such as a group II source and a dopant source, the pressure downstream of the dopant PCVS was considerably higher. For the MKS 1150 mass flow controller, the pressure at the outlet is 0.1 torr. Accounting for pressure drops at the shut-off valve and the tubing, the downstream pressure at the exit orifice of the dopant PCVS can be as high as .01 torr. As high downstream pressures can alter the flow characteristics of the dopant PCVS, flow characteristics were also determined as a function of downstream pressure.

For these calibrations nitrogen gas was admitted to the upstream volume through the control valve, which was then closed. The decrease in upstream pressure with time was measured for a fixed orifice opening, as the downstream pressure was systematically varied between $2x10^{-4}$ - $2x10^{-1}$ torr. The downstream pressure was controlled by admitting N_2 gas through another variable leak valve into the vacuum chamber. For $P_{dn} \leq 1x10^{-3}$ torr, $\log(P_{up})$ decreased linearly with time, thus, choked flow was observed. Fig. 3 shows the variation of $\log(P_{up})$ with time for the same orifice opening, measured for $P_{dn} \geq 1x10^{-3}$ torr. Once again, only every tenth data point is shown in the figure. The dashed straight line in Fig. 3 represents the theory as predicted for choked flow (Eq.10) and indicates choked flow for $P_{dn} = 1x10^{-3}$ torr. However as P_{dn} is increased, a plot of $\log(P_{up})$ with time is not linear, thus the flow pattern departs from choked flow conditions.

Eq.11 which describes molecular flow, approximates to the expression for choked flow when $P_{up} >> P_{dn}$. The theory of molecular flow, described by Eq. 11 was used to model the behavior of the PCVS for $P_{dn} > 1x10^{-2}$. Least square fits to the data were made using the expression for molecular flow with the conductance of the orifice, C_1, being the only variable parameter. As shown in Fig. 3, the agreement between experiment and the theory of molecular flow (represented by continuous straight lines) is excellent. As the orifice opening is narrow, gas-wall interactions dominate the flow across the orifice, even when P_{up} is as high as 1 torr. For these special conditions when $P_{up} \geq 50\ P_{dn}$, the expression for choked flow adequately describes the flow characteristics. However, for this PCVS system, when $P_{up} < 50\ P_{dn}$ the expression for molecular flow was found to best describe the flow characteristics. In this study, the PCVS was operated under choked flow conditions for the iodine doping of CdTe.

Reproducibility and stability of flow rate

For the commercial leak valve used in this study the conductance changes by three orders of magnitude for only two full turns of the valve, up to a valve setting of 20. Hence, even after eliminating any backlash error in the gear and counter assembly, it was found that the reproducibility in conductance for the same valve opening as indicated by the mechanical counter was typically ±10%. However, by measuring the conductance and adjusting the upstream pressure, flow rate can be adjusted to within ± 1% of a desired value. In this manner, the counter in the leak valve was used only as a rough guide of the conductance, and throughput measurements were used to fine tune the flow rate of the system.

For nitrogen, the stability in the flow rate was better than ±0.5%. When a condensible gas was used, all the vacuum tubing was maintained at a temperature >60°C to minimize condensation on the walls. The capacitance manometer has its own temperature controlled enclosure and was maintained at 45°C. For ethyliodide, the dopant used in this study, the flow rate stability was typically ±5%, for flow rates from 10^{-5} to 10^{-3} sccm. This error is attributed to the inherent stability of the variable leak valve, at the low flow rates. A fixed-conductance orifice is currently being developed to control flow rates < 10^{-5} sccm, with improved stability.

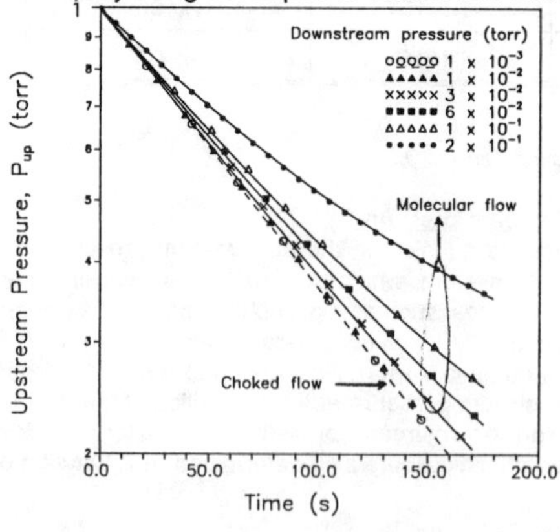

Figure 3. The variation of P_{up} with time for choked and molecular flow. The points represent experimental data. The dashed line represents the theory of choked flow, and the continuous lines were calculated from the theory of molecular flow.

IODINE DOPING OF CdTe

Low temperature growth of Hg-based alloys is required to minimize diffusion and to increase the sticking coefficient of Hg. Growth conditions rich in group II elements are required for the low temperature (< 160°C) epitaxial growth of (Hg,Cd)Te alloys [4]. Furthermore, as the sticking coefficient of the group II elements is less than that of Te, a metal rich flux is required to grow stoichiometric compounds. The inherent advantages associated with group II rich growth conditions also favor the incorporation of dopants on the Te sublattice, at V_{Te} sites. Thus the development of a group VII dopant source is compatible with the materials science issues of the MBE growth of (Hg,Cd)Te. Furthermore, Cd-rich growth of CdTe suppresses the formation of V_{Cd} acceptor defects and $[V_{Cd}I_{Te}]$ acceptor complexes which are know to produce compensation [5]. Indium has been the most widely used donor impurity in (Hg,Cd)Te and occupies the group II sublattice. Highly conductive CdTe:In films have been grown by MBE [6-8]. We show in this work that high n-type conductivity can also be achieved in iodine doped CdTe. Besides being compatible with group II rich growth conditions, iodine diffuses slower than indium as iodine occupies the Te sublattice.

In order to dope CdTe n-type, ethyliodide was used as the dopant source material and flow rates were controlled by the PCVS described above. Gas source iodine doped (001) CdTe layers were grown using conventional solid CdTe and Cd in a Varian GEN II system modified for the CBE growth of (Hg,Cd)Te alloys. (001) CdTe:I layers were grown on (001) CdTe substrates at 230°C at a growth rate of 0.5 μm/hr. The substrate temperature was calibrated by a tellurium condensation technique described elsewhere [9]. For these films, the CdTe and Cd beam equivalent pressures were 8×10^{-7} and 3×10^{-7} torr, respectively. Under these conditions the RHEED pattern exhibited mixed (2x1) Te-stabilized and C(2x2) Cd-stabilized surface reconstructions [10]. CdTe:I layers were grown for flow rates of ethyliodide between 10^{-5} to 10^{-2} sccm. Ethyliodide incident on the substrate was uncracked and decomposition of the compound occurred on the growth surface by a catalytic process. The electron concentration and mobility were determined by resistivity and Hall effect measurements. The variation of the room temperature electron concentration, n, with dopant flow is shown in Fig. 4. A least square fit to the data indicated that n varied as the square root of the dopant flow rate. For electron concentrations up to 3×10^{18} cm^{-3}, n did not saturate with increase in dopant flow. This indicates that even higher electron concentrations can be achieved by iodine doping. Layers with electron concentrations of 8.2×10^{16} and 3.1×10^{18} cm^{-3} had room temperature electron mobilities of 630 and 460 cm^{2}V^{-1}s^{-1}, respectively, and compare favorably with the properties reported for CdTe:In layers grown by MBE.

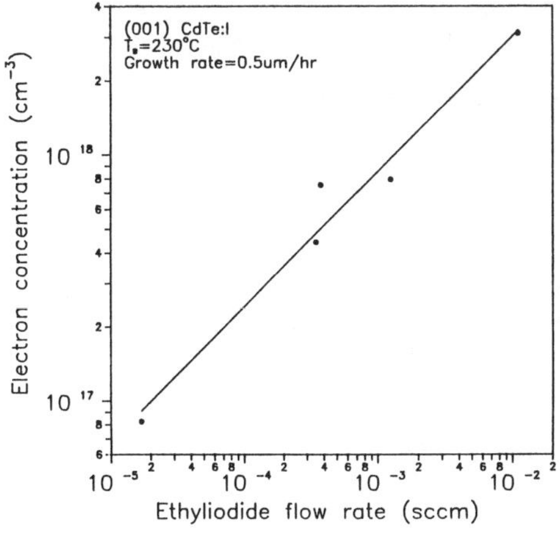

Figure 4. Room temperature electron concentration in CdTe:I plotted versus the dopant flow rate.

SUMMARY

A pressure controlled vapor source was developed to control the flow of dopant vapor in the 10^{-6} to 10^{-2} sccm range. The flow characteristics were studied and modelled based on the theory of choked and molecular flow. In general the flow across the exit orifice is molecular. However, when $P_{up} \geq 50\ P_{dn}$, the expression for choked flow adequately describes the flow pattern. For choked flow conditions, throughput across the orifice is independent of P_{dn} and directly proportional to P_{up}. Thus desired flow rate can be achieved by simply controlling P_{up}. The PCVS was used to control the flow rate of ethyliodide for the gas source iodide doping of CdTe. Highly conductive n-type CdTe epitaxial layers were achieved by iodine doping, and electron concentrations varying between $8x10^{16}$ and $3x10^{18}$ cm^{-3} were measured in these films. The electron mobility compared favorably with CdTe:In films grown by MBE. A comprehensive study of the optical, electrical and structural properties of CdTe:I films is currently underway, and will be reported later.

ACKNOWLEDGMENT

The authors wish to thank Don Swank for welding the PCVS components. This work was supported by SERI (contract XH-9-19056-1), WRDC (contract F33615-C-1066) and the Georgia Tech Research Institute, and is thankfully acknowledged.

REFERENCES

1. B.K. Wagner, R.G. Benz and C.J. Summers, J. Vac. Sci. Technol. **A 7,** 295 (1989).
2. J.F. O'Hanlon, A Users Guide to Vacuum Technology, (John Wiley and Sons, New York, 1980) p. 20.
3. D.J. Santeler, J. Vac. Sci. Technol. **A 4,** 348 (1986).
4. J.M. Arias, S.H. Shin, D.E. Cooper, M. Zandian, J.G. Pasko, E.R. Gertner and R.E. DeWames, J. Vac. Sci. Technol. **A 8,** 1025 (1990).
5. Y. Marfaing, Prog. Crystal Growth Charact. Vol.4, 317 (1981).
6. R.N. Bicknell-Tassius, A. Waag. Y.S. Wu, T.A. Kuhn and W. Ossau, J. Crystal Growth **101,** 33 (1990).
7. S. Hwang, R.L. Harper, K.A. Harris, N.C. Giles, R.N. Bicknell, J.W. Cook and J.F. Schetzina, J. Vac. Sci. Technol. **A 6,** 2821 (1988).
8. N.R. Taskar, V. Natarajan, I.B. Bhat and S.K. Gandhi, J. Crystal Growth, **86,** 228 (1988).
9. D. Rajavel, F. Mueller, J.D. Benson, B.K. Wagner, R.G. Benz and C.J. Summers, J. Vac. Sci. Technol. **A 8,** 1002 (1990).
10. J.D. Benson, B.K. Wagner, A. Torabi and C.J. Summers, Appl. Phys. Lett. **49,** 1034 (1986).

FORMATION OF CdTe AND GaAs BY ELECTROCHEMICAL ATOMIC LAYER EPITAXY (ECALE)

D. Wayne Suggs, Ignacio Villegas, Brian W. Gregory and John L. Stickney
School of Chemical Sciences, University of Georgia, Athens, Georgia 30602

ABSTRACT

The principles of Atomic Layer Epitaxy (ALE) have been applied to the formation of compound semiconductors by an electrochemical technique, referred to as Electrochemical Atomic Layer Epitaxy (ECALE). Atomic layers of the component elements are alternately electrodeposited at underpotential (UPD) from separate solutions and at separate potentials. Results are presented concerning the structures of both CdTe and GaAs deposits formed by ECALE. Studies were performed using single-crystalline Au electrodes in a UHV surface analysis instrument coupled directly with an electrochemical cell. This instrument was used in order to prevent corruption by contact with air during transfer to the surface analysis environment.

INTRODUCTION

The method of Electrochemical Atomic Layer Epitaxy (ECALE), described here, is the electrochemical analog of ALE [1]. It is based on the alternated electrodeposition of atomic layers of elements at underpotential in order to form a compound.

It is anticipated that ECALE can be developed into a reliable method for the formation of thin films of compound semiconductors. There are certain inherent advantages to an electrochemical method of deposition, such as the low temperatures at which deposits can be formed; in the present case, for example, deposition was performed at room temperature. Equilibrium conditions are controlled by the electrode potential instead of temperature. Optimal growth is facilitated by the use of deposition potentials where an atomic layer of an element deposits and is stable on the surface, while further deposition is energetically unfavorable, a condition referred to as underpotential deposition (UPD). Another advantage of ECALE is the ease with which the controlling variable, potential, is changed from one element to the next. Generally with ALE the controlling variable, temperature, remains constant for the deposition of both elements, and is thus chosen as a compromise between the ideal temperatures for atomic layer formation with the two reactants.

Underpotential deposition (UPD) refers to the phenomenon whereby the potential necessary to deposit one element onto a second element occurs before (under) that necessary to deposit the element on itself. This is a natural consequence of the free energy for compound formation. Classically, it has referred to the reductive deposition of a less noble metal onto a more noble metal, resulting in the formation of a two-dimensional bimetallic compound [2] (reductive UPD). It also occurs in the formation of other compounds, such as the adsorption of hydrogen on a metal surface (e.g., the hydrogen waves on Pt [3]). Anodic processes, such as the initial stages of oxide formation on copper surfaces [4] or the oxidative deposition of I atoms on a metal surface from an I^- solution [5], can also be described as underpotential deposition (oxidative UPD).

Most compound semiconductors consist of both a metal and a main group element such as Te, Se, S, As, Sb or Bi. Metals are generally soluble as cations or their complexes, and may be deposited by reduction. In ECALE deposition these metals are deposited by reductive UPD. The main group elements listed above form soluble oxidized species, as do metals; however, they can also form soluble reduced species such as Te^{-2}, Se^{-2}, S^{-2}, As^{-3}, Sb^{-3}, and Bi^{-3}. These species are amenable to oxidative deposition of the element and thus to oxidative UPD. It is the formation of these reduced main group species and their oxidative UPD which makes ECALE formation of a compound possible.

Reductive UPD of both elements can be used to form the first monolayer of CdTe. An atomic layer of Te can be deposited at -0.2 V, Figure 1. An atomic layer of Cd can then be deposited at -0.6 V. Subsequent reductive UPD of Te to form a second atomic layer of Te at -0.2 V would result in the spontaneous dissolution of the previously deposited Cd. For this reason, oxidative UPD of one of the elements is a requirement for ECALE deposition. By alternating between the potential for oxidative UPD of Te, -1.2 V, and reductive UPD of Cd, -0.6 V, the product CdTe remains stable.

Codeposition of both elements simultaneously is presently the most frequently used method for electrodepositing compound semiconductors [1]. Most of the previous codeposition studies have been modeled on those of Kroger, et al. [6]. Their procedure, and many subsequent modifications of it, consisted of codepositing both elements from the same solution. Optimal stoichiometries for the deposits were achieved by adjusting the deposition potentials so that the metal cation, Cd^{+2} for example, deposited at underpotential on the main group element, Te for example. To insure quantitative reaction of the depositing Te, low concentrations of its precursor, TeO_2, were used with a large excess of $CdSO_4$. The rate of CdTe deposition was thus limited by transport of Te to the surface, since TeO_2 was the limiting reagent, and Cd only deposited on previously deposited Te.

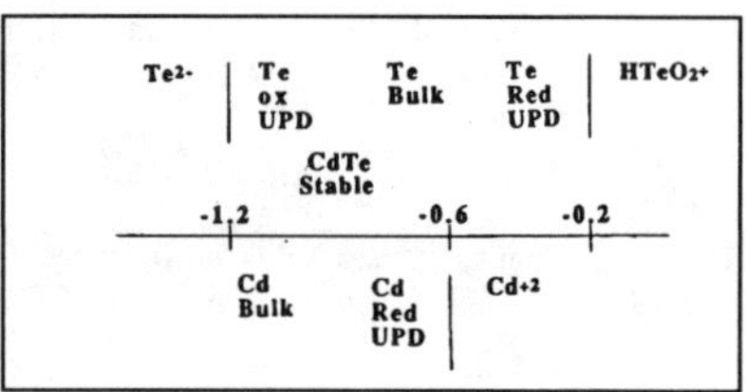

Figure 1: Potential diagram for CdTe deposition.

The results in every case for codeposition were the formation of polycrystalline deposits [1], sometimes described as "cauliflower" deposits due to their convoluted morphology [7]. These morphologies are at least partially the result of excess formation of three-dimensional nuclei during deposition. Other problems resulting in polycrystallinity were the ubiquitous use of polycrystalline substrates and the presence of contaminants.

ECALE deviates from previous methods of compound semiconductor electrodeposition in its use of separate solutions and potentials for each reactant. In this way, transport control of the deposition rate, present in the codeposition of compounds, is avoided. Instead, solutions containing a single reactant are exposed to the deposit surface for a time sufficient for equilibration, and at a potential where UPD of the element occurs. After deposition of an atomic layer of the first element, the solution and potential are changed and an atomic layer of the second element is deposited, completing one cycle. Thin films are then formed by repeating this cycle.

Initial studies in the development of ECALE involved the use of polycrystalline thin-layer electrodes (TLEs) to determine the potentials for UPD of Cd and Te on Cu, Pt and Au electrodes [8]. Those results indicated that Au was the most suitable metal for these studies, from an electrochemical perspective. That is, the stability range of Au in the aqueous solutions used in these studies is by far the most convenient. Subsequently, the first ECALE depositions were performed with CdTe on a Au TLE [1]. Studies of the structures formed by ECALE of CdTe on the low-index planes of Au single crystals have been carried out and are briefly described in this paper.

Studies of the surface chemistry of the CdTe(111) planes have also been performed as they relate to its preparation as a substrate for ECALE deposition [9]. Those studies resulted in formulation of a procedure where single crystals of CdTe can be etched and then electrochemically reduced to remove contamination and disordered material, forming a clean non-reconstructed ordered surface. Previous treatments described in the literature either left a disordered layer of Te or involved ion bombardment and subsequent annealing which resulted in reconstruction of the CdTe surfaces.

EXPERIMENTAL

The experiments described below were performed using a three-sided Au single crystal electrode, with each side oriented and polished to one of the low-index planes: (111), (110) and (100). The electrode was used in a UHV surface analysis instrument equipped with an integral electrochemical cell [10]. This configuration allowed the electrode to be prepared and examined in UHV prior to each experiment, and then to be transferred to the electrochemical cell without exposure to air. After deposition, the electrode was transferred back to the analysis chamber where the resulting deposits were examined. Instrumentation available includes optics for low energy electron diffraction (LEED), Auger electron spectroscopy (AES), and thermal desorption spectroscopy (TDS). Also available are facilities for ion bombardment and annealing, which are used to prepare electrode surfaces prior to each experiment.

Solutions were prepared with high purity, research grade chemicals and pyrolytically triply-distilled water [11]. Electrochemistry was performed with a three-electrode potentiostat based on conventional op-amp circuitry. Potentials are reported vs. a Ag/AgCl(1M NaCl) reference electrode.

RESULTS AND DISCUSSION

As stated in the introduction, Au is an ideal electrode material for these studies. It is not the ideal substrate for the epitaxial electrodeposition of CdTe. A comparison of the lattice constant for CdTe and twice the lattice constant of Au reveals an 11% mismatch (Table 1). The mismatch with GaAs, on the other hand, is only about 2%. As mentioned in the introduction, work on CdTe substrates is underway and will facilitate the homoepitaxial electrodeposition of CdTe.

Table I: Substrate lattice matching possibilities: Lattice constants for various small unit cells (A).

	1	$\sqrt{2}$	$\sqrt{3}$	2	$\sqrt{5}$
Ni	2.50	3.54	4.33	5.00	5.59
Cu	2.56	3.62	4.43	5.12	5.72
Zn	2.78	3.93	4.81	5.56	6.22
Pt	2.78	3.93	4.81	5.56	6.22
Au	2.88	4.07	4.99	5.76	6.44
Ag	2.90	4.10	5.02	5.80	6.48
Ti	2.92	4.13	5.06	5.84	6.53
Cd	3.14	4.44	5.44	6.28	7.02
In	3.32	4.70	5.75	6.64	7.42
InP	5.07	7.17	8.78	10.14	11.34
GaP	5.45	7.70	9.44	10.90	12.19
GaAs	5.65	7.99	9.79	11.30	12.63
InAs	6.06	8.57	10.50	12.12	13.55
GaSb	6.10	8.63	10.57	12.20	13.64
InSb	6.48	9.16	11.22	12.96	14.49
CdTe	6.48	9.16	11.22	12.96	14.49

Studies of ECALE deposit structure began with investigations of the first monolayer of CdTe. Experiments were performed by immersion of the clean and ordered Au tri-crystal into reactant solutions. These solutions were contained in a Pyrex electrochemical cell, housed in an antechamber of the surface analysis instrument. Examples of the resulting voltammetry are shown in Figure 2 for the deposition of Cd, Te, Ga and As. The experimental procedure involved emersion (withdrawal) of the

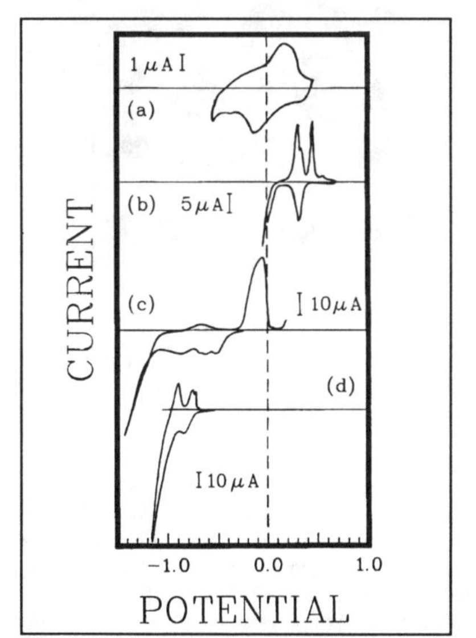

Figure 2: I-V curves for various metals on the Au tri-crystal: a) 1 mM $CdSO_4$; b) 0.4 mM TeO_2; c) 1 mM $HAsO_2$; d) 0.5 mM Ga_2SO_4.

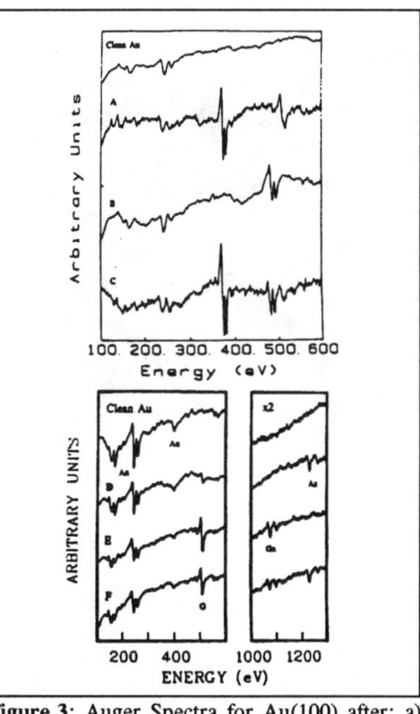

Figure 3: Auger Spectra for Au(100) after: a) Cd UPD; b) Te UPD; c) one monolayer of CdTe; d) Ga UPD; e) As UPD; f) one monolayer of GaAs. Beam currents: (a-c) 0.5 μA and (d-f) 10.0 μA.

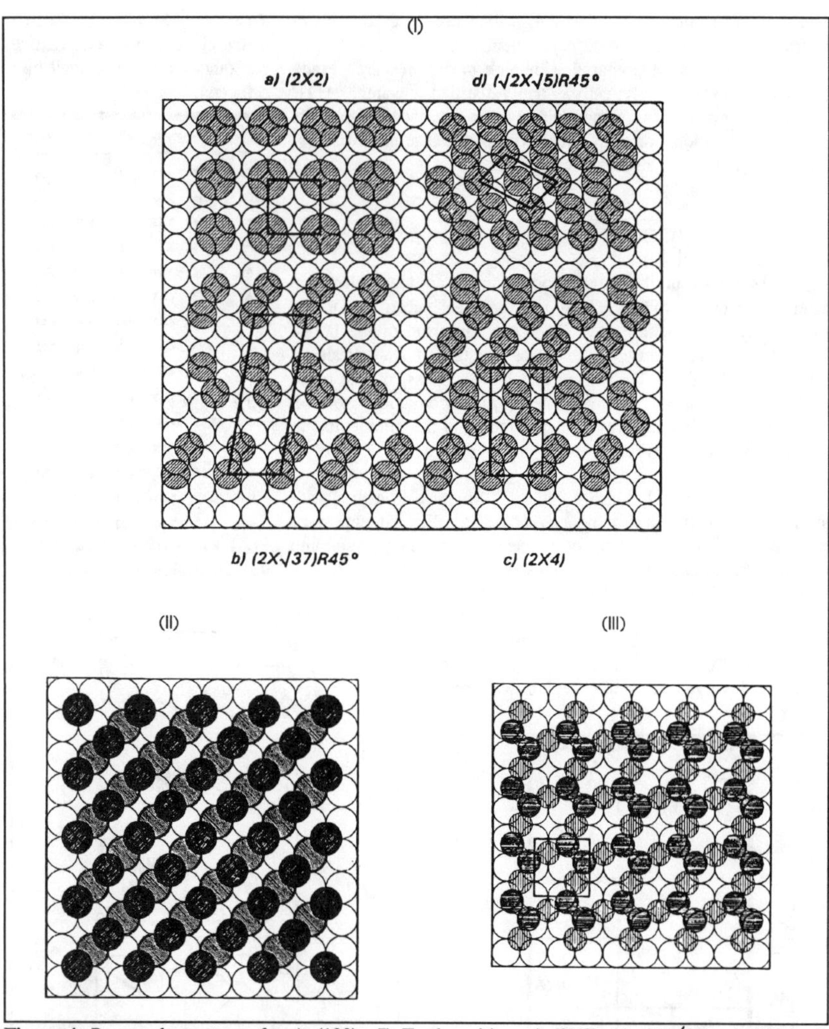

Figure 4: Proposed structures for Au(100): (I) Te deposition: a) (2X2); b) (2X√37); c) (2X4); d) (√2X√5)R45°; (II) one monolayer of CdTe, (√2X√2)R45°; (III) one monolayer of GaAs, (2X2).

electrode at a potential of interest and its transfer to the surface analysis chamber. The resulting surfaces where examined using both LEED and AES. Initial studies were performed by depositing Cd from a Cd^{+2} solution. A UPD peak for Cd is clearly visible (Figure 2a). However, emersion of the electrode at any potential below the onset of Cd deposition and examination by LEED and AES indicated the presence of a disordered layer of cadmium oxide or hydroxide (Figure 3a). This is a natural consequence of the emersion process for less noble metals. As the electrode is withdrawn from solution, potential control is lost; the result is Cd metal in contact with solution at open circuit. Without potential control, Cd metal is thermodynamically unstable in the presence of aqueous solutions, resulting, for example, in the formation of Cd^{+2} or $Cd(OH)_2$ [12]. As dissolution of the Cd metal in solution probably results in the formation of H_2, the pH in the solution layer emersed with the electrode will rise, driving the formation of cadmium hydroxide. The emersion techniques used in the present studies make it difficult to examine any structures actually formed in solution by the depositing Cd. The spontaneous dissolution of less noble metals at open circuit is a fairly general process, and has been observed, for instance, in studies of the deposition of Pb on Pt [13].

Te on the other hand is stable in water [12]. Te deposition from an aqueous solution of TeO_2 (Figure 2b) takes place in two clearly observable UPD processes: an initial reduction peak followed by a shoulder on the bulk deposition wave. An extensive study of the deposition of Te on all three low-index planes as a function of potential has been carried out and is to be published elsewhere [14]. As an example of the surface chemistry observed, some of the structures formed on Au(100) will be described here, Figure 4Ia-d. Deposition through the first UPD feature in Figure 2b results in a clear (2X2) LEED pattern and 1/4 coverage of Te, determined by AES and coulometry. No oxygen was evident. We propose the formation of a p(2X2) structure such as that diagramed in Figure 4(Ia). This figure is drawn with isolated Te atoms at their Van der Waals diameter and set in high coordinate sites. Subsequent Te deposition does not occur for another 200-300 mV, until initiation of the shoulder (Figure 2b) at which point a (2X$\sqrt{37}$) LEED pattern is observed. Subsequent deposition quickly reveals two other structures: a (2X4) and a ($\sqrt{2}$X$\sqrt{5}$). The ($\sqrt{2}$X$\sqrt{5}$) persists as bulk Te is formed. Proposed structures, consistent with the observed LEED patterns and coverages from both coulometry and AES, are also shown in Figure 4Ib-d. These structure are drawn with dimers, as structures based on isolated atoms were not workable. Other evidence for dimer formation is offered by STM where studies of S on Re showed the initial formation of a p(2X2) at 1/4 coverage and subsequent formation of dimers and trimers at higher coverages [15]. The hiatus in deposition between the first UPD peak and the shoulder in Figure 2b indicates that a fundamental chemical change\phase transition has taken place on the surface. We propose that this hiatus corresponds to the energy needed to convert isolated Te atoms on Au into dimers on Au. Similar structures and transitions were observed on the other two Au low-index planes [14].

Subsequent Cd deposition has been studied as a function of the initial Te coverage, as has the deposition of Te as a function of the initial Cd coverage. Both studies resulted in the formation of structures with equivalent LEED patterns and Te and Cd coverages. Evidently, which element is deposited first is not very significant. Also of interest is the resulting CdTe structure's lack of dependence on the initial Te layer structure. In general, a c(2X2) LEED pattern was observed irrespective of the initial Te structure, as long as the coverage was relatively close to 1/2. Optimal LEED pattern clarity was observed for 1/2 monolayer coverages of both Cd and Te. Some variations in the Cd and Te coverages, near 1/2, also resulted in the same c(2X2) patterns, but with increased diffuse intensity. Evident in the AES spectra for the layer of CdTe is the minimal amount of oxygen present (Figure 3c) indicating that the Cd is bonded to the Te and is thus less readily oxidized during the emersion process.

A simple c(2X2) structure is diagramed in Figure 4II. This is the structure that would be expected if a (100) layer of CdTe were superimposed on the Au(100) surface, with the required 11% compression in the CdTe lattice. This strained layer is not expected to persist for long before the critical thickness is exceeded. From Table 1 it might be anticipated that a ($\sqrt{5}$X$\sqrt{5}$)R26.6° unit-cell would form, not the c(2X2), since the mismatch for the ($\sqrt{5}$X$\sqrt{5}$) would have been only 0.6%.

Similar investigations were performed with the GaAs system. The stability of GaAs as a function of pH and potential in aqueous solutions is considerably less than that for CdTe, which adds to the difficulty of forming GaAs by the ECALE method. Initial investigations, though, proved encouraging. As with the Cd deposition, deposited Ga spontaneously oxidizes upon emersion from solution (Figure 3d). Arsenic deposition resulted in formation of a p(2X2) at 1/4 coverages, as observed for Te. Unlike Te deposition, subsequent As deposition resulted in disordered LEED patterns. A structure equivalent to the p(2X2) Te structure (Figure 4Ia) is proposed for the As p(2X2).

Alternated deposition of As and Ga appeared to be optimal for 1/2 coverage of each, as it was for CdTe above. Actual coverage measurements were more difficult than for Cd and Te due to decreased sensitivity to the Ga and As Auger transitions, and due to increased interference from hydrogen evolution in coulometric measurements. The p(2X2) LEED pattern observed for the GaAs structure differed significantly from the c(2X2) formed for a monolayer of CdTe on the Au(100) surface. The unit cell for the GaAs is twice that for the CdTe. It is also twice what would be expected if a GaAs(100) slice were superimposed on the Au(100) surface, with the 2% expansion. As with

CdTe, a c(2X2) should result. We are proposing a structure in which the As atoms form an alternated array of dimers on the Ga atoms (Figure 4III) thus accounting for the larger unit cell. This structure is consistent with previous STM studies of the surfaces of GaAs(100) studied in vacuum, where dimerization of the surface is normally observed [16].

ACKNOWLEDGEMENTS

Acknowledgement is made to the Donors of The Petroleum Research Fund, administered by the American Chemical Society, for partial support of this research. Acknowledgement is also given to the National Science Foundation for partial support of this work under grant Nº DMR-9017431, and to the Office of Naval Research. I.V. would like to thank INTEVEP S.A. (Caracas, Venezuela) for financial support.

REFERENCES

1. B.W. Gregory and J.L. Stickney, J. Electroanal. Chem. 300, 543 (1991).
2. D.M. Kolb, Advances in Electrochemistry and Electrochemical Engineering, Vol. 11, Eds. H. Gerischer and C.W. Tobias (John Wiley: New York, 1978), p. 125; K. Juttner and W.J. Lorenz, Z. Phys. Chem. N.F., 122, 163 (1980).
3. P.N. Ross, Jr., Surf. Sci., 102, 463 (1981).
4. J.M.M. Droog and B. Schlenter, J. Electroanal. Chem., 112, 387 (1980).
5. B.G. Bravo, S.L. Michelhaugh, M.P. Soriaga, I. Villegas, D.W. Suggs, and J.L. Stickney, J. Phys. Chem., in press.
6. M.P.R. Panicker, M. Knaster, and F.A. Kroger, J. Electrochem. Soc., 125, 566 (1978).
7. M. Tomkiewicz, I. Ling, and W.S. Parsons, J. Electrochem. Soc., 129, 2016 (1982).
8. B.W. Gregory, M.L. Norton, and J.L. Stickney, J. Electroanal. Chem., 293, 85 (1990).
9. I. Villegas and J.L. Stickney, J. Electrochem. Soc., 138, 1310 (1991).
10. J.L. Stickney, C.B. Ehlers, and B.W. Gregory, Electrochemical Surface Science: Molecular Phenomena at Electrode Surfaces, ACS Symposium Series, no. 378, Ed. M.P. Soriaga (American Chemical Society: Washington, D.C., 1988), p. 99.
11. B.E. Conway, H. Angerstein-Kozlowska, W.B.A. Sharp, and E.E. Criddle, Anal. Chem., 45, 1331 (1973).
12. M. Pourbaix, Atlas of Electrochemical Equilibria in Aqueous Solutions (Pergamon Press: Oxford, 1966).
13. B.C. Schardt, J.L. Stickney, D.A. Stern, A. Wieckowski, D.C. Zapien, and A.T. Hubbard, Surf. Sci., 175, 520 (1986).
14. D.W. Suggs and J.L. Stickney, J. Phys. Chem., submitted.
15. M.D. Pashley, K.W. Haberern and J.M. Woodall, J. Vac. Sci. Technol. B 6 (1988) 1468; D.K. Biegelsen, L.-E. Swartz, and R.D. Bringans, J. Vac. Sci. Technol. A 8 280 (1990).
16. D.F. Ogletree, R.Q. Hwang, D.M. Zeglinski, A.L. Vazquez-de-Parga, G. A. Somorjai, M. Salmeron, J. Vac. Sci. Tech. B, 9, 886 (1991).

PART VI

High Tc Superconductors and Other Materials

LOW TEMPERATURE EPITAXIAL GROWTH OF HIGH TEMPERATURE SUPERCONDUCTORS: Bi-Sr-Ca-Cu-O.

Maki KAWAI*, Masami MORI**, Shunji WATABE**, Ziyuan LIU**, Yasunori TABIRA** and Nobuo ISHIZAWA**
*The Institute of Physical and Chemical Research (RIKEN), Surface Chemistry Lab., 2-1 Hirosawa, Wako-shi, Saitama 351-01 and **Tokyo Institute of Technology, RLEM, 4259 Nagatsuta, Midori-ku, Yokohama, 227 Japan.

ABSTRACT

Molecular beam epitaxy of ultra thin films of $Bi_2Sr_2CuO_6$- (2201 phase) is realized on the surface of $SrTiO_3$(100) and $LaAlO_3$(100) at the substrate temperature of 573 K, using 10^{-5}Pa of NO_2 as an oxidant. The film epitaxially grown from the surface of the substrate has identical in-plane lattice constant to the substrate itself. Such a growth can only be obtained on the substrate with similar lattice constant to those of the material to be formed. The crystallinity of the film strongly depended on the sequence of the metal depositions and the oxidation process. In the case of the Bi system, the elementary unit of the epitaxial growth has proved to be the subunit of the perovskite structure (Sr-Cu-Sr). The structure of the film grown on a substrate with large mismatch (MgO) is also discussed.

INTRODUCTION

In the film formation of the high temperature superconducting materials the layer by layer synthesis is not only useful from the quality point of view but also for the material synthesis itself. An example is Bi-Sr-Ca-Cu-O system. In the synthesis of bulk materials, the control of the single phase synthesis is mainly done by the careful control of the sintering temperature and atmosphere. In such a complicated material synthesis, the layer by layer formation of films has proved to be one of the most powerful technique[1-4] . Especially, in realizing the low temperature epitaxy, the design of the surface reaction is indispensable[5].

The mechanism of the layer by layer growth of cuprate superconductors has been the focus of much attention. In connection with whether the precise control of the material synthesis can be done by atomic layer or not, unitcell epitaxy has been shown for $YBa_2Cu_3O_{7-x}$ film by Terashima et al. using reactive evaporation method [6]. Kanai et al. have shown that the elementary unit of epitaxial growth depends on the experimental conditions and that by the careful control of the element supply, even an atomic layer control is possible [7].

Various techniques have been applied to form ultra thin films of cuprate superconductors. However, the techniques reported so far have quite high processing temperatures (800-1000K). In the case of the epitaxial growth of oxides, the process to incorporate oxygen into the structure is said to be the most difficult step. An enormously low temperature (600K) and low pressure (10^{-5}Pa) process to possess the epitaxial growth of the ultra thin film of cuprates has been realized using NO_2 as an oxidant [8,9]. In this paper, I will report on the structure of the epitaxial grown ultra thin film of Bi-Sr-Ca-Cu-O on various oxide substrates below 600K and at 10^{-5}Pa of NO_2.

EXPERIMENTAL

The MBE is carried out in an ultra high vacuum (UHV) chambers, with the base pressure of 10^{-8}Pa. Bi, Sr, Ca and Cu are supplied by metals from effusion cells. The rate of the metals deposited on the surface of the substrates is estimated by the quartz thickness monitor prior to the film formation. The nitric dioxide, NO_2, is supplied from a doser through a variable leak valve. Substrate, $SrTiO_3$(100), $LaAlO_3$(100) and MgO(100) (Earth Jewelry Co. Ltd.), are introduced into the UHV chamber followed by surface cleaning[10]. The amount of carbon contamination on the surface of the substrates is monitored by Auger electron spectroscopy (AES), and is completely removed by cleaning treatment. Every epitaxial growth described in this paper was carried out starting from the surface thus cleaned.

In order to determine the detailed structure of the ultra

thin film, x-ray diffraction (XRD) measurement was carried out using a four-circle diffractometer (RIGAKU AFC-5) using Cu $K\alpha$ radiation. The films were mounted so as to align the c axis of the film with the ϕ axis of the diffractometer.

RESULTS AND DISCUSSION

Low temperature epitaxial growth of $Bi_2Sr_2CuO_x$.

On the $SrTiO_3$(100) substrate, the clean surface of which is obtained as described in the experimental section, each elements are supplied from the metal source using standard effusion cells. Starting from the Sr, each elements is supplied with the amount which corresponds to one monolayer. First unit is started from Sr, then Cu and Sr (Sr-Cu-Sr) are deposited on the surface at room temperature. Then the oxidant NO_2 is introduced into the chamber through a variable leak valve. The temperature of the substrate is then increased gradually to 573K and kept at that temperature for 10-20 min in 10^{-5}Pa of NO_2, until a clear pattern of streaks is observed by the reflection high energy electron diffraction-(RHEED). After the formation of the first layer, second layer formation is started from the double monolayers of Bi, a monolayer of Sr, Cu and Sr (Bi-Bi-Sr-Cu-Sr) deposition at room temperature again. Following oxidation with NO_2 at 573K, a pattern of streaks due to the epitaxially grown phase appears. The procedure of this second layer formation is repeated 4 times. The X-ray diffraction pattern of the ultra thin film thus formed on various substrates are shown in Fig. 1. Besides the strong diffraction pattern of (100) and (200) of the $SrTiO_3$, the diffraction patterns due to the well known $Bi_2Sr_2Cu_1O_6$ are observed.[11] Only the diffractions indexed 00l (l=6,8,10,16) are observed indicating that the c axis of the film is perpendicular to the surface of the substrate. As the number of cells in the direction of c axis are limited, each diffraction peaks has broad profile. The thickness of the film can be estimated from the broadening of the (008) diffraction peak. Using the Scherrer's equation,[12] the thickness of the film is calculated to be 5.7nm, which is in good correspondence with the film of 5 x 1/2 unit cells in c axis direction, where the length for half unit cell is 1.2nm.

Structure of the epitaxially grown films.

The lattice constant for the c axis of the film on $SrTiO_3$(100), calculated from the (006) and the (008) diffraction peaks is 2.41nm. This value is shorter than the value known for the 2201 phase of the bulk material, which is 2.46nm[11]. The shortening of the c axis can be understood by the fact that due to the epitaxial growth from the surface of the substrate $SrTiO_3$(100), the lattice constant of the a and b axes are lengthened for 3%, as a result the c axis of the ultra thin film was shortened.

A similar phenomenon is observed for the epitaxial growth of Bi-Sr-Ca-Cu-O on the $LaAlO_3$(100). The a and b axes of the $LaAlO_3$(100) are 0.379nm, which is 0.4% shorter than the known lattice constants for the bulk Bi-Sr-Ca-Cu-O compounds. In the case of the film grown on $LaAlO_3$(100), the observed value for the c axis for $Bi_2Sr_2CuO_6$ is 2.48nm. The observed value of the c axis is lengthened for the substrate, $LaAlO_3$(100), with shorter a and b axes.

On the contrary, with large lattice mismatch of 10%, MgO(100), the structure of the grown film was identical to those for the bulk materials (Fig. 1).

The lattice constants of the ultra thin film formed on above three substrates are summarized in Fig. 2. The value of the c axis are plotted against the a and b axes of the substrates. The value for the bulk materials are also indicated in the figure. It is clearly shown that for the ultra thin film formed on the substrate with a small lattice mismatch, films are epitaxially grown from the surface and as a result, the lattice constant of the ultra thin film formed are affected by this value. However, on the substrate with too large mismatch, that is on MgO, the epitaxial growth from the surface of the substrate can no more be realized and the film begin to show bulk properties.

In order to determine the lattice constants for the ultra thin film formed on various substrates, more precise study is carried out using a four-circle diffractometer. The a and b axes of the ultra thin films formed on $SrTiO_3$(100) , $LaAlO_3$-

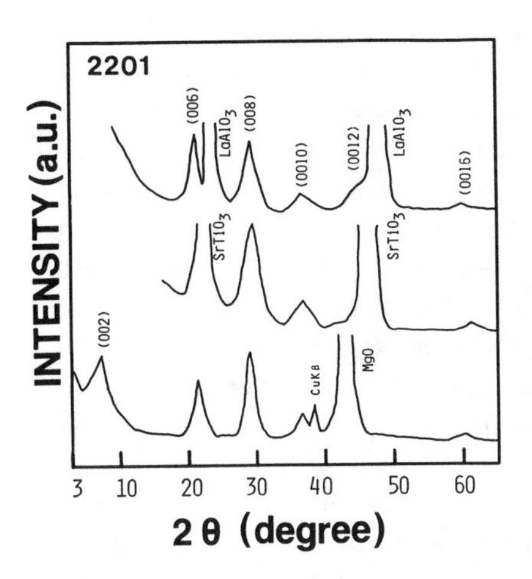

Fig. 1. X-ray diffraction patterns of the ultra thin films of $Bi_2Sr_2CuO_6$. Substrates are $SrTiO_3$(100), $LaAlO_3$-(100) and MgO(100).

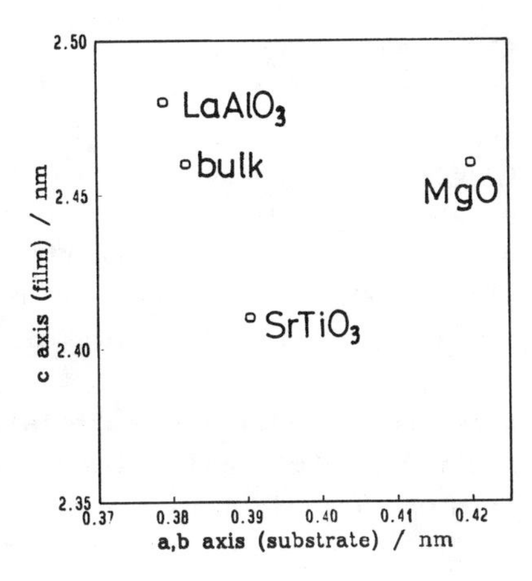

Fig. 2. The observed value of the c axis of the $Bi_2Sr_2CuO_6$ film formed and the lattice constant(a and b) for the substrates. The value for the bulk materials are also shown.

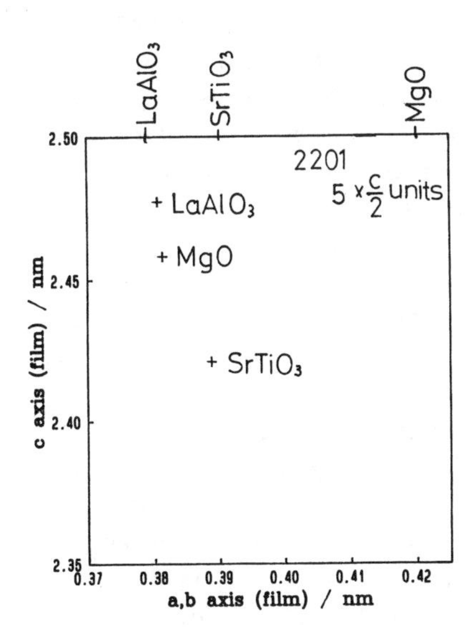

Fig. 3. The observed lattice constants of the ultra thin films on $Bi_2Sr_2CuO_x$ formed on $SrTiO_3$(100), $LaAlO_3$(100) and MgO(100). The lattice constant for c axis is plotted against that of a and b axes. The substrates are indicated in the figure. Thickness of the films are 1/2c x 5 units, which corresponds to ca. 1.2nm. The in-plane lattice constant for the substrates are indicated in the upper part of the figure.

(100) and MgO(100) are examined. Over 20 diffraction peaks observed, such as (002), (008), (115),(208), (028) and so on, are used in calculating the a, b, and c axes of the film. The observed a,b and c axes are summarized in Fig. 3. Lattice constants for the substrates are also sown in the upper part of the figure. For the film formed on $SrTiO_3$(100) and on $LaAlO_3$(100), the observed a, b axes are similar to those for the substrates. The mismatch of the lattice constants of these substrates are less than 3% from the lattice constants known for Bi 2201 bulk materials. As for the films formed on MgO(100), with large mismatch, the lattice constant of the film was identical to those for the bulk materials.

It is of great importance to know the epitaxial relationship between the film formed and the substrate. Changes in the intensity of the 115 diffraction was examined by rotating the film on the substrate in the ϕ axis, where the c axis of the sample is aligned with. The 2201 ultra thin film formed on $SrTiO_3$(100) is shown in Fig. 4. As is clearly shown in this figure, the 115 diffraction peak is observed in [11] direction of the $SrTiO_3$(100) surface, telling that the 2201 ultra thin film is formed normal to the surface of the substrate taking lattice match in this direction. No other orientation was observed on $SrTiO_3$(100) and on $LaAlO_3$(100).

On the other hand ϕ scan of the 115 peak on MgO(100) gave entirely different feature. The strong 115 peak was observed in the [10] direction of MgO(100) and side peaks were also observed in the direction of ± 16 to the [10] direction as well (Fig. 5). In order to have better idea of the structure of film on the MgO(100), the reflection contour map in h-k reciprocal plane at l=5 is shown in Fig. 6, where the hkl indices are defined using the observed lattice constants of the film on MgO (a=0.539nm, b=0.539nm and c=2.458nm). Four intense peaks are observed in the figure. The intense peak at (0.98, 1.00) is originated from the phase observed in [10] direction of the MgO(100). Two peaks at ± 16 are the side peaks observed in Fig. 5. These peaks has maximum at (0.66, 1.22) and (1.22, 0.68) in the (h, k) space, telling that these domains have ca. 1% different in-plane lattice constant to those observed in the [10] direction. As the l is fixed to l=5, the observed lattice constants only tells about the in-plane structure, however, domains at $\pm 16^{\circ}$ and 0° seems to

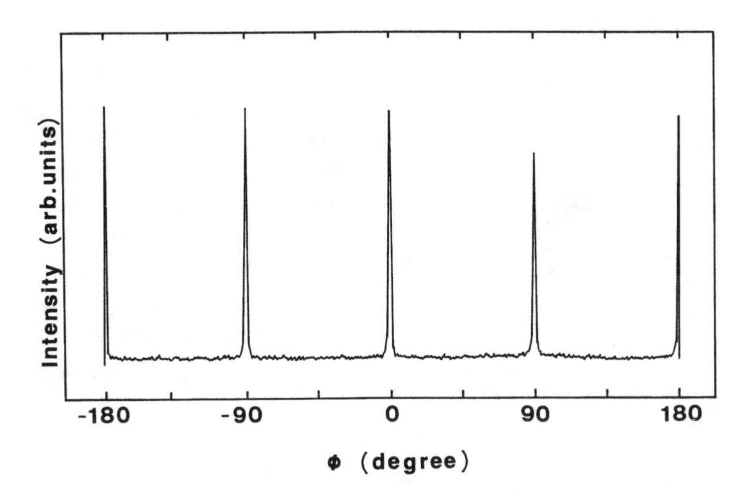

Fig. 4. ϕ scan of the 115 diffraction peak of $Bi_2Sr_2CuO_x$ ultra thin film on $SrTiO_3$(100), the thicknessof which is 1/2 c x 5 units.

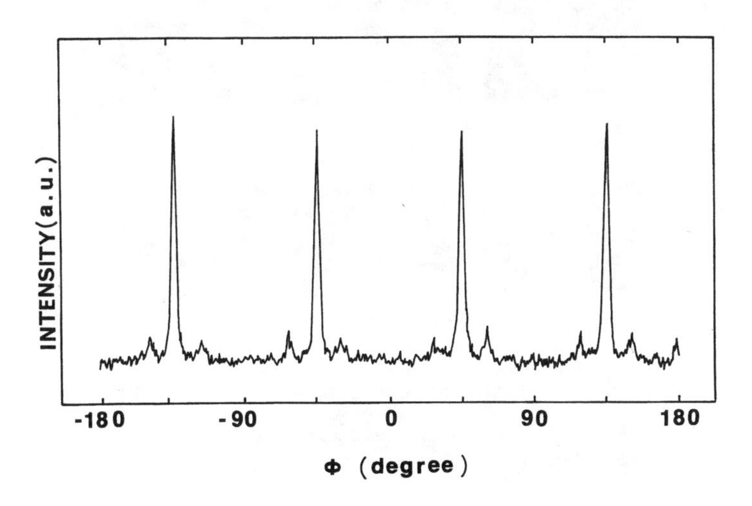

Fig. 5. ϕ scan of the 115 diffraction peak of $Bi_2Sr_2CuO_x$ ultra thin film on MgO100), the thickness of which is 1/2 c x 5 units.

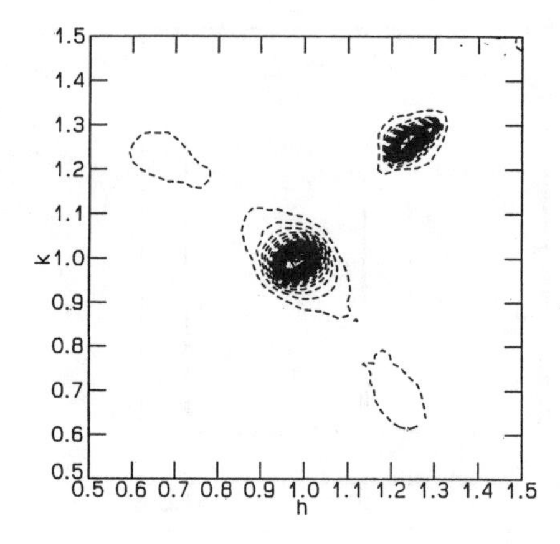

Fig. 6. A contour map of the intensity of reflection in h-k reciprocal plane for Bi 2201 ultra thin film on MgO(100), with 1/2 c x5 units thickness. The hkl indices are defined using the observed averaged lattice constants of the film on MgO(100) which are a=b=-0.539 nm and c=2.458 nm.

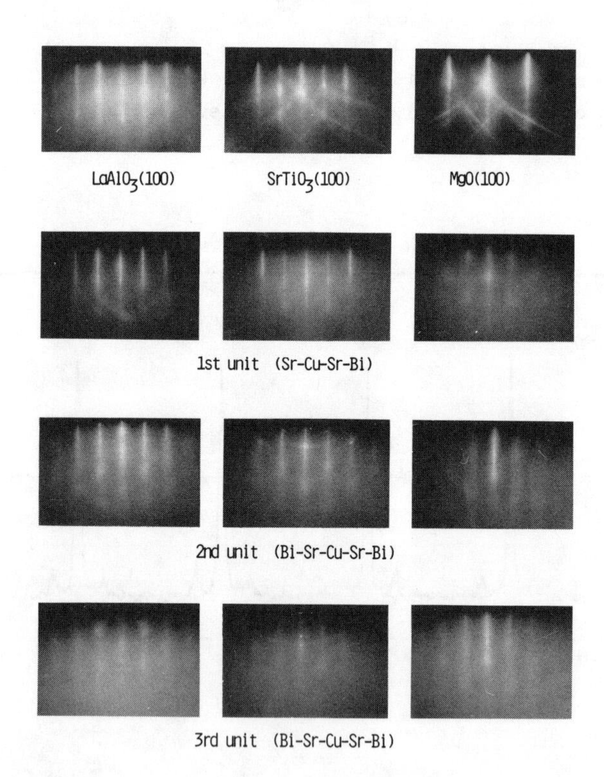

Fig. 7. RHEED pattern after the every layer formation. (a)on $SrTiO_3$(100), on (b) $LaAlO_3$(100) and on (c)MgO(100).

have different lattice constants.

For the Bi 2201 and 2212 phase formed on MgO(100) at higher temperature, Schlom et al[13] has observed similar multiple domain formation. In their observation the angle between these domains seems to be flexible. They have observed $\pm 11°$, $\pm 13°$, etc. Fujita et. al have reported Bi 2212 phase formed by ion beam sputtering on cleaved MgO(100), where the similar multiple domain formation in ± 13 direction is observed [14].

The reason why these domains are formed in certain directions to the surface of MgO is not clarified, yet. However, from our observation, it became clear that the structure of the films with different domains has different lattice constants. The MgO(100) surface is formed with Mg^{+2} and O^{-2} ions, Giving ordered change in the ionic potential. The periodical change in the ionic potential may play a role in determining the angle between the film domain and the substrate. More precise study has to be carried out in order to determine the exact reason for the domain formation.

Here, it is clearly shown that the film grown at low temperature as 573K has similar crystallinity to those formed at higher temperature of 1000K.

These structural feature of the film has been realized from the very 1st layer formation. The RHEED patterns after the formation of each layer in 2201 ultra thin film formation are shown in Fig. 7. Starting from the clean surface of the substrates, a pattern of streaks appears after the every layer formation. It is clearly shown that the spacing between the streaks for the grown films are identical to those for the substrate in the case of $SrTiO_3(100)$ and $LaAlO_{32}(100)$. This indicate that the 2201 layers formed have an identical lattice constant in a and b axes directions. Changes in the RHEED pattern during the film formation on MgO(100) is quite different. Here, the lattice constant of the film formed had different value from that of the substrate from the very 1 st layer. These fact suggests that on the substrate with large lattice mismatch, the epitaxial growth does not obey the atomic position of the surface of the substrate, and the film formed on such substrate has the property of the bulk material. However, due to the interaction between the substrate and the film at the interface results in a growth in a certain direction.

CONCLUSION

Ultra thin film of Bi-Sr-Ca-Cu-O are formed on various oxide substrates by a low temperature and low pressure MBE. On the substrate with a small lattice mismatch, the film formed has an identical in-plane lattice constant to those of the substrate. Ultra thin films formed on such substrates has proved to possess single domain. However, on the substrate with large mismatch, the film with bulk structural property was formed. And the multiple domain with different lattice constants are observed. The reason for the certain domain formation has to be further examined.

REFERENCES

1. T.Kawai, Y.Egami, H.Tabata and S.Kawai, Nature 349, 200(1991).
2. H.Adachi, S.Kohiki, K.Setsune, T.Mitsuyu and K.Wasa, Jpn.J.Appl.Phys. 27, L1883(1988).
3. K.Nakamura, J.Sato, M.Kaise and K.Ogawa, Jpn.J.Appl.Phys. 29, L437 (1990).
4. J.Fujita, T.Tatsumi, T.Yoshitake and H.Igarashi, Appl. Phys.Lett. 54, 2364 (1989).
5. M.Kawai, S.Watanabe andT.Hanada, J.Crys.Growth (1991) in press.
6. T.Terashima, Y.Bando, K.Iijima, K.Yamamoto, K.Hirata, K.Hayashi, K.Kamigaki and H.Terauchi, Phys.Rev.Lett. 65, 2684 (1990).
7. M.Kanai, T.Kawai and S.Kawai, Appl.Phys.Lett. 58, (1991).
8. S.Watanabe, M.Kawai and T.Hanada, Jpn.J.Appl.Phys. 29, L1111(1990).
9. M.Kawai, S. Watanabe and T.Hanada, J. Vac. Sci. & Technol.A 8 , 4140(1990).

10. S. Watanabe, M. Kawai and T. Hanada, J. Vac. Sci. & Technol. (1991) in press.
11. J. M. Tarascon, Y. Le Page, P. Barboux, B. G. Bagley, L. H. Greene, M. R. Mckinnon, G. W. Hull, M. Giroud and D. M. Hwang, Phys. Rev. B 37 , 9382 (1988).
12. B. D. Cullity: Elements of X-ray Diffraction, (Addison-Wesley, Massachusetts, 1977)
13. D.G.Schlom, Thesis, Stanford University (1990)
14. J.Fujita, T.Yoshitake, T.Satoh and H.Igarashi, Appl. Phys. Lett.58, 1092(1991)

PLASMA ENHANCED OMCVD ATOMIC LAYERED GROWTH OF Y-BA-CU OXIDE FILMS

S. J. DURAY, D. B. BUCHHOLZ, S.N. SONG, D. S. RICHESON, J. B. KETTERSON, T. J. MARKS, AND R. P. H. CHANG
NSF Science and Technology Center for Superconductivity and Materials Research Center, Northwestern University, Evanston, Illinois, 60208

ABSTRACT

We report the results of a pulsed organo-metallic beam epitaxy (POMBE) process for growing complex oxide films at low background gas pressure (10^{-4} - 10^{-2} torr) and low substrate temperature (600 to 700 C) using organo-metallic precursors in an oxygen plasma environment. Our results show that POMBE can extend the capability of organo-metallic chemical vapor deposition to growing complex oxide films with high precision both in composition and structure without the need for post-deposition oxidation and heat treatments. The growth of phase-pure, highly oriented Y-Ba-Cu-O superconducting oxide films {[T_c (R=0)=90.5K] and J_c (77K, 50K gauss)=1.1×10^5 A/cm^2} is given as an example. Similar to the pulsed laser deposition process, the POMBE method has the potential for in-situ processing of multilayer structures (e.g. junctions).

INTRODUCTION

Organo-metallic chemical vapor deposition (OMCVD) is an attractive crystal growth method for compound semiconductors. Recently this method has been used for oxide film growth, with high T_c superconductors being a particularly successful example [1]. In fact, it has been shown that Y-Ba-Cu-O superconducting oxide films grown by OMCVD or plasma enhanced OMCVD have qualities comparable or superior to those achieved by other deposition methods [2]. Some investigators also believe that OMCVD could be the preferred method for large scale processing of superconducting oxide films. In this paper we describe how complex oxide films can be formed at low oxygen background pressures (10^{-4}-10^{-2} torr) using a new pulsed organo-metallic beam epitaxy (POMBE) technique. This method controls the oxide film growth with atomic level accuracy. We present the results of high-quality Y-Ba-Cu-O superconducting oxide film growth as an example.

Recent studies on the growth of Y-Ba-Cu-O films indicate that the superconducting properties of these films are a strong function of the oxide composition and structure. Carlson et al [3] have shown that stoichiometric deviations greater than $\pm$ 1% in the epitaxial growth of Y-Ba-Cu-O films on $LaAlO_3$ can cause: 1. depression of the critical temperature, T_c ; 2. broadening of the superconducting transition width; 3. decrease in the critical current density, J_c ; 4. deterioration of the surface morphology; and 5. decrease in the crystallinity of the films. In addition, it has been shown that atomic oxygen species and oxygen plasmas present during growth play a significant role in determining the ultimate electrical properties of the Y-Ba-Cu-O film [4-6]. These recent results point to the need for precise control of stoichiometry during oxide growth.

For device applications it is highly desirable to have the capabilities for growing heteroepitaxial junctions, such as superconducting-insulating-superconducting layered structures. For high temperature superconductors with short coherence lengths, high quality insulating films of a few tens of angstroms are needed. This implies the need to grow very sharp junctions with minimum defects. As in the case of heterostructures for semiconductors, the availability of a low temperature process will significantly reduce unwanted interdiffusion between the layers.

EXPERIMENTAL SETUP

The pulsed organo-metallic beam epitaxy (POMBE) technique is designed to meet the need for a low temperature "digital" process for complex oxide film growth with atomic precision. The present system can accommodate both solid and/or liquid organo-metallic precursors. Helium is used throughout as the carrier gas for the precursors. Each precursor is pulsed onto the substrate via a pneumatic valve. All the precursors are injected through a single nozzle facing the substrate surface. A computer is used to control all the valves, which can be turned on and off in sequence or in parallel (and any other combination the software dictates). Atomic oxygen is generated via a microwave excited plasma discharge which can also be pulsed on and off at will. The flux impinging on the substrate surface is calibrated via a quartz crystal monitor.

Figure 1 shows a simplified schematic of the POMBE growth system. The sample and the substrate holder face the injection nozzle while the plasma is produced by a microwave generator. A turbomolecular pump is used to evacuate the system. With such a system, one can grow oxide films one atomic layer at a time. For instance, in the case of Y-Ba-Cu-O, we have used the following growth process, which is the layered structure of the oxide: Y pulse--oxidation period--2x[Cu pulse--oxidation period--Ba pulse--oxidation period]--Cu pulse--oxidation period--repeat. Each oxidation period can be adjusted differently (to allow sufficient oxidation time), depending on the precursor. By using this pulsed procedure, films can be epitaxially grown without post-oxidation annealing and at lower substrate temperatures. Films of Y-Ba-Cu-O are deposited from the following volatile organo-metallic sources, $Y(dpm)_2$, $Ba(hfa)_2$(tetraglyme), and $Cu(acac)_2$ [7]. The barium source is a second-generation precursor that exhibits significantly improved thermal stability and volatility over previously used compounds. These characteristics have provided better control of film stoichiometry [7]. To increase the production of atomic oxygen species and increase the cracking (hydrolysis) efficiency of barium fluoride, we introduced 2% water vapor into the oxygen plasma discharge. The discharge also consists of about 50% helium derived from the precursor carrier gas.

Figure 1
A simplified schematic diagram of the pulsed deposition system.

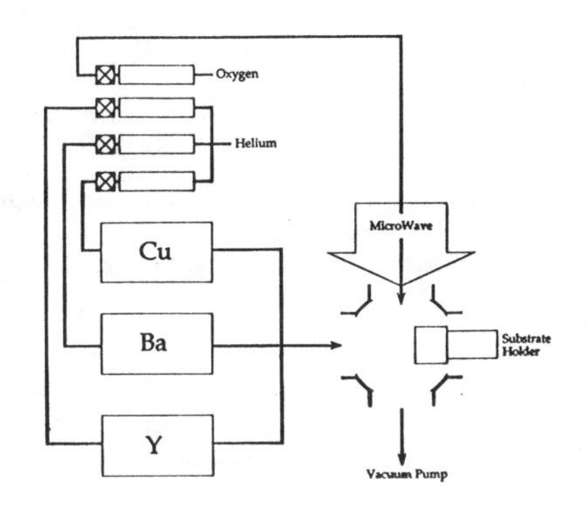

Our initial operating parameters for growing Y-Ba-Cu-O are as follows: precursor pulse width, 15 sec.; oxidation period, 15 sec.; substrate temperature, 600 C - 700 C; oxygen partial pressure, 10^{-3} torr. At this low pressure the mean free path is longer than the nozzle orifice - substrate surface separation. Thus most if not all of the oxidation of the precursor takes place on the substrate surface. This avoids any gas phase reactions which might deposit unwanted oxide phases in the film. It should also reduce defect density from gas phase particle inclusion. Under these deposition conditions, the typical growth rate is in the range of 10 - 20 Å per minute. Unlike previous plasma enhanced metal-organic chemical vapor deposition methods, the POMBE technique does not require post-deposition soaking in an oxygen ambient.

RESULTS

Figure 2 shows a typical θ-2θ X-ray diffraction scan of a POMBE grown Y-Ba-Cu-O film (about 700Å thick) on a $LaAlO_3$ substrate at 600 C. This figure shows that the film has highly C-axis oriented growth, in agreement with previous results[8]. The rocking curve (figure 2 inset) data of the Y-Ba-Cu-O (005) peak yields a full width at half maximum of 0.58 degree. Within the sensitivity of our instrument, Auger measurements of these oxide films indicate that there is no measurable carbon contamination. Scanning electron microscopy shows that these films have a very fine surface texture and are very uniform. Figure 3 shows four-probe variable temperature electrical resistance data for typical films grown at different temperatures (600 C and 680 C) and having thicknesses at 600 Å and 1,500 Å respectively. The as-deposited (680C) film (curve 1) shows metallic behavior in the normal state and a sharp superconducting transition with a T_c (R=0) of 90.5K and a transition width of about 1.5 K; the as-deposited (600 C) film (curve 2) has a T_c (R=0) of 83K. The critical current, J_c, was measured on a strip (112 μm wide, 2 mm long, and 1500Å thick) etched by a laser beam. The current density measured at 77K and in a magnetic field of 50 K gauss is $(1.0 \pm 0.1) \times 10^5$ A/cm^2. These results are comparable to if not better than Y-Ba-Cu-O films produced by laser ablation with a similar substrate and substrate temperature but much higher (by as much as two orders of magnitude) ambient oxygen gas pressure [9,10]. These favorable synthesis conditions should allow us to grow multilayer heterostructures with the POMBE technique.

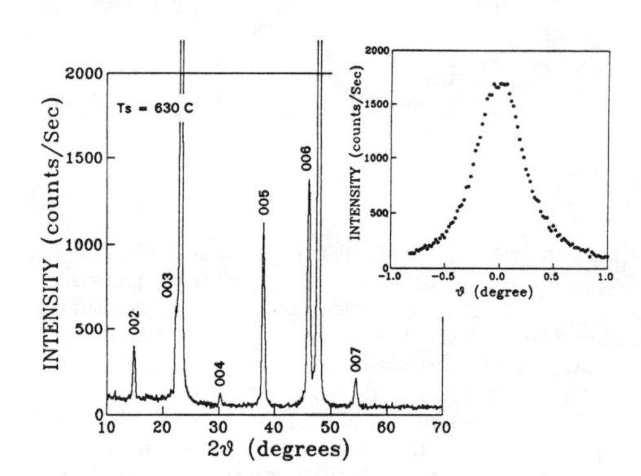

Figure 2. θ-2θ X-ray diffraction data for a 700 Å thick Y-Ba-Cu-O film grown on $LaAlO_3$ at 600 C by POMBE.; inset shows the rocking curve of the film.

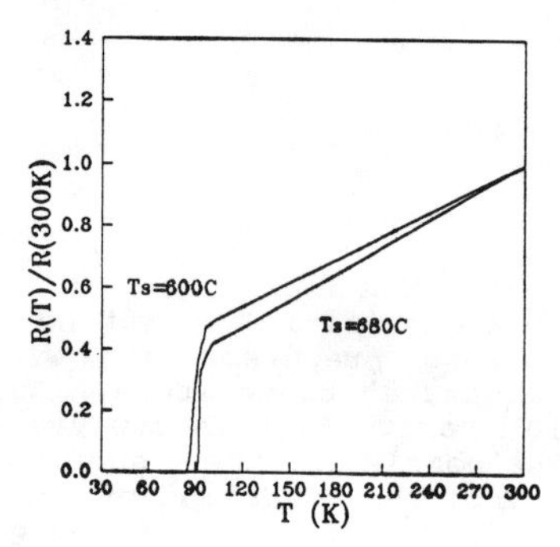

Figure 3. Resistance vs. temperature curve of an as-deposited Y-Ba-Cu-O film grown on $LaAlO_3$ at 680 C (curve 1) and at 600 C (curve 2) by POMBE.

ACKNOWLEDGEMENT

This research has been supported by NSF Science and Technology Center for Superconductivity under grant number DMR 88-09854 and Northwestern University Materials Research Center under grant number DMR 88-21571.

References

1. J. Zhao. H. K. Dahman, H. O. March, L. M. Tonge, T. J. Marks, B. W. Wessels, and C. R. Kannewurf, Appl. Phys. Lett. **53**, 1750 (1988).

2. Y. Q. Li, J. Zhao, C. S. Chern, W. Huang, G. A. Kulesha, P. Lu, B. Gallois, P. Norris, B. Kear, and F. Cosandey, Appl. Phys. Lett. **58** 648 (1991).

3. D. J. Carlson, M. P. Siegal, J. M. Phillips, T.H. Tiefel, and J. H. Marshall, J. Mater. Res., Vol 5, 2797 (1990).

4. N. G. Chew, S. W. Goodyear, J. A. Edwards, J. S. Satchell, S. E. Blenkinsop, and R. G. Humphreys, Appl. Phys. Lett. **57** 2016 (1990).

5. T. Yoshitake, S. Miura, J. Fujita, N. Shohata, H. Igarashi, T. Satoh, A. Sekiguichi, and K. Katoh, Appl. Phys. Lett. **56** 575 (1990).

6. K. Yamamoto, B. M. Lairson, C. B. Eom, R. H. Hammond, J. C. Bravman, and T. H. Geballe, Appl. Phys. Lett. **57** 1936 (1990).

7. G. Malandrino, D.S. Richeson, T.J. Marks, D.C. DeGroot, J.L. Schindler, C.R. Kannewurf, Appl. Phys. Lett. **58** 182 (1991).

8. C. S. Chern, J. Zhao, Y. Q. Li, P. Norris, B. Kear, B. Gallois and Z. Kalman, Appl. Phys. Lett. **58** 185 (1990).

9. T. Venkatesan, A. Inam, B. Dutta, R. Ramesh, M. S. Hegde, X.D. Wu, L. Nazar, C.C. Chang, J. B. Barner, D. M. Hwang, and C. T. Rogers, Appl. Phys. Lett. **56** 391 (1990).

10. A. Inam, X.D. Wu, L. Nazar, M.S. Hegde, C.T. Rogers, T. Venkatesan, R. W. Simon, K. Daly, H. Padamsee, J. Kirchgessner, D. Moffat, D. Rubin, Q.S. Shu, D. Kalokitis, A. Fathy, V. Pendrick, R. Bjrown, B. Brycki, E. Belohoubek, L. Drabeck, G. Gruner, R. Hammond, F. Gamble, B. M. Lairson, and J.C. Bravman, Appl. Phys. Lett. **56** 1178 (1990).

ATOMIC LAYER AND UNIT-CELL LAYER GROWTH OF $(Ca,Sr)CuO_2$ AND $Bi_2Sr_2Ca_{n-1}Cu_nO_{2n+4}$ THIN FILMS BY LASER MOLECULAR BEAM EPITAXY

MASAKI KANAI, TOMOJI KAWAI, TAKUYA MATSUMOTO AND SHICHIO KAWAI
The Institute of Scientific and Industrial Research, Osaka University, 8-1, Mihogaoka, Ibaraki, Osaka 567, Japan

ABSTRACT

Thin films of $(Ca,Sr)CuO_2$ and $Bi_2Sr_2Ca_{n-1}Cu_nO_{2n+4}$ are formed by laser molecular beam epitaxy with in-situ reflection high energy electron diffraction observation. The diffraction pattern shows that these materials are formed with layer-by-layer growth. The change of the diffraction intensity as well as the analysis of the total diffraction pattern makes it possible to control the grown of the atomic layer or the unit-cell layer.

INTRODUCTION

The crystal structures of high Tc superconductors, Bi_2Sr_2-$Ca_{n-1}Cu_n$-O_{2n+4} can be constructed by the insertion of Bi_2O_2/Sr_2O_2 layers into the structure of the parent material of $(Ca,Sr)CuO_2$, which is the infinite stacking of CuO_2 planes[1]. (see Fig.1.) Based on this concept, we have formed $Bi_2Sr_2Ca_{n-1}Cu_nO_{2n+4}$ thin films (n=1 to 4) by the successive stacking of atomic or sub-unit cell layer taking advantage of the two dimensional layer structure. In-situ reflection high energy electron diffraction (RHEED) observation shows that both the parent material and $Bi_2Sr_2Ca_{n-1}Cu_nO_{2n+4}$ can be formed with layer-by-layer growth. The process of atomic or sub-unit cell layer growth is controlled using the change of the RHEED intensity. This method makes it possible to control the number of CuO_2 planes (n) in $Bi_2Sr_2Ca_{n-1}Cu_nO_{2n+4}$ thin films.

EXPERIMENTAL

The films were formed on $SrTiO_3(100)$ substrates by computer controlled laser ablation method under molecular beam epitaxial condition[2]. The light source is an ArF excimer laser. The thin films of $(Ca,Sr)CuO_2$ were formed by the ablation of $(Ca_{0.86}Sr_{0.14})CuO_2$ target

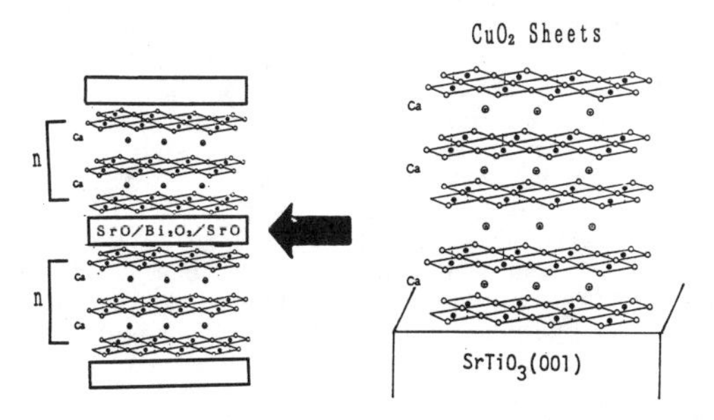

Fig.1 The schematic diagram of the $Bi_2Sr_2Ca_{n-1}Cu_nO_{2n+4}$ formation by the insertion of Bi_2O_2/Sr_2O_2 layer into the parent structure.

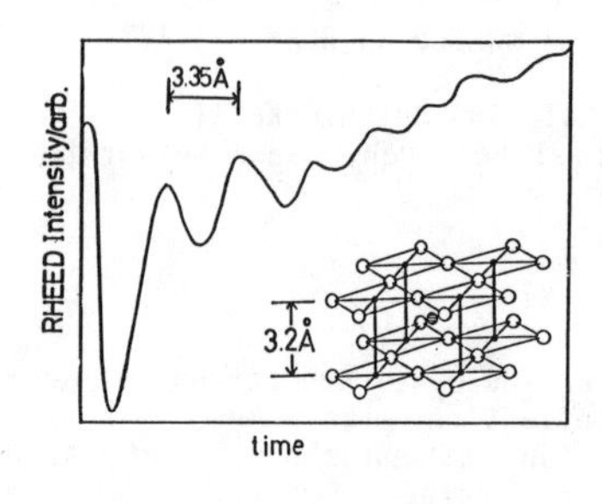

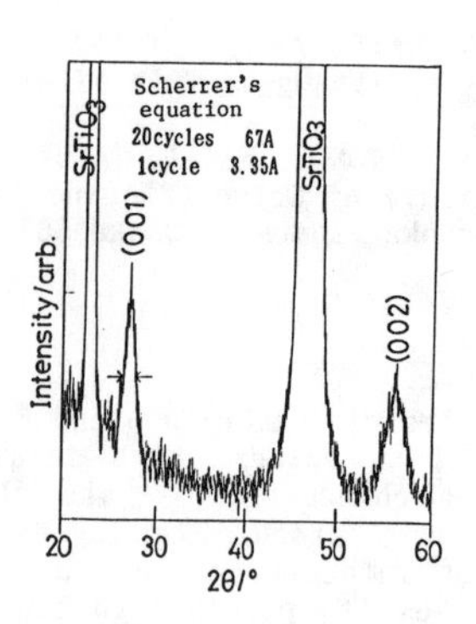

Fig.2 Variation of RHEED intensity with time during the film formation by the ablation of $Ca_{0.86}Sr_{0.14}CuO_2$ target in NO_2 of $1x10^{-5}$mbar and XRD pattern of the film.

or successive ablation of Ca, Sr and Cu metal targets. The films of $Bi_2Sr_2Ca_{n-1}Cu_nO_{2n+4}$ were formed by the successive ablation of Bi_2O_3, $SrCuO_x$, $(Ca_{0.86}Sr_{0.14})CuO_2$ and Sr metal targets. During the depositions, the substrate was heated at 650°C and $1x10^{-5}$mbar of NO_2 was dosed into the formation chamber. The growth mechanism was investigated with in-situ RHEED observation and Auger electron spectroscopy.

RESULTS AND DISCUSSION

Streak pattern of RHEED during the growth and X-ray diffraction (XRD) patterns of the (00ℓ) orientation show that both $(Ca,Sr)CuO_2$ and $Bi_2Sr_2Ca_{n-1}Cu_nO_{2n+4}$ are grown epitaxially. Fig.2 shows the variation of RHEED intensity during the formation of $(Ca,Sr)CuO_2$ films by the ablation of $(Ca_{0.86}Sr_{0.14})CuO_2$ target and XRD pattern of formed films. The oscillation of RHEED intensity shows the two-dimensional layer growth of the material, similar to the growth observed in semiconducting material. The thickness with one cycle of the intensity oscillation is 3.35A, which is calculated from the width of the (001) peak using Scherrer's equation. The thickness corresponds to the lattice constant c of this material. This result indicates that $(Ca,Sr)CuO_2$ can be formed with two dimensional unit cell layer growth, which corresponds to the sub-unit cell layer growth in $Bi_2Sr_2Ca_{n-1}Cu_n$-O_{2n+4}. When the elements are supplied successively according to the layered crystal structure, different growth mechanism is observed. Fig.3 shows the data for the successive ablation of Ca (Sr)and Cu metal targets. The analysis of the data shows that two cycles of oscillation (1 cycle for deposition of alkaline earth metal and 1 cycle for Cu deposition) corresponds to the growth of one unit cell layer. Namely, each oscillation corresponds to the growth of one atomic layer of Ca (Sr) and CuO_2, respectively.[2]

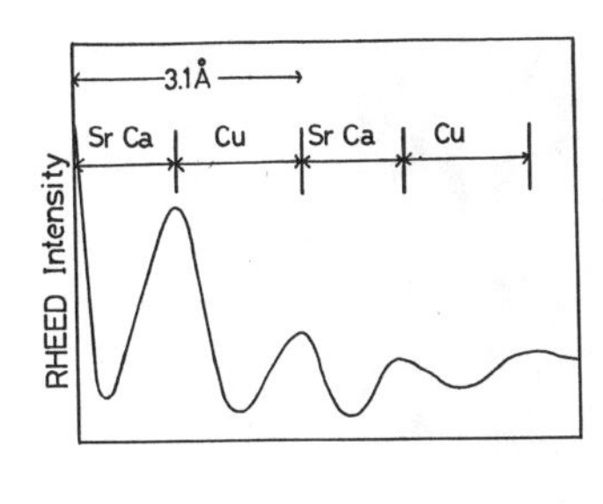

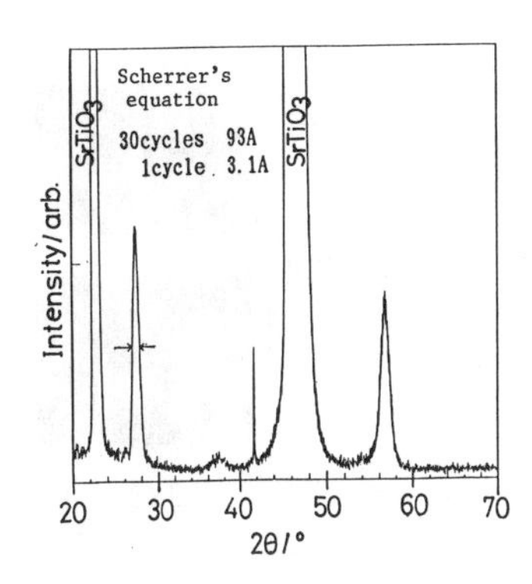

Fig.3 Variation of RHEED intensity with time during the $(Ca,Sr)CuO_2$ formation by the successive ablation of Ca (Sr) and Cu targets in NO_2 of $1x10^{-5}$mbar and XRD patron of the film.

Based on the formation of the $(Ca,Sr)CuO_2$ parent material, $Bi_2Sr_2Ca_{n-1}Cu_nO_{2n+4}$ is formed by the insertion of Bi_2O_2/Sr_2O_2 layers. First, $Bi_2Sr_2CuO_6$ layer was formed as a buffer layer on $SrTiO_3$ substrate by the successive ablation of Sr, $SrCuO_x$ and Bi_2O_3 targets. Fig.4 shows the RHEED pattern of the surface of the strontium deposited substrate. The change of the RHEED pattern and intensity during the first $Bi_2Sr_2CuO_6$ layer deposition is shown in Fig.5. In each step, the oscillation or the unique change of the RHEED intensity is observed and the deposition process of one layer can be controlled using this change of the diffraction intensity. The change of the RHEED pattern shows that the first deposition of Sr (that is SrO) monolayer makes cleaner surface than bare substrate surface. (see Fig.4) The deposition of $SrCuO_x$ on the clearer surface leads to the 2x2

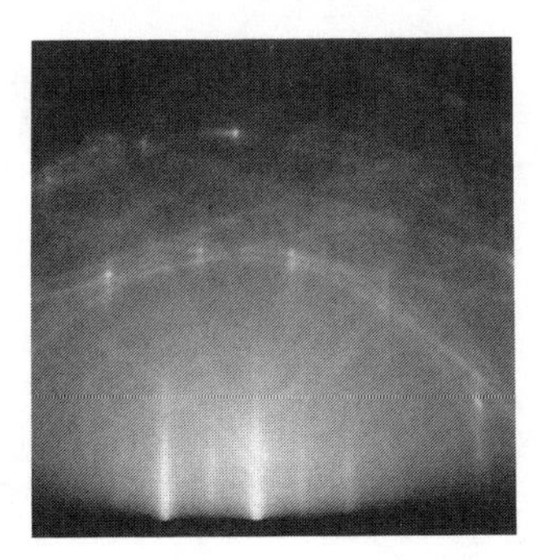

Fig.4 RHEED pattern of the strontium deposited $SrTiO_3(100)$ substrate. Even the diffraction spots of higher order are observed.

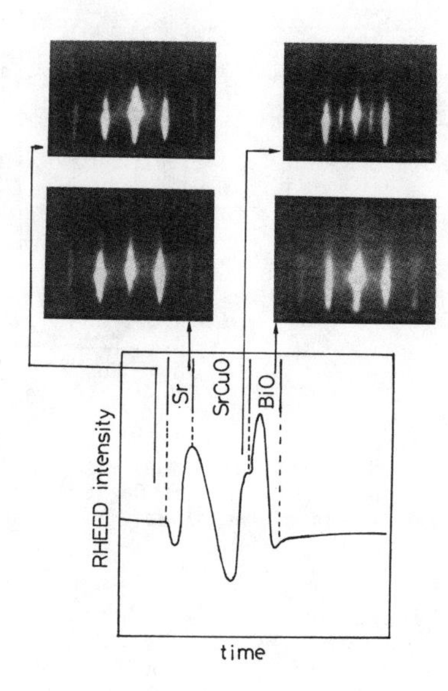

Fig.5 The variation of the RHEED pattern and intensity during the successive formation of the first $Bi_2Sr_2CuO_6$ buffer layer.

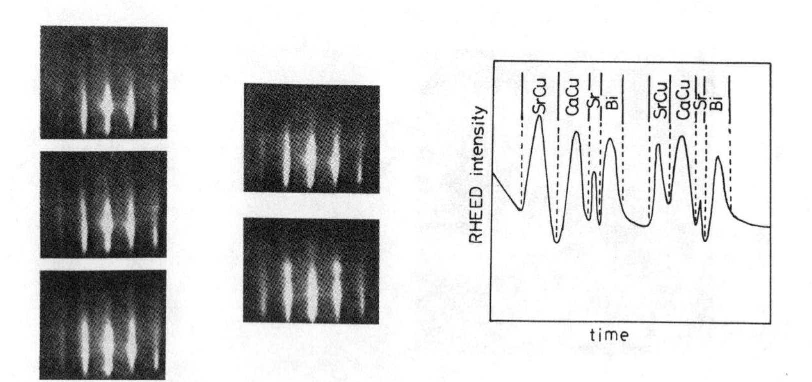

Fig.6 The variation of the RHEED pattern and intensity during the successive formation of the $Bi_2Sr_2CaCu_2O_8$ structure.

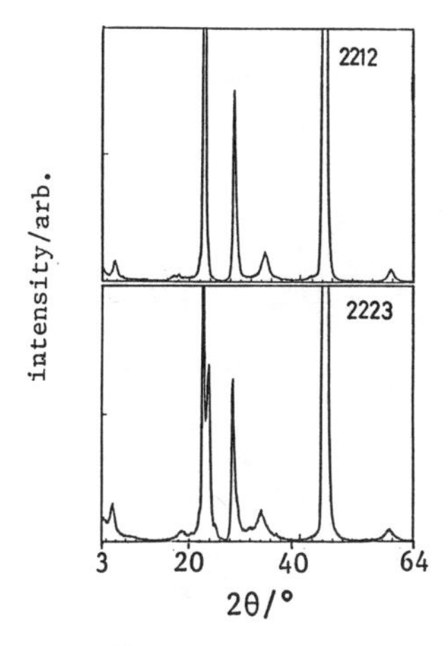

Fig.7 XRD patterns of $Bi_2Sr_2CaCu_2O_8$ and $Bi_2Sr_2Ca_2Cu_3O_{10}$ thin films.

superstructure at the surface. We consider that this superstructure may be oxygen deficient surface structure to satisfy the charge neutrality. This superstructure disappears with Bi deposition. On the $Bi_2Sr_2CuO_6$ buffer layer, $Bi_2Sr_2Ca_{n-1}Cu_nO_{2n+4}$ film was formed by the successive ablation of $SrCuO_x$, $(Ca,Sr)CuO_2$, Sr and Bi_2O_3 targets. Fig.6 shows the change of the RHEED pattern and intensity during the $Bi_2Sr_2CaCu_2O_8$ film formation. Similar to the growth of $Bi_2Sr_2CuO_6$ buffer layer, the change of RHEED intensity is observed in each step keeping the streak pattern and the formation of one layer can be controlled using this change. The data shows that the all steps in the $Bi_2Sr_2Ca_{n-1}Cu_nO_{2n+4}$ film formation proceed with layer-by-layer growth mechanism in contrast to the previous report[3], in which the particles or or amorphous state is formed during the deposition of sub-unit cell layer. The XRD patterns of $Bi_2Sr_2Ca_1Cu_2O_8$ and $Bi_2Sr_2Ca_2Cu_3O_{10}$ films formed by this method are shown in Fig.7. These patterns indicate that the as-grown crystallization of these materials is possible and the structure, namely the number of CuO_2 plane, can be controlled by this method.

In summary, The parent material of cuprate superconductor, $(Ca,Sr)CuO_2$, and $Bi_2Sr_2Ca_{n-1}Cu_nO_{2n+4}$ thin films have been formed by laser molecular beam epitaxy method with in-situ RHEED observation. The variations of RHEED pattern and intensity show that both materials can be formed with layer-by-layer growth in all steps of the formation. Furthermore, the oscillation of the RHEED intensity is observed and the deposition of one atomic layer can be controlled using the intensity variation. By this method, the number of CuO_2 plane in $Bi_2Sr_2Ca_{n-1}Cu_nO_{2n+4}$ thin films have been successfully controlled.

REFERENCES

1)T.Siegrist, S.M.Zahurak, D.W.Murphy and R.S.Roth, Nature **334**, 231 (1988).

2)M.Kanai, T.Kawai and S.Kawai, Appl.Phys.Lett.**58**, 771 (1991).

3)D.G.Shlom, J.N.Eckstein, I.Bozovic, Z.J.Chen, A.F.Marshall, K.E. von Dessonneck and J.S.Harris, SPIE proceeding, **1285**, 234 (1990)

GROWTH OF SrS THIN FILMS BY ATOMIC LAYER EPITAXY

M. LESKELÄ*, L. NIINISTÖ,** E. NYKÄNEN,** P. SOININEN** AND M. TIITTA**
*Department of Chemistry, University of Helsinki, SF-00100 Helsinki, Finland
**Laboratory of Inorganic and Analytical Chemistry, Helsinki University of Technology, SF-02150 Espoo, Finland

ABSTRACT

The growth of strontium sulfide thin films in a flow-type Atomic Layer Epitaxy reactor from $Sr(thd)_2$ (thd = 2,2,6,6-tetramethyl-3,5-heptanedione) and H_2S has been studied. The growth is independent on flow rate and duration of the purge gas (N_2) pulse and it does not depend on the $Sr(thd)_2$ and H_2S pulses either provided their amounts are sufficient to saturate the surface. The variables significantly affecting the growth rate are the substrate temperature and source temperature for $Sr(thd)_2$. The observed lower than one monolayer growth rate is mainly due to the large size of the $Sr(thd)_2$ molecule.

INTRODUCTION

Currently, only the yellow-emitting ZnS:Mn AC operated thin film electroluminescent (ACTFEL) devices are commercially available. Full color ACTFEL devices are needed, however, to display more information. Rare earth doped alkaline earth sulfides are extensively studied as potential candidates for full color ACTFEL phosphors [1].

Strontium sulfide is a well-known phosphor matrix material which can be doped with cerium to give bluish-green EL [2]. The EL properties can be improved by codoping with F^- and K^+ ions [3,4]. Recently, very interesting results have been reported for SrS doped with CeF_3 and $CuBr_2$ [5]. Although SrS:Ce is probably the most promising blue EL phosphor available as thin film its brightness values and efficiency are not high enough for practical applications. On the other hand, strontium sulfide can also act as a matrix for white EL phosphors. SrS doped with Pr [6], Nd or Ho [7] and double doped with Eu and Ce [8], Pr and Ce [9], Sm and Ce as well as with Nd and Ho [10] have been reported to yield white emission.

SrS thin films have usually been deposited by the electron beam (EB) evaporation method. This method, however, requires that the substrate is heated to at least 450 °C and that sulfur is coevaporated in order to obtain stoichiometric SrS. The cracking of sulfur molecules to smaller species has been reported to lower the substrate temperature and sulfur consumption and to improve luminance [11]. Sputtering methods, both diode rf [12,13] and magnetron [14], have also been developed to lower the processing temperature. Succesful deposition of SrS films has also been made by a multi source (MSD) method from Sr and S [15].

Crystallinity of SrS affects the luminance. According to Shoki and Yamaguchi [14] poor crystallinity and low luminance are caused by shortage of sulfur or by oxygen inclusions. The crystallinity can be improved by annealing in an inert atmosphere and consequently the EL characteristics of SrS:Ce will improve [16]. The orientation of the SrS films depends on the experimental conditions used but the effect of orientation on the luminance is unclear. In EB deposited films Tanaka et al. obtained at first non-oriented [17] and later (200) oriented films [16]. In sputter deposited films higher temperature favoured the (220) orientation and lower temperature the (200) one [14]. In MSD deposited films the orientation changed from (111) to (200) and (220) along with the increase of the S/Sr ratio [15].

SrS is rather sensitive for oxygen and moisture. Generally it is believed that oxygen is harmful for the luminance [18]. Recently, Onisawa et al. [19] have shown that a slight oxidation of SrS to $SrSO_3$ may improve the EL characteristics, however.

The present paper is a part of our investigation into ALE deposited thin films for ACTFEL devices. The ALE process offers an attractive way to process SrS:Ce thin films of high luminance without codoping. The EL properties of the devices fabricated will be discussed elsewhere [20].

EXPERIMENTAL

The SrS films were deposited by the ALE method in a flow-type reactor. The source materials were $Sr(thd)_2$ (thd = 2,2,6,6-tetramethyl-3,5-heptanedione) and H_2S [21]. The thd-chelates are among the few alkaline earth compounds exhibiting sufficient vapor pressure and stability under the thin film growth conditions [22,23]. The crucibles for the solid source materials were either glass or titanium. The reactant pulses were separated by N_2 gas purging. The substrates used were aluminium oxide coated soda lime glass (5 x 5 cm^2). The film growth was studied by varying the temperatures of the substrate (260-510 $^{\circ}C$) and source furnace (190-250 $^{\circ}C$) as well as the durations of the reactant and purge gas pulses (0.2-1 s). Typical pulse durations were 0.6, 0.4 and 0.4 s for $Sr(thd)_2$, H_2S and N_2, respectively and the H_2S flow rate was 15 sccm. The total pressure, measured after the reaction chamber, was about 2.9 mbar. Typically 3000 or 6000 cycles were grown; the average thickness was around 200 nm.

The crystallinity and preferential orientation in the SrS films were studied by X-ray diffraction. Thicknesses of the films were measured by profilometer and X-ray fluorescence [24]. XRF, however, gave a mean thickness without taking into account the presence of some thickness profile which can be estimated from profilometer measurements to be within ± 15 %.

RESULTS AND DISCUSSION

First the sublimation of $Sr(thd)_2$ was studied by varying the source temperature and keeping the reactor temperature at 450 $^{\circ}C$ (Fig. 1). The growth rate of SrS films was increasing with the source temperature up to 210 $^{\circ}C$ after which saturation to a rate of 0.4 Å/cycle occurred. With the higher source temperatures (> 230 $^{\circ}C$) the reproducibility in growth rates was poor. This

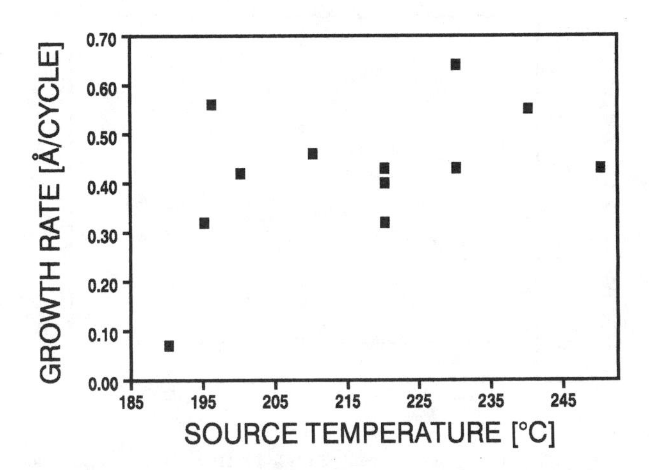

Fig. 1. The growth rate of SrS thin film *vs*. $Sr(thd)_2$ source temperature. The pulse duration is 0.6 s and the substrate temperature 450 °C.

indicates that $Sr(thd)_2$ begins to decompose in these temperatures which can also be seen as darkening of the source tube. The sublimation behaviour of $Sr(thd)_2$ is more complicated than that of $Ca(thd)_2$ and it may also depend on the preparation method and the amount of free Hthd ligand left in the product. The corresponding Ba chelate appears to behave in a even more complicated way because the solid phase and the sublimating molecule is an oligomer with high molecular weight [25].

The growth rate of SrS depends on the reactor temperature in an unexpected way. The rate increases up to 0.6 Å/cycle along with the increase of reactor temperature from 260 to 400 °C (Fig. 2). Above this temperature a competing mechanism with a lower growth rate appears with a value of 0.3 Å/cycle at 420-450 °C. Between 450 and 510 °C both growth rates again increases with the temperature indicating possibly a CVD-type decomposition growth. For the overall reaction $Sr(thd)_2$ + H_2S -> SrS + 2Hthd it is difficult to find an "ALE-window" where a self-saturated growth would occur independent on experimental conditions. There appears, however, to be a constant growth rate at 410-450 °C but the growth conditions are difficult to control due to the competing reaction (cf. Fig. 2). Decomposition and recombination of the $Sr(thd)_2$ molecules in the gas phase [23] obviously lead to a complicated situation where several mechanisms are possible. Futhermore, the nucleation of differently oriented grains may have an effect on the growth rate as, for instance, with the (111) orientation the rate is highest (d = 3.479 Å) and much smaller with the (220) (d = 2.129 Å) orientation. In the nucleation the choice of the orientation may be very sensitive for small changes in the experimental conditions. In our samples the low growth rate appears to favor the (200) orientation while

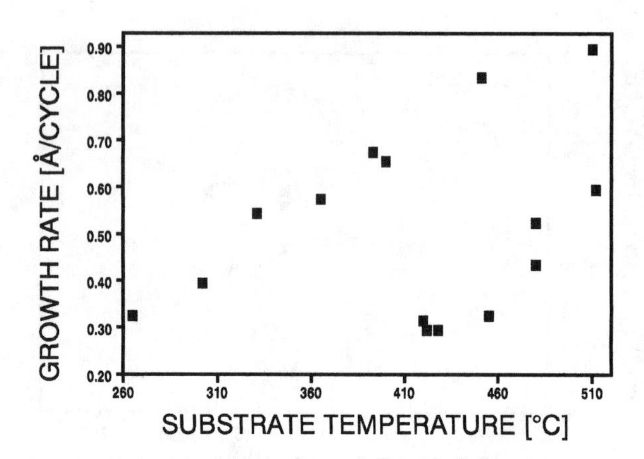

Fig. 2. The growth rate of SrS thin film vs. substrate temperature. Source temperature is 230 °C and pulse duration 0.6 s.

the (111) preferred orientation is dominant when growth range is 0.45-0.65 Å/cycle. At higher rate the orientation was random.

The next variable studied was the duration of the $Sr(thd)_2$ pulse at 400 °C. The growth rate increased slightly when the pulse was lengthened from 0.2 s to 0.4 s. With longer durations the growth rate was constant (0.65 Å/cycle) (Fig. 3). This indicates that a certain minimum pulse duration is needed to get

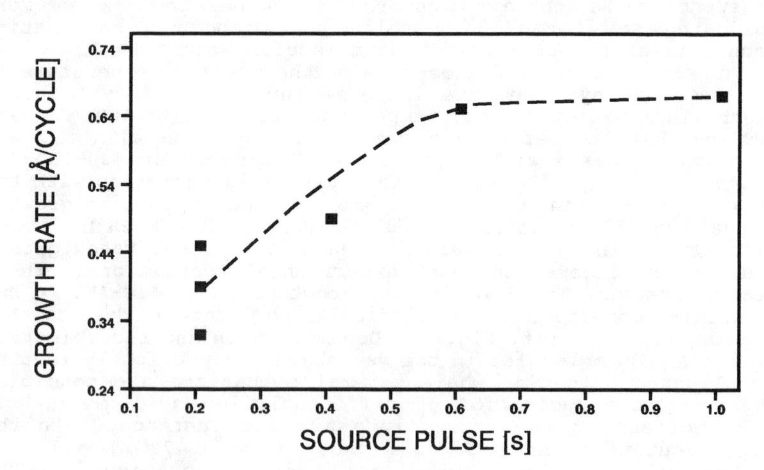

Fig. 3. The growth rate of SrS thin films vs. $Sr(thd)_2$ pulse duration. The source and substrate temperatures are 225 and 400 °C, respectively.

a full coverage of the substrate. When the full coverage is reached the situation is stable. Variation of the H_2S pulse duration gave a similar picture.

The duration of the purge gas pulses did not affect the growth rate of SrS. The small variations observed are within the limits of accuracy of the thickness measurements.

CONCLUSION

The results show that, although chemically good SrS films can be grown in a flow-type ALE reactor from $Sr(thd)_2$ and H_2S, the complicated sublimation behaviour and possible partial decomposition of $Sr(thd)_2$ make the mechanism of the growth reaction very complicated and highly sensitive for the source and reactor temperatures. Therefore the films may easily contain thickness profile and different orientations within the same substrate. It is possible to obtain films with single preferred orientation only by a very careful control of the film processing parameters.

REFERENCES

1. M. Leskelä and M. Tammenmaa, Mater. Chem. Phys. 16, 349 (1987).
2. W.A. Barrow, R.E. Coovert and C.N. King, SID 84 Digest 1984, 249.
3. S. Tanaka, V. Shanker, M. Shiiki, H. Deguchi and H. Kobayashi, SID 85 Digest 1985, 218.
4. M. Ogawa, T. Simouna, S. Nakada and T. Yoshioka, Jpn. J. Appl. Phys. 24, 168 (1985).
5. R. Mach, G.O. Muller, E. Schneuerer, R. Selle and H. Ohnishi, Acta Polytechn. Scand. Appl. Phys. Ser. Ph 170, 197 (1990).
6. S. Tanaka, H. Yoshiyama, J. Nishiura, S. Ohshio, H. Kawakami and H. Kobayashi, SID 88 Digest 1988, 293.
7. S. Okamoto, E. Nakazawa and Y. Tsuchiya, Jpn. J. Appl. Phys. 28, 406 (1989).
8. S. Tanaka, H. Yoshiyama, J. Nishiura, S. Ohshio, H. Kawakami and H. Kobayashi, Proc. SID 29/4 1988, 305.
9. Y. Abe, K. Onisawa, K. Tamura, T. Nakayama, M. Hanazono and Y.A. Ono, Jpn. J. Appl. Phys. 28, 1373 (1989).
10. S. Okamoto, E. Nakazawa and Y. Tsuchiya, Jpn. J. Appl. Phys. 29, 1987 (1990).
11. S. Okamoto, E. Nakazawa, T. Kuki and Y. Tsuchiya, Acta Polytechn. Scand. Appl. Phys. Ser. Ph 170, 203 (1990).
12. C. Gonzales, Conf. Record 1987 Inter. Display Res. Conf., 1987, 21.
13. H. Ohnishi and T. Okuda, SID 89 Digest 1989, 317.
14. T. Shoki and Y. Yamaguchi, Acta Polytechn. Scand. Appl. Phys. Ser. Ph 170, 207 (1990).
15. S. Tanada, A. Miyakoshi and T. Nire, Springer Proceedings in Physics 38, 180 (1990).
16. S. Tanaka, K. Nakamura, H. Morita, S. Wada and H. Kobayashi, Acta Polytechn. Scand. Appl. Phys. Ser. Ph 170, 211 (1990).
17. S. Tanaka, H. Deguchi, Y. Mikami, M. Shiiki and H. Kobayashi, SID 86 Digest 1986, 29.

18. K. Okamoto and K. Hanaoka, Jpn. J. Appl. Phys. 27, L1923 (1988).
19. K. Onisawa, Y. Abe, K. Tamura, T. Nakayama, M. Hanazono and Y.A. Ono, J. Electrochem. Soc. 138, 599 (1991).
20. M. Leppänen, M. Leskelä, L. Niinistö, E. Nykänen, P. Soininen and M. Tiitta, SID 91 Digest (1991), in press.
21. M. Tammenmaa, H. Antson, M. Asplund, L. Hiltunen, M. Leskelä and L. Niinistö, J. Cryst. Growth 84, 151 (1987).
22. M. Leskelä and L. Niinistö, in Atomic Layer Epitaxy edited by T. Suntola and M. Simpson (Blackie & Sons, Glascow 1990) p. 1.
23. M. Leskelä, L. Niinistö, E. Nykänen, P. Soininen and M. Tiitta, Thermochim. Acta 175, 91 (1991).
24. M. Tammenmaa, I. Yliruokanen, M. Leskelä and L. Niinistö, Anal. Chim. Acta 195, 351 (1987).
25. S.B. Turnipseed, R.M. Barkley and R.W. Sievers, Inorg. Chem. 30, 1164 (1991).

STUDY OF THE GROWTH MECHANISMS OF NbS_2 SCALES USING MARKER EXPERIMENTS

CHUXIN ZHOU AND L. W. HOBBS
Department of Materials Science & Engineering, MIT, Cambridge, MA 02139

ABSTRACT Au thin films were deposited using electron beam evaporation onto the surface of Nb metal to serve as thin film markers to study the growth mechanisms of NbS_2 scales. Scanning Auger microscopy (SAM) was used to measure the depth profiles for the compositions of Au, Nb, S, Si and O across the sulfide scale. Three other marker experiments were also studied. All four marker experiments indicated that the inward diffusion of sulfur is the dominant process responsible for the growth of NbS_2 scales.

INTRODUCTION

NbS_2, and other transition metal sulfides, have attracted recent attention because of their unique properties in several new applications. In energy conversion systems, all conventional corrosion-resistant metallic materials fail due to sulfidation at high temperatures. Nb (and Mo and W as well), by comparison, show substantial stability and slow sulfidizing kinetics in severe sulfidizing environments owing to the diffusion-barrier properties of their sulfides[1, 2]; the generation of these sulfides in scales formed on several useful alloys could convey substantial protection[3, 9], provided the source of the resistance they convey can be understood. Other applications of these solids derive from their strong anisotropy in optical, mechanical and electrical behavior as well as from their intercalation properties.

The growth mechanisms, growth orientations and rate controlling factors of NbS_2 films are important factors to document. Little study has been made of these features. The present work has employed thermogravimetric measurements of scaling rate and four different markers (microscopic and thin film markers) to study the kinetics and growth mechanisms of NbS_2 sulfidation films. Only the results from Au thin film marker experiments are reported here. The other three marker results are reported elsewhere [9].

EXPERIMENTS

Elemental Nb was utilized in the experiments. Polycrystalline Nb of 99.8$wt.\%$ purity, (0.2$wt.\%$Ta, minor impurities are Si, Al) was obtained from Johnson–Matthey, Seabrook, NH. The as-purchased niobium rod, 12.7 mm diameter, was cut into discs 0.5 $\sim$ 1mm thick. Pure niobium metal is very soft, softer than aluminum. A 1μm surface polish with diamond paste or 0.3 μm polish with alumina powder gave a greyish surface. The final step of the surface finish was carried out with Buehler Mastermet (Buehler $No.$40−6370−064), a colloidal silica polishing suspension, which provided a shiny surface. The specimens were ultrasonically agitated and cleaned in distilled water, acetone and methanol to remove the residues of the colloidal suspension.

The SAM used (Perkin Elmer 660 System) employs Auger electron spectroscopy (AES) and a finely focused, computer–controlled electron beam, with deflection and rastering capability, to provide point, line or area compositional analysis of the surface to a depth of less than 1nm. SAM is a preferred technique for many applications because it combines high speed and surface sensitivity with beam scanning to determine the spatial distribution of elements at the surface. The electron gun is run at 15kV in the photographic mode and at 3–5kV in the AES mode for elemental analysis. The Ar ion gun works at

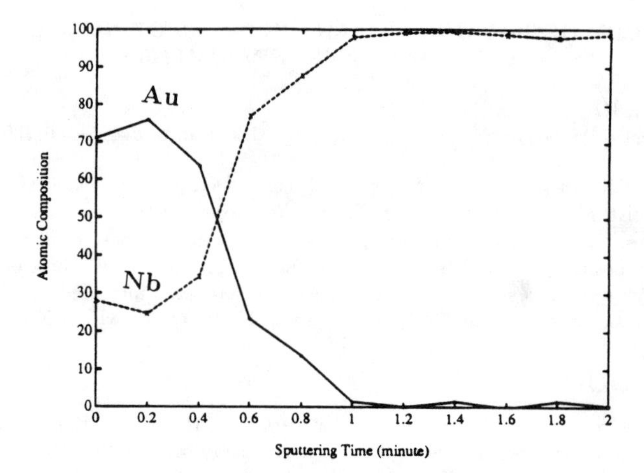

Figure 1: Auger depth profiles of Au thin film deposited onto a Nb polycrystalline wafer, showing the distributions of Au and Nb. Data were collected for every 12 seconds sputtering with an argon ion gun. The sputtering rate was 0.5nm/s.

$<$ 3kV with an emission current of 25mA. The sputtering ion current was $30\mu A$ and was carefully measured by using a Faraday cup placed at one of the sample stage positions. The sputtering area is about $1.5mm \times 1.5mm$.

Au–THIN FILM MARKER RESULTS

Continuous gold thin films were deposited onto the surface of elemental Nb wafers using vacuum evaporation. The gold film thickness was approximately 20 nm. Figure 1 shows the composition profile across the as-deposited thin film and indicates a reasonable thickness for AES measurements.

The Nb wafers were then sulfidized in sulfur vapor at 1173K for 1 minute, and the reaction products were characterized by glancing angle X-ray diffraction and electron diffraction, as well as by high resolution transmission electron microscopy (HREM) techniques[9]. The products were identified as rhombohedral NbS_2. Figure 2 presents a composition profile across the overall scale. The original peak–to–peak energy profiles were computed from the raw spectra and transformed to atomic concentrations which were further calibrated with standard NbS_2 scales formed on pure Nb metal wafers. This NbS_2 standard scale was analyzed by SAM and by electron probe microanalysis (EPMA) in order to calibrate the depth profile data collected by SAM. Figure 2 shows that the Au thin film marker remains on the surface of the NbS_2 scale. This suggests that sulfur inward diffusion may be the key factor for the growth of NbS_2 scale.

Consistent results were also observed using alternative markers, such as Pt microscopic markers using a photolithographic technique, Si atomic markers using ion implantation and Si thin film markers using CVD[9].

DISCUSSION

All four marker experiments, using discontinuous Pt markers, continuous thin films of Au or Si and an implanted Si–Nb layer, indicate that new sulfide is being formed within

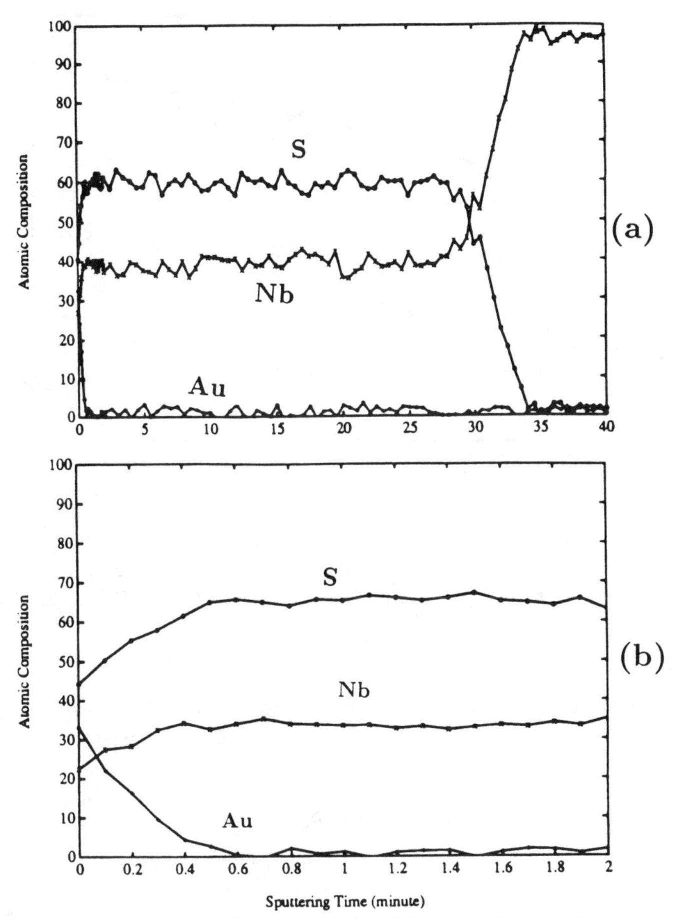

Figure 2: **(a)** Auger depth profiles of Au, Nb and S across the overall NbS_2 scale grown on Nb wafer with a Au film deposited onto the surface. Data were collected for every 30s of sputtering (every 6s for the initial 2 minutes.) The sulfide scale is about $1\mu m$ thick, and the profiles establish the location of Au marker at the outer surface of the NbS_2 scale. **(b)** Data collected for every 6 seconds of ion sputtering to show the details of the location of the Au marker film. The sputtering rate was 0.5nm/sec.

the scale or at the Nb–NbS_2 interface, beneath the markers. We suggest on the basis of this evidence that the growth of the NbS_2 scale is occurring by the inward diffusion of sulfur. Such a conclusion has also been alluded to for sulfidation of other refractory metals by Gerlach and Hamel[6, 8] . Assuming sulfur is the much more mobile species in the NbS_2 scale, we still do not know whether sulfur diffuses as gaseous molecules, or by interstitial $[S_i^{-z}]$ or vacancy $[V_S^{+z}]$ lattice or boundary diffusion within the sulfide scale. No diffusion measurements have been reported. We nonetheless evaluate evidence for the form of sulfur lattice defects, assuming them to be implicated, arising from the present marker experiments and our kinetic measurements[5] of niobium sulfidation.

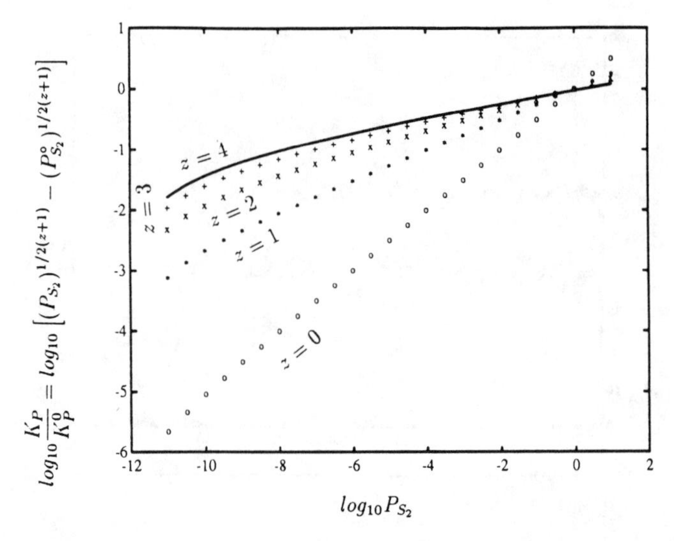

Figure 3: Assuming sulfur interstitials $[S_i^{-z}]$ as the dominant defects, $log_{10}(K_P/K_P^0)$ is plotted against $log_{10}P_{S_2}$ ($P_{S_2} = 10^{-12}$ atm was used for this computation) and shows strong dependence of K_P on P_{S_2} in the external environment.

$[S_i^{-z}]$ Dominant

If sulfur interstitials are dominant, we must evaluate the following processes:

$$S_S^{\times} = S_i^{-z} + z \cdot h^{\cdot} \tag{1}$$

$$\frac{1}{2}S_2 = S_S^{\times} \tag{2}$$

$$\frac{1}{2}S_2 = S_i^{-z} + z \cdot h^{\cdot} \tag{3}$$

where $S_S^{\times}$ represents a sulfur ion occupying a normal sulfur site, electrically neutral with respect to the sulfur sublattice; S_i^{-z} an interstitial sulfur ion with effective charge $-z$; and $h^{\cdot}$ an electron hole. Reaction (1) describes a sulfur lattice ion leaving its normal sublattice site to form an interstitial, while reaction (2) is the thermodynamic equilibrium between the gas phase and the sulfide phase. Reaction (3) describes sulfur atoms entering to form interstitials directly from the gas phase. All three reactions are interrelated; at thermodynamic equilibrium, however, only two of the three reactions are independent. The relationship between $[S_i^{-z}]$ and P_{S_2} is:

$$[S_i^{-z}] \propto P_{S_2}^{1/2(z+1)} \tag{4}$$

so that the parabolic rate constant varies with P_{S_2} as

$$K_P \propto K_P^0 \cdot \left[(P_{S_2})^{1/2(z+1)} - (P_{S_2}^{o})^{1/2(z+1)}\right], \tag{5}$$

where P_{S_2} is the sulfur partial pressure in the environment, $P_{S_2}^{o}$ is the equilibrium sulfur partial pressure at the sulfide/metal interface, and K_P^0 is a proportionality parameter.

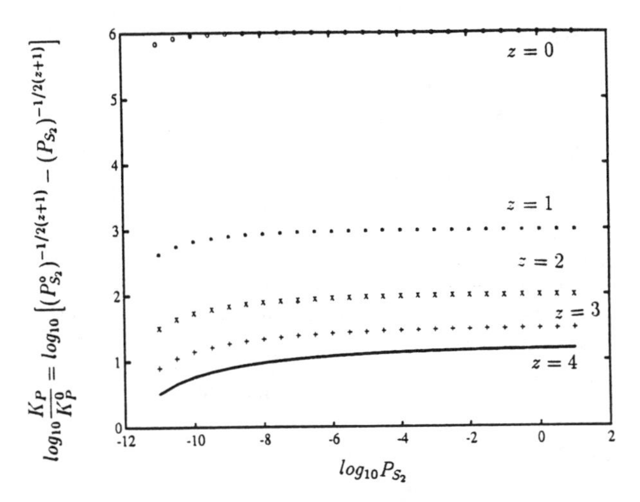

Figure 4: Plot of $log_{10}(K_P/K_P^0)$ against $log_{10}P_{S_2}$, assuming sulfur vacancies $[V_S^{+z}]$ to be the dominant defects ($P_{S_2} = 10^{-12}$ atm was used for this computation).

Under our experimental conditions,

$$P_{S_2} \gg P^o_{S_2}$$

Therefore, K_P in equation (5) is strongly dependent on P_{S_2} in the environment. This behavior was computed and is illustrated in Figure 3. Our kinetic results[5,9] showed, however, little dependence of K_P on P_{S_2}. Consequently, the assumption that $[S_i^{-z}]$ is dominant is incorrect.

$[V_S^{+z}]$ Dominant

If sulfur vacancies are the dominant lattice defects in the NbS_2 scale, we must evaluate the following reaction:

$$S_S^{\times} = \frac{1}{2}S_2 + V_S^{+z} + z \cdot e \tag{6}$$

where V_S^{+z} represents a vacant sulfur site with an effective charge $+z$ in respect to its normal lattice site. This reaction implies that a sulfur atom leaves its normal lattice site to enter the gas phase, and a positively-charged sulfur vacancy and free electrons are created to maintain electronic neutrality. At thermodynamic equilibrium, we derive the relationship between $[V_S^{+z}]$ and P_{S_2} as

$$[V_S^{+z}] \propto P_{S_2}^{-1/2(z+1)}. \tag{7}$$

This dependence was found in qualitative agreement with that of Tatùruki *et al* [4]. The corresponding parabolic rate constant is related to P_{S_2} by

$$K_P = K_P^0 \cdot \left[(P^o_{S_2})^{-1/2(z+1)} - (P_{S_2})^{-1/2(z+1)}\right]. \tag{8}$$

Under our experimental conditions, $P_{S_2} \gg P^o_{S_2}$, and K_P in equation 8 is almost independent of the P_{S_2} in the environment. This behavior was computed and is illustrated in Figure 4, in agreement with the our kinetic results. Consequently, the assumption that $[V_S^{+z}]$ defects are dominant is consistent with experiments.

Calculation of Diffusion Coefficient

For diffusion–controlled growth of thin scales or films, the growth rate can be approximately expressed as

$$x(t) = 2\gamma\sqrt{D \cdot t} \tag{9}$$

where γ is a dimensionless parameter defined in the function[7]

$$g(\gamma) = \frac{C(x=0)}{\overline{C}} = \sqrt{\pi}\gamma e^{\gamma^2} erf(\gamma) \tag{10}$$

where $C(x = 0)$ is the composition of sulfur at the gas/sulfide scale interface and $\overline{C}$ is the average composition of sulfur across the sulfide scale. For the non-stoichiometric compound $Nb_{1+\alpha}S_2$,

$$g(\gamma) = \frac{C(x=0)}{\overline{C}} = \frac{1+\overline{\alpha}}{1+\alpha(x=0)} \approx 1 \tag{11}$$

We can easily solve equation (11) to yield $\gamma \approx 0.62$. The growth kinetics measured previously imply $K_P \approx 3 \times 10^{-16} m^2/sec.$[5,9]. The calculated diffusion coefficient of sulfur at 1173K is

$$D_S = \frac{K_P}{4\gamma^2} = 2 \times 10^{-16} m^2/sec. \tag{12}$$

CONCLUSIONS

Four marker experiments indicate that new sulfide NbS_2 is formed within the existing sulfide scale or at the scale/substrate interface, beneath the markers, consistent with a conclusion that sulfur inward diffusion controls the growth of the sulfide scales formed by sulfidation of Nb at 1173K. The measured P_{S_2} dependence of the parabolic rate constant in sulfidation suggests that sulfur vacancy diffusion is the dominant mechanism for the inward sulfur transport.

References

[1] M. S. Kovachenko, W. W. Syczew, D, Z, Jurczenko and I. G. Tkaczenko, *Izv. Akad. Nauk. USSR. Metall.* **5**(1974)221

[2] K. N. Strafford and J. R. Bird, *J. Less-common Metals*, **68**(1979)223-228

[3] Ge Wang, R. Carter and D. L. Douglass, Oxidation of Metals, **31**(1989)273.

[4] K. Tatsuki, M. Wakihara and M. Taniguchi, *J. Less-Common Metals*, **68**(1978)183.

[5] Chuxin Zhou, L. W. Hobbs and G. J. Yurek, *"Growth Kinetics and Structure of NbS_2 Sulfidation Scales"*, to be published by J. Electrochem. Soc., 1991.

[6] J. Gerlach and H. J. Hamel, Metall., **23**(1969)1006.

[7] J. Crank, *The Mathematics of Diffusion*, Oxford University Press, 1976, p. 305.

[8] J. Gerlach and H. J. Hamel, Metall., **24**(1970)488.

[9] Chuxin Zhou, *"Growth, Structures and Properties of Cr_2O_3 and NbS_2 Corrosion Scales"* Sc.D thesis, MIT, Cambridge, MA (1991).

Low-Temperature Growth of Thin Films of Al_2O_3 with Trimethylaluminum and Hydrogen Peroxide

J. F. FAN, K. SUGIOKA AND K. TOYODA
The Institute of Physical and Chemical Research (RIKEN), Wako, Saitama 351-01, Japan.

ABSTRACT

Thin films of Al_2O_3 were prepared by sequential surface chemical reaction of trimethylaluminum and hydrogen peroxide at low temperatures. It has been found that hydrogen peroxide reacts very easily with trimethylaluminum, resulting in growth of Al_2O_3 at the temperature as low as the room temperature. Another favorable feature of the technique is that the growth of excellent Al_2O_3 occurs identically wherever the reactants reach, making it possible to completely coat the surface of the sample with arbitrary shape.

INTRODUCTION

High quality thin films of Al_2O_3 are necessary not only for fabrication of electronic devices also for coatings of optical and mechanical components. To prepare such films, low temperature and damage-free process is preferred in many cases. R.Solanki et al studied a photo-chemical vapor deposition of Al_2O_3 using trimethylaluminum (TMA) and N_2O.[1] Recently, G.S.Higashi et al reported on a sequential surface chemical reaction limited growth technique using TMA and water (H_2O) vapors.[2] By the technique, growth of thin films of Al_2O_3 could be achieved at low temperatures between 100°C and 500°C. The electrical resistivity and breakdown field of the film grown at 450°C was found to be excellent. However, low temperature growth behavoir was not studied. Also the excellent electrical properties were demonstrated only in the low field region (below 2 MV/cm), i.e., the high field behaviors remain unclear.

In this study, we carry out Al_2O_3 growth at low temperatures and find that the growth rate decreases markedly. In order to overcome this difficulty, we grow Al_2O_3 using TMA and a new oxidant, hydrogen peroxide (H_2O_2).

EXPERIMENTAL

Figure 1 shows the schematic diagram of our growth apparatus. A computer-controlled gate valve was added in comparison with the apparatus developed by G.S.Higashi et al.[2] High purity (7N) TMA, supplied from

Shinetsu Chemicals, and H_2O_2 prepared by vacuum-distillation of the commercially available EL-grade H_2O_2 aqueous solution(30%) were used as the reactants. Wafers of various materials such as silicon, gallium arsenide, glass, stainless steel and etc were used as substrates. The substrates were first ultrasonically cleaned in conventional organic solvents, then, blown dry with N_2 gas and immediately loaded into the chamber.

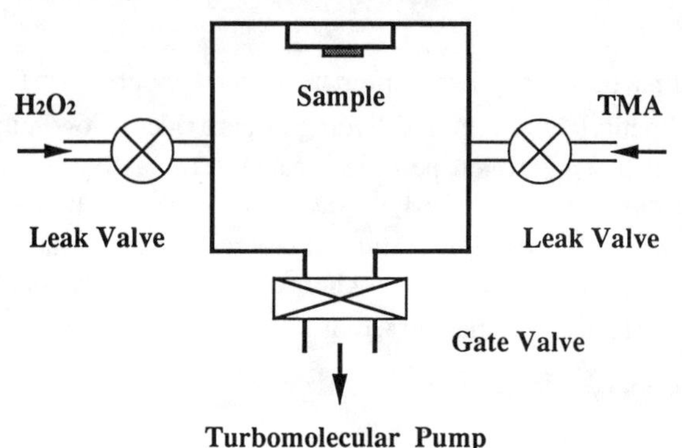

Fig.1. Schematic diagram of the growth apparatus.

The chamber, made of aluminum alloy, was evacuated by a turbo molecular pump (TMP) to a pressure below 10^{-5} Torr. Prior to the growth, initialization was done by purging the chamber with either TMA or H_2O_2 vapor at the pressure of 20 mTorr for 10 s. In the growth cycle, the reactants were sequentially introduced into the chamber through two computer-controlled leak valves, held for a period, and then exhausted out from the gate valve. Surface chemical reaction and adsorption occur during the holding period. Typical chamber pressures and the hold times were calibrated to 20~30 mTorr for 1 s for the holding period. The gate valve closes during the holding period, which markedly reduces the doses of the source reactants, resulting in both marked reduction in the evacuation time and increase in the source efficiency. During the growth, the substrate was kept at the growth temperature.

The films obtained were evaluated in the following ways. Thickness of the films was measured by both stylus surface profiler and ellipsometer. To determine the chemical composition, X-ray photoelectron spectroscopy (XPS) and Rutherford back-scattering spectroscopy (RBS) were carried out. An α-sapphire (0001) wafer was used as the standard. Dielectric properties were evaluated by the conventional current-voltage (I-V) and capacitance-voltage (C-V) characterizations with $Ag/Al_2O_3/n^+$-GaAs MIS structures. Reflection high energy electron diffraction (RHEED) and scanning electron microscope (SEM) were used to determine the crystal and interfacial structures of the films.

RESULTS AND DISCUSSIONS

We found that the growth of films was nearly independent of the substrate. This is reasonable because different adsorption coefficients and different surface structures associated with different substrates would only affect the growth of initial few layers.

Figure 2 shows the dependence of film-thickness on the growth cycles for 150°C growth. Also shown are our results for growth using H_2O as the oxidant. The thickness of film is precisely proportional to the number of growth cycles as expected from the self-limiting nature of surface adsorption and reaction. The growth rate, determined by dividing the film-thickness by the number of the growth cycles, was found to be 1.13 Å/cycle for 150°C growth using H_2O_2, which is about the same as that for growth at 450°C using H_2O.[2] It decreased, however, to 0.74 Å/cycle for growth using H_2O at the same growth temperature. This indicates that the reaction of TMA and H_2O to produce Al_2O_3 becomes incomplete at low temperature.

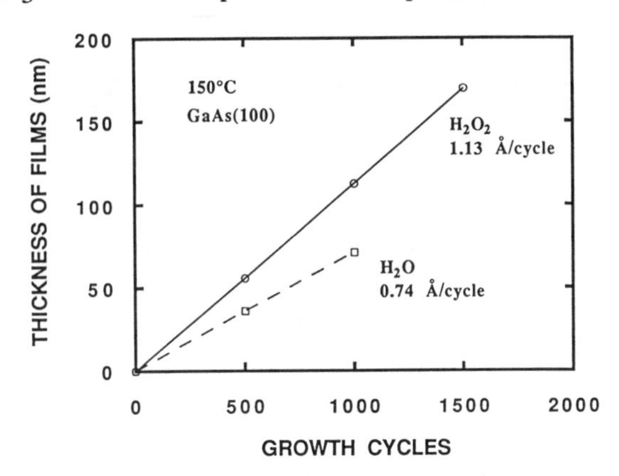

Fig.2. Film-thickness versus the number of growth cycles.

Figure 3 shows the RBS and XPS spectra of the film grown at 150°C. The sample for RBS measurement was prepared on the substrate of beryllium. No impurity in the film was detected in the sensitivity level of RBS (1%). Theoretical analysis gives the results of Al 40% and O 60%, indicating that the films are of stoichiometric composition. In the XPS spectra, the peak position of Al 2p indicates that bonds of Al-O are similar with those of sapphire. From the areas of peaks Al 2p and O 1s, values of the composition, Al 39.7% and O 60.3%, were obtained which well agree with the RBS results.

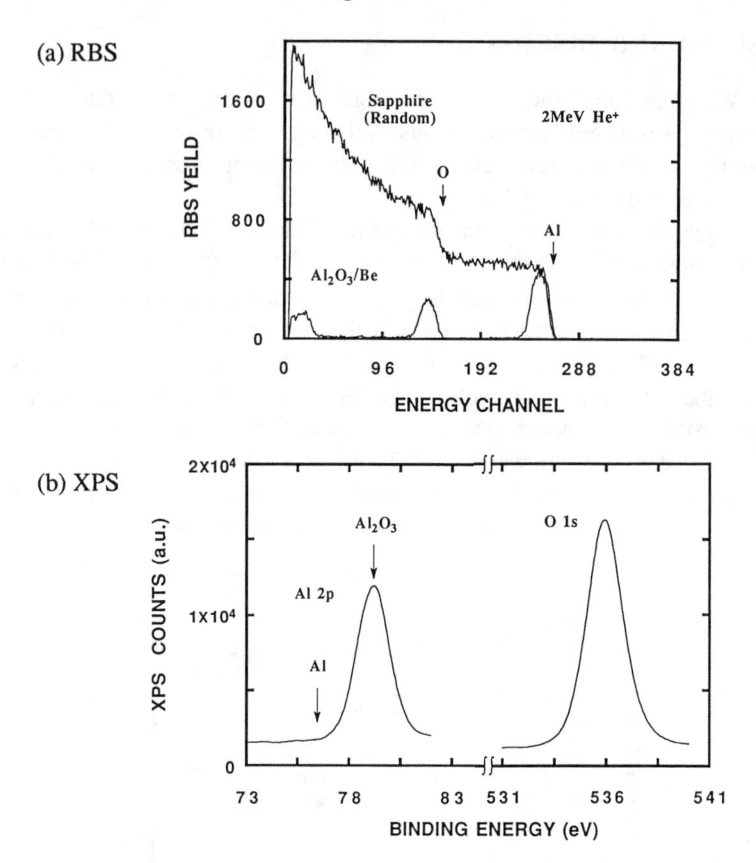

Fig.3. RBS(a) and XPS(b) spectra of Al_2O_3 films grown at 150℃.

RHEED measurement gave a halo pattern, indicating that the films are amorphous. Interestingly, the growth of Al_2O_3 obtained per cycle (1.13 Å) is just half atomic layer of α-Al_2O_3 (0001). Molecular size, adsorption coefficient, and desorption are factors to dominate the growth.[3]

The I-V measurement gave the results as shown in Fig. 4. The electrical resistivity was calculated to be 1.2×10^{16} Ωcm, and the catastrophic breakdown field was found to be 6~7 MV/cm. The defect-related leakage begins at ~3 MV/cm, which is about 3 times larger than that of the film grown at 450°C using H_2O,[2] suggesting that Al_2O_3 prepared in our study is of fewer defects.

The dielectric constant ε_r of the film was calculated from the accumulation capacitance of the $Ag/Al_2O_3/GaAs$ structure and the thickness of the film. As the result, a value of 7.1 for frequency of 1 MHz was obtained.

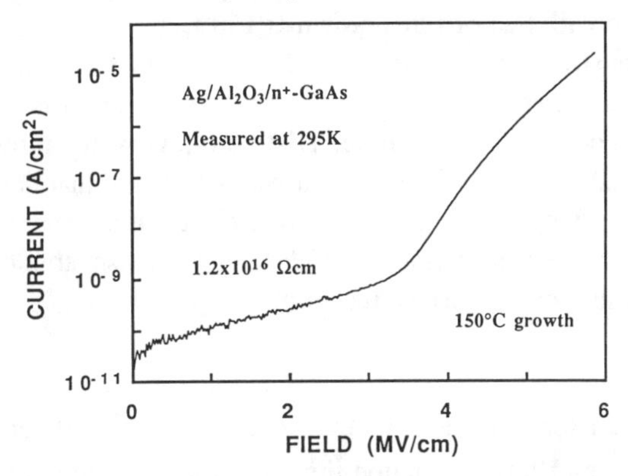

Fig.4. I-V characteristics of $Ag/Al_2O_3/n^+$-GaAs structures.

The refractive index n determined by ellipsometry was found to be 1.61 for wavelength of 632.8 nm. As noted, both the dielectric constant and the refractive index are smaller than those for sapphire (ε_r: 9.8, and n: 1.76), indicating that the films are less dense in comparison with sapphire. In fact, reduction in the film thickness by post-growth thermal annealing was observed.

Above results were obtained for films grown at 150°C. For growth carried out at the room temperature (21°C), growth of 1000 cycles resulted in a film of ~2270 Å thickness. However, the thickness decreased to ~1200 Å by an thermal annealing at 150°C for 15 min in flowing N_2. Because no noticeable

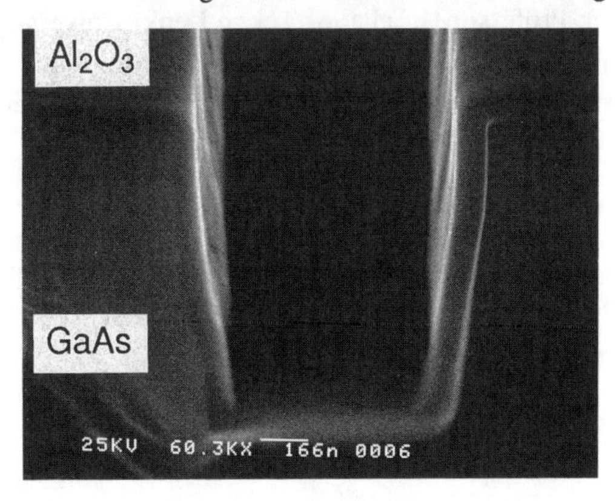

Fig.5. SEM image of the cross section of the Al_2O_3/GaAs structure.

change in the chemical composition while increase in the refractive index were observed, it was concluded that the film grown at the room temperature was porous but easy to be densified. Interestingly, the quality of the annealed films was comparable with that of films grown at 150°C.

To demonstrate another feature of the growth technique, growth was carried out on a trench-shaped substrate of GaAs. Figure 5 is the SEM image of the cross section of the sample. Distinctly, growth of the film occurs identically on the surface of the trench. It was also found that deviation in film-thickness over 5 samples located at different places in the chamber was within the error limit of the measurements. Both facts are reasonable considering the self-limiting nature of the surface reaction.

SUMMARY

We studied the behaviors of low temperature growth of Al_2O_3 in the sequential surface chemical reaction limited growth and found that the growth rate decreased significantly in the case using H_2O and TMA as the reactants. The problem was successfully overcome by substituting H_2O with H_2O_2, which reacted with TMA very easily and completely at the temperature as low as the room temperature. As the result, growth of thin films of Al_2O_3 was achieved at low temperature. The growth was independent of the substrate, and the film was ideally uniform. Also the quality of films was excellent. Thus, it has been concluded that the technique is capable of preparing high quality thin films for electrical insulation, for optical and mechanical coatings, and etc.

ACKNOWLEDGEMENTS

We would like to thank Professor H.Takai of Tokyo Denki University for RBS analysis and Dr. T.Takahashi of ETL for ellipsometry measurement. This work was supported in part by the system of Special Researchers, Basic Science Program, the Agency of Science and Technology of Japan.

REFERENCES

[1] R.Solanki, W.H.Ritchie, and G.J.Collins: Appl. Phys. Lett. 43(1983) 454.

[2] G.S. Higashi and C.G.Fleming: Appl. Phys. Lett. 55(1989) 1963.

[3] S.Iwai, A.Doi, Y.Aoyagi and S.Namba: Symp. GaAs and Related Compounds, Heraklion, Greece, 1987, p191.

SELF TERMINATING REACTION OF DIPIVALOYLMETHANATE COMPLEXES WITH HYDROXYL GROUPS ON OXIDE SURFACE

RIKA SEKINE*, MAKI KAWAI*, KIYOTAKA ASAKURA**
AND YASUHIRO IWASAWA**
*Research Laboratory of Engineering Materials, Tokyo Institute of Technology, 4259, Nagatsuta, Midori-ku Yokohama, 227, Japan and Institute of Physical and Chemical Research, Wako-shi, 351-01, Japan
**Department of Chemistry, Faculty of Science, The University of Tokyo, 7-3-1, Hongo, Bunkyo-ku, Tokyo, 113, Japan

ABSTRACT

We have already reported that copper and calcium dipivaloylmethanates [$Cu(DPM)_2$ and $Ca(DPM)_2$] reacts selectively and stoichiometrically with surface hydroxyl groups (OH) on SiO_2. In order to clarify the structure of the adsorbed species and the origin of the reaction between $M(DPM)_2$ (M=Cu and Ca) and OH groups, the surface adsorbed species are studied by infrared spectroscopy (IR), X-ray photoelectron spectroscopy (XPS), and the extended X-ray absorption fine structure (EXAFS). As a result, it was found that H from surface OH has moved into $M(DPM)_2$ after the adsorption, where the four oxygen coordinated structure around Cu still exists in the adsorbed $Cu(DPM)_2$. Introducing water vapor at 673 K to this surface results in the removal of ligand DPM from the adsorbed $Cu(DPM)_2$. At 673 K, Cu atoms decomposed from the adsorbates aggregated on the surface. This fact supports that the interaction between the adsorbed $Cu(DPM)_2$ and SiO_2 surface is originated from that between the ligands and the surface.

INTRODUCTION

Layered oxides have drawn much attraction for these several years. An example is high-Tc cuprate superconductive oxides with perovskite type layered structure. In order to realize the epitaxial growth of these oxide, amount of stacking atoms have to be controlled precisely in each layers. At the same time, low temperature is favorable in avoid mixing between different phases of high-Tc super conductors. Self-terminating reaction is one of the promising way to realize above condition. Because: (1) The amount of metal loaded can be controlled by the self-terminating reaction, independent to the amount of materials supplied, and (2) making use of chemical compounds which have often higher vapor pressure than its simple substance enable to reduce the process temperature. The self-terminating reaction system, i.e. the atomic layer epitaxy (ALE) system have already been established for III-V or II-VI semiconductors but not many are known for oxide system [1]. In order to consider the possibility of the ALE system for layered complex oxide system, we examined a reaction between surface functional groups on oxides and compounds includ-

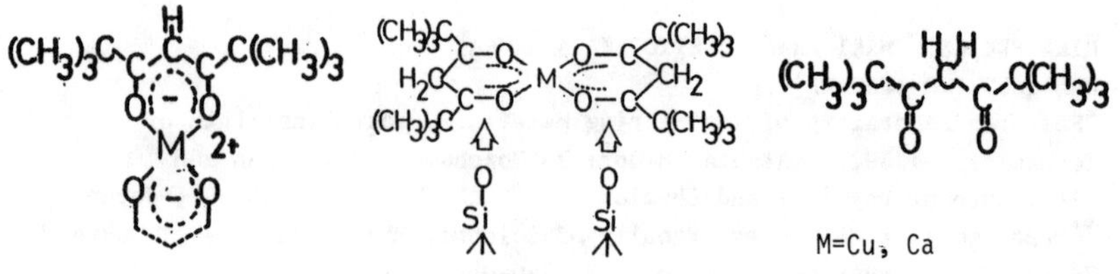

Fig.1 Structures of (a) $M(DPM)_2$ and (b) the adsorption model of $M(DPM)_2$ on SiO_2. The bonding state between Si-O and $M(DPM)_2$ is ambiguous. (c) DPM.

ing the object metal, such as complexes, organometallics, or halides. We have already found that the hydroxyls (OH) on oxide surface reacted stoichiometrically with copper or calcium dipivaloylmethanates [$Cu(DPM)_2$ or $Ca(DPM)_2$, see Fig. 1(a)], a kind of β -diketonate complexes [2,3]. Reaction between H_2O and the adsorbed $M(DPM)_2$ (M=Cu,Ca) has also been discussed.

In this paper, we examined minutely the structure of the adsorbed species by infrared spectroscopy (IR), X-ray photoelectron spectroscopy (XPS), and the extended X-ray absorption fine structure (EXAFS). We also investigated structure of metal atoms after the removal of ligands by above spectroscopies and the transmission electron microscope (TEM).

EXPERIMENTAL

The SiO_2 sample was prepared by spraying a suspension of SiO_2 (supplied by Cab-O-Sil, HS-5) in ethanol onto the surface of NaCl crystal. The NaCl with SiO_2 layers thus prepared was placed in a glass infrared (IR) reaction cell and the treatment was carried out at 733 K under $1x10^3$ Pa oxygen, for 2 h, followed by evacuation at room temperature. After this pretreatment, the DPM complexes (supplied by Toso Akuzo Co. and Trichemical Lab.) was evaporated at 353 K [$Cu(DPM)_2$] or 483 K [$Ca(DPM)_2$]. The amount of surface OH on SiO_2 and that of adsorbed $Cu(DPM)_2$ or $Ca(DPM)_2$ were determined by IR transmission spectroscopy (NICOLET FT-IR 510), together with the structure of adsorbed species. The electronic states of Cu and Ca on SiO_2 were investigated by XPS (JEOL JPS-80). The binding energy of emission peaks in XPS were calibrated by assuming the Si 2p in SiO_2 to be 103.4 eV and C 1s in CH to be 285.1 eV [4]. The local structure around Cu atom was determined by the EXAFS. The EXAFS was measured at BL 10B and BL 7C of Photon Factory in National Laboratory for High-Energy Physics. The morphology of Cu metal after the removal of ligand by H_2O was observed by an electron microscope operated at acceleration energy of 200 keV.

RESULTS AND DISCUSSIONS

After the pretreatment of substrate, SiO_2, at 733 K, only isolated OH's are remained on the surface. The isolated OH gives a sharp vibrational absorption due to stretching mode at 3750 cm^{-1} in IR spectrum [2]. When $Cu(DPM)_2$ was supplied from gas phase onto this surface, the peak assigned to the OH stretching disappears, and that of CH stretching mode, which originates from the ligands of the complexes increases. The successive evacuation at the same temperature hardly affected the amount of OH and CH adsorbed on the surface of SiO_2. This indicates that $Cu(DPM)_2$ was chemisorbed on SiO_2.

The stoichiometric ratio of surface OH and adsorbed $Cu(DPM)_2$ was estimated from the intensity of IR absorbance of OH and CH's. Figure 2 shows that the decrease in the amount of surface OH on SiO_2 and increase in the amount of CH_3 groups of adsorbed $Cu(DPM)_2$ (solid line) are proportional, suggesting that they reacted stoichiometrically. The stoichiometry between the initial surface OH and adsorbed $Cu(DPM)_2$ is estimated to be (2-3):1 [2].

The structure of adsorbed $Cu(DPM)_2$ can be figured out as shown in Fig. 1(b), which was determined from following consideration. At first, as mentioned in above paragraph, stoichiometric ratio of reacted $Cu(DPM)_2$ and OH is approximately 1:2. The model shown in Fig. 1(b) satisfies this condition. Secondly, the adsorbed $Cu(DPM)_2$ no longer has))>CH structure but $>CH_2$ structure. The evidence of this change was obtained by the IR spectrum in CH's region, where absorption due to))>CH disappeared and that due to $>CH_2$ increased [2]. Thus, the model where H has moved from surface OH to the adsorbed $Cu(DPM)_2$ is suitable. Thirdly, the structure around Cu does not change after adsorption. This model is confirmed from the IR spectrum in C-C and C-O stretching region and from the EXAFS. The IR absorption spectrum of adsorbed $Cu(DPM)_2$ is very similar to that of isolated $Cu(DPM)_2$ (Fig. 1(a)) and not to that of liquid DPM (Fig. 1(c)) [2,3]. Therefore, we propose a model where the coordination around the metal by the ligand DPM does not change after the adsorption. This model is also supported by the analysis of the Cu K-edge EXAFS spectra for isolated (A̲) and adsorbed (B̲) $Cu(DPM)_2$.

Figure 3 shows the Fourier transformation of the EXAFS of sample A̲ and B̲. They are quite similar. The peak between 0.1-0.2 nm is assigned to Cu-O bond and that between 0.2-0.3 nm to the second coordinated Cu ··· C. The clear existence of Cu ··· C coordination in adsorbed species (B̲) indicates that the ligand is still coordinated to Cu atom after the adsorption. The distance between Cu and the first coordinated atoms, O, and that between Cu and the second coordinated atoms, C, were obtained by a curve fitting analysis. The coordination number of Cu with O or B was also obtained. Results for $Cu(DPM)_2/SiO_2$ (B̲) are displayed in Table I, where $Cu(DPM)_2$ (A̲) was chosen as a model compound. Since the structure of the $Cu(DPM)_2$ has not been determined precisely yet, the bond length around Cu atom for model compound (A̲) was taken from copper acetylacetonate [$Cu(acac)_2$] [5]. Comparing A̲ with B̲, it was found that the bond length between Cu and the first coordinated atoms, and also those between the second coordinated atoms, are identical. In addition, the coordinated numbers were identical as well. Therefore, it can be concluded that after the adsorption, Cu-O bond length and the ring structure of the ligand remains unchanged. [Fig. 1(b)].

Reaction between water vapor and $Cu(DPM)_2$ adsorbed on SiO_2 is also studied. After the treatment with H_2O vapor at 673 K, the IR absorption

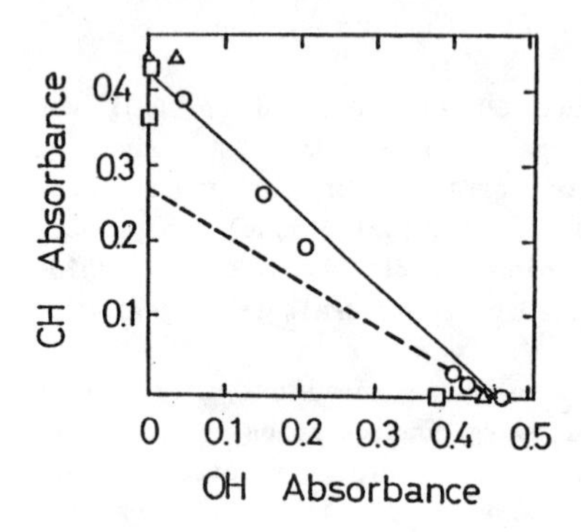

Fig.2. The relation between the absorbance of CH_3 (2969 cm^{-1}) in adsorbed $Cu(DPM)_2$ and that of νOH on SiO_2 (solid line). The broken line indicates the similar relation observed for the adsorbed $Ca(DPM)_2$ and OH on SiO_2.

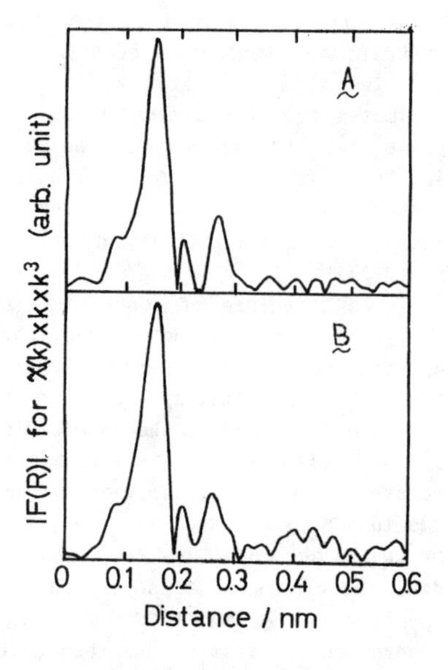

Fig.3 Fourier transformation of the EXAFS. (A):$Cu(DPM)_2$, (B):$Cu(DPM)_2$/ SiO_2

Table I. Structure and the coordination number around Cu atom for $Cu(DPM)_2$ complex (A) and $Cu(DPM)_2/SiO_2$ (B).

	Cu-O (first coordination)		Cu-C (second coordination)	
	distance /nm	coordination number	distance /nm	coordination number
A[1)	1.914	4.0	2.84	4.0
B	1.92(1)	3.6(4)	2.87(3)	3.6(6)

[1] Taken from the structure of $Cu(acac)_2$ [5].

bands due to CH stretching almost disappeared and those for OH reproduced to similar amount of the initial value [2]. This fact shows that the adsorbed $Cu(DPM)_2$ was decomposed by H_2O and the ligand DPM desorbed by this treatment. The emission peak due to the remained Cu from $2p_{2/3}$ and $2p_{1/2}$ are observed at 934 and 954 eV. The absence of the satellite structure indicates that Cu is not in the +2 oxidized state. From this photoelectron spectra it is hard to determine whether the Cu is in +1 states or is metallic, though previously we have concluded it to be in Cu_2O [2]. In order to clarify the local structure around Cu atom, EXAFS was observed. The observed spectrum in the region of absorption near edge structure (XANES), was similar to that for the Cu metal. By the curve fitting analysis of the EXAFS, Cu-Cu bond length is determined to be 0.255(1) nm and coordination number to be 11.3(7). These values are identical to those for the Cu foil, where Cu-Cu distance is 0.2556 nm and coordination number was 12.0. Therefore it can be concluded that remained Cu atoms in a form of metal particle on the SiO_2 surface. To estimate the dimensions of the copper particle, we have observed the sample by a transmission electron microscope (TEM). The micrograph image is displayed in Fig. 4. Copper metals were proved to form particles with diameter of ~10 nm, surrounded by Cu_2O layers. These phases, Cu metal and surrounding layers of Cu_2O were determined from the electron diffraction. It can be concluded that after the decomposition Cu metal aggregated to form particles with diameter in ~10 nm, where the surface of which is covered with Cu_2O.

As mentioned above, when the ligand was removed from the adsorbed $Cu(DPM)_2$, Cu atom begins to diffuse on the surface and finally forms metal aggregates. When the ligands are coordinated to Cu atoms, four coordinated oxygen structure around Cu atom is held even though the H atom has moved from surface OH to ligand. These facts show that the interaction between $Cu(DPM)_2$ and the surface OH works between the ligand and the surface. The ligand in the adsorbed species becomes cationic after accepting H from the OH and the surface after the removal of H becomes ionic, $Si\text{-}O^{\delta -}$ (Fig. 1(b)). Therefore, the existence of the ligand is indis-

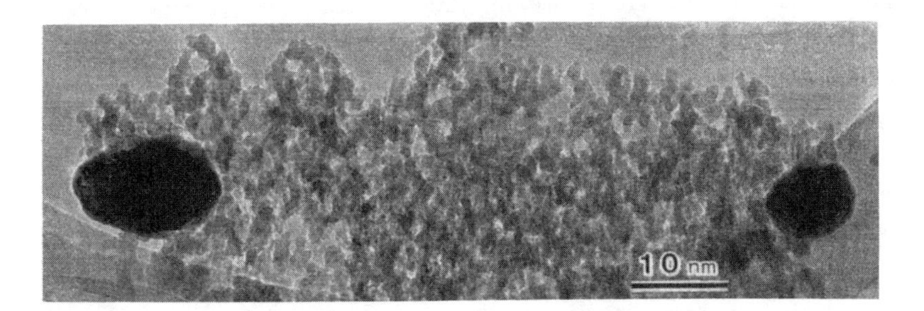

Fig.4 Electron micrograph for the remained species after the decomposition of $Cu(DPM)_2/SiO_2$ with H_2O at 673 K. The spheres ~10 nm was proved to be Cu metal particle.

pensable for stabilizing the adsorption of $Cu(DPM)_2$. Because of this reason, Cu atoms begins to move on the surface to form Cu metal aggregated when the ligand was removed.

Similar reaction between $Ca(DPM)_2$ and surface OH was examined. The results obtained are quite similar to those for $Cu(DPM)_2$ [3]. The results are: (1) The adsorbed $Ca(DPM)_2$ on SiO_2 gave a quite similar IR spectrum to that of $Cu(DPM)_2$ on SiO_2, (2) the reaction between OH and $Ca(DPM)_2$ is also stoichiometric, and (3) the adsorbed $Ca(DPM)_2$ is decomposed by H_2O. Here, only the relation between the amount of surface OH reacted and adsorbed CH is shown (broken line in Fig. 2). The increment of the two lines in Fig. 2 are almost the same. Here we can conclude that $Ca(DPM)_2$ reacts with surface OH on SiO_2 with a same ratio to $Cu(DPM)_2$. The model for adsorbed $Ca(DPM)_2$ can also be described as shown in Fig. 1(b), same as in the case of $Cu(DPM)_2$. In spite of the difference in the center metal atoms, $Cu(DPM)_2$ and $Ca(DPM)_2$ showed similar interaction with the surface OH on SiO_2. These facts support that the interaction between $M(DPM)_2$ and surface OH described above, i.e., the adsorption of $M(DPM)_2$ is originated from the interaction between ligand and OH, and is independent to the center metal.

CONCLUSION

The structure of adsorbate $M(DPM)_2$ (M=Cu,Ca), which adsorbs stoichiometrically and selectively on surface OH of SiO_2, was determined. The fact that (1) the adsorptions of quite different metal DPM complexes (i.e., $Cu(DPM)_2$ and $Ca(DPM)_2$) are similar, and that (2) in the case of $Cu(DPM)_2$, Cu atom was stable only when the ligand is coordinated. Thus it is concluded that the adsorption should originate from the interaction between the ligand and the surface OH, and not strongly related to the kind of the center metal atom.

ACKNOWLEDGMENT

A part of this work was supported by a Grant-in Aid for Scientific Research on Chemistry of New Superconductors, from the Ministry of Education, Science and Culture of Japan. This work has been performed under the approval of the Photon Factory Program Advisory Committee (Proposal Nos. 88109 and 90143). The authors are grateful to Mr. Gonda and Professor Koinuma of the Tokyo Institute of Technology for the XPS measurements. The authors are also grateful to Professor Hirotsu of Nagaoka University of Technology for the TEM measurement.

References

1. T Kawai, T. Choda an S. Kawai, Mat. Res. Soc. Symp. Proc., 75, 289 (1987).
2. R. Sekine and M. Kawai, Appl. Phys. Lett. 56, 1466 (1990).
3. R. Sekine, M. Kawai, T. Hikita and T. Hanada, Surface Sci., 242, 508 (1991).
4. C. D. Wagner, W. M. Riggs, L. E. Davis, J. F. Moulder and G. E. Muilenberg, in Handbook of X-ray Photoelectron Spectroscopy (Perkin-Elmer, Eden Prairie, MN, 1978).
5. S. Shibata, T. Sasase, and M. Ohta, J. Mol. Struct. 96 347 (1983).

RHEED INTENSITY OSCILLATION DURING THE EPITAXIAL GROWTH OF SILVER AND GOLD FILMS

KAZUKI MAE, KENTARO KYUNO, TAKEO KANEKO, AND RYOICHI YAMAMOTO
Research Center for Advanced Science and Technology, University of Tokyo, Komaba 4-6-1, Meguro-ku, Tokyo 153, Japan

ABSTRACT

The intensity oscillation of reflection high-energy electron diffraction (RHEED) during the autoepitaxy of Au on Au(001) and Ag on Ag(001) and during the heteroepitaxy of Au on Ag(001) and Ag on Au(001) were measured at 310 and 330 K. The morphologies of the growing surfaces are discussed in terms of the surface energies and the surface reconstruction. RHEED intensity oscillation during Au deposition on Ag(001) at 330 K rapidly decayed at the 6th period when reconstruction of the Au(001) surface appeared. Such a rapid decay was not observed at 310 K. On the other hand, the reconstruction entirely disappeared at the first minimum of the RHEED intensity oscillation in the case of Ag deposition on Au.

INTRODUCTION

RHEED intensity oscillation is considered to be induced by the layer-by-layer growth and now employed to control the film thickness and surface flatness. It has been applied to synthesize not only for semiconductor superlattices but also for metallic superlattices[1,5]. The phase effect and the transient effect, however, make it difficult to control the film thickness precisely, i.e. the maxima of the oscillation do not always corespond to the monolayer completion. Joyce et al. claim that the transient effect is associated with a change in reconstruction [7]. Detailed studies of the initial stages of metallic epitaxy are needed to understand the phenomenon of RHEED intensity oscillation and to make it more powerful method to control the growth of thin films.

The purpose of the present study is to measure RHEED intensity oscillations during homo and heteroepitaxy of Au and Ag. Au and Ag have very similar lattice constants, 4.078Å and 4.086Å, respectively, and epitaxially grow on the (001) plane with few mismatch dislocations. Au(001) has reconstructed structures such as c(26x68), but Ag does not. The topmost layer of Au(001) contracts within the layer about 20%, and becomes hexagonal-close-packed arrangement to reduce the surface area per atom [6]. The difference of the electron scattering factors between Au and Ag also seems to have an influence on the average RHEED intensity. According to Doyle and Turner, the electron scattering factors of Au and Ag are 9.251 and 7.267, respectively, when the glancing angle of the electron beam is 0.5 degrees [8]. We are interested in the effects of the reconstruction on the growing surfaces and the RHEED intensity oscillation and the difference in the RHEED intensity oscillation during heteroepitaxies, Au/Ag and Ag/Au. Since Au has a larger surface energy than Ag, Ag can be considered to grow on Au in a layer-by-layer mode [Frank-Van der Merwe (FM) mode] and, on the contrary, Au grows on Ag forming 3D islands [Stranski-Krastanov (SK) mode]. The epitaxial growth of Au and Ag on Au(111) surface were studied by scanning tunneling microscopy (STM) [2,3]. The growth front of Au has more roughness and smaller size of islands than Ag. Almost entire surface is covered by Ag and no evidence of the reconstructed structure of Au substrate can be seen at the coverage of 3/2 ML.

EXPERIMENTAL

The base pressure was about 6×10^{-10} Torr and the pressure during depositions was 2×10^{-9} Torr. The substrate of MgO(001) single crystal was cleaned by heating at 1100 K. Ag was deposited on the substrate at room temperature to prepare the buffer layers of 2000 Å. The films were annealed at 750 K for about 5 minutes [4]. The buffer layer of Au was further deposited on the Ag layer in the cases of Au/Au and Ag/Au. The thickness of Au buffer layer was chosen as 300 Å and were anealed at 500 K for about 2 minutes [5]. RHEED intensity measurements were performed during depositions of Au and Ag at room temperature. The deposition rate was chosen as about 0.1 Å/sec. The electron beam with the energy of 20 keV impinged on the film surface at a glancing angle of between 0.2 and 0.5 degrees. The incident azimuths were the [100] and [110] directions. The RHEED image on the fluorescent screen was taken by CCD TV camrera and the image was recorded by VTR. The image was also digitized for later analysys.

RESULTS

Fig.1 shows the RHEED intensity oscillations of Au/Au. The oscillation was not observed at 373 K. We can recongnize the oscillation of several periods at 330 K and clearly see a long lived oscillation at 310 K. The reconstructed pattern of Au(001) was observed during the deposition. Long lived oscillations were observed for Ag on Ag(001) at 330 K as shown in Fig.2.

For Au on Ag(001), clear and bright Kikuchi-lines disappeared when depositing Au. The observed oscillations both at 330 K and 310 K are shown in Fig.3. The oscillation at 330 K rapidly decayed at the sixth period when RHEED pattern of the reconstructed Au(001) surface appeared. Such a rapid decay was not observed at 310 K and the reconstructed RHEED pattern was not so clear.

In the case of Ag on Au(001), the reconstructed pattern of Au(001) substrate entirely disappeared at the first minimum of the RHEED intensity oscillation and Kikuchi-lines gradually appeared.

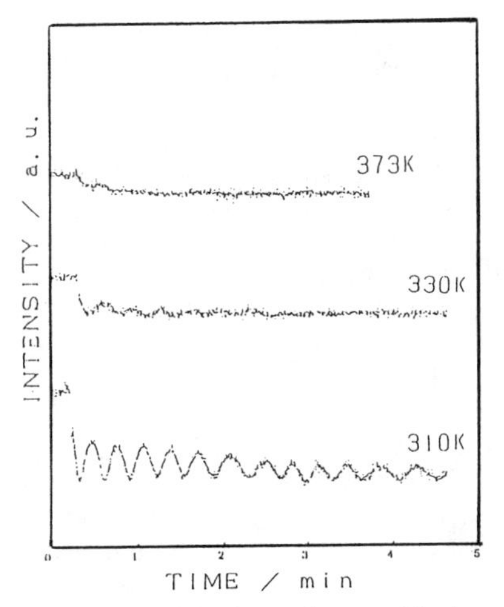

Fig.1 RHEED intensity oscillations during the deposition of Au on Au(001) at 373 K, 330 K, and 310 K

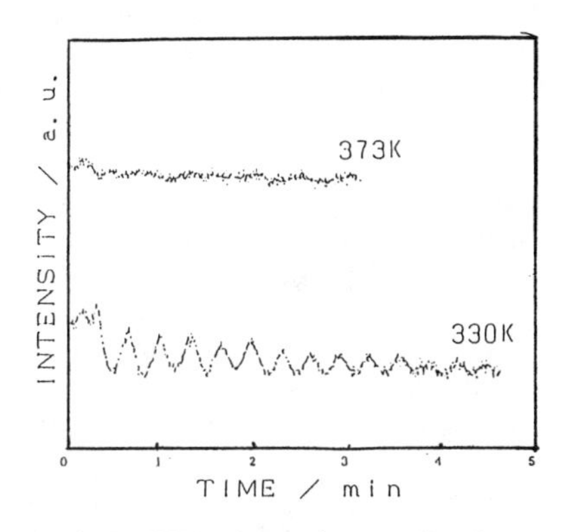

Fig.2 RHEED intensity oscillations during the deposition of Ag on Ag(001) at 373 K and 330 K.

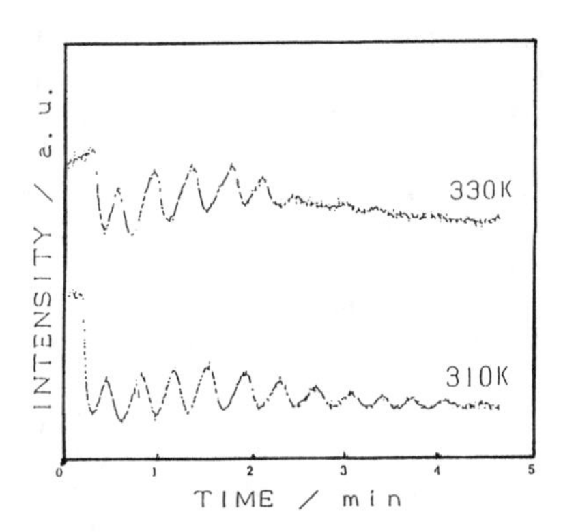

Fig.3 RHEED intensity oscillations during the deposition of Au on Ag(001) at 330 K and 310 K.

DISCUSSION

The upper limit of the temperature that RHEED intensity oscillations are observable is considered to be determined by the transition temperature from 2D nucleation growth mode to step propagation growth mode. According to Flynn [9], the upper limit temperature T for a RHEED intensity oscillation is written by

$$T/T_m < 3/10$$

where Tm is the melting temperature of a depositing metal. Since the melting temperatures of Au and Ag are 1336 K and 1234 K, the upper limit temperatures for Au and Ag are 401 K and 370 K, respectively. These results are consistent with the present experiments of Au/Au(001) and Ag/Ag(001). The decay of the RHEED intensity oscillation is more rapid in Au/Au than in Ag/Ag because the surface diffusion constant of Au is smaller than that of Ag.

Fig.4 shows the intensity evolutions at various points along the (00) streak in the case of Ag/Au at 330 K. The increase of the average intensities at some measuring points is due to the effect of the Kikuchi-line. The changes of the intensity distribution on the (00) streak is depicted in Fig.5. The RHEED pattern of Ag surface has clear and bright Kikuchi-lines. On the other hand, Kikuchi-lines of Au surface was weak. The intensity change of a Kikuchi-line alone was too small to detect, but it enhanced the diffuse streaks at their crossings and had not a little influence on the RHEED intensity oscillation. The intensities at the crossings of Kikuchi-lines and streaks increased as depositing Ag on Au.

The intensity reduction of the specular beam due to the surface reconstruction is explained by the basic scattering geometry. Meyer-Ehmsen et al. adopted so called two-layer system to consider the diffraction from disordered surfaces [10].

In addition, the boundary of the reconstructed domains can become additional surface defects when terraces are larger than the reconstructed domains. The reconstruction did not seem to work as an additional defect for Au/Ag because of the small size of the islands due to the small surface diffusion length of Au atoms at 310 K.

The size of islands seem to have another effect on the reconstruction, i.e. small islands make it difficult to form the reconstructed structure. The step edges of an island can be a barrier to form the reconstructed structure. The driving force of forming reconstructed structures in noble metals such as Au and Pt is the atomic interactions within the surface layer which tend to reduce the surface energy, though reconstructions of semiconductors, e.g. Si(001)-2x1, are drived by the local interactions which combine the surface dangling bonds with each other. It seems that long-range interactions within the surface layer are interrupted at the step edges of an island and that the reduction of the surface energy by the transformation of the topmost layer is not superior to the energetical increase by the mismatch between the top layer and the second layer.

The difference between the RHEED intensity oscillations at 330 K and 310 K for Au/Ag seems to be induced by the effects of the reconstruction described above.

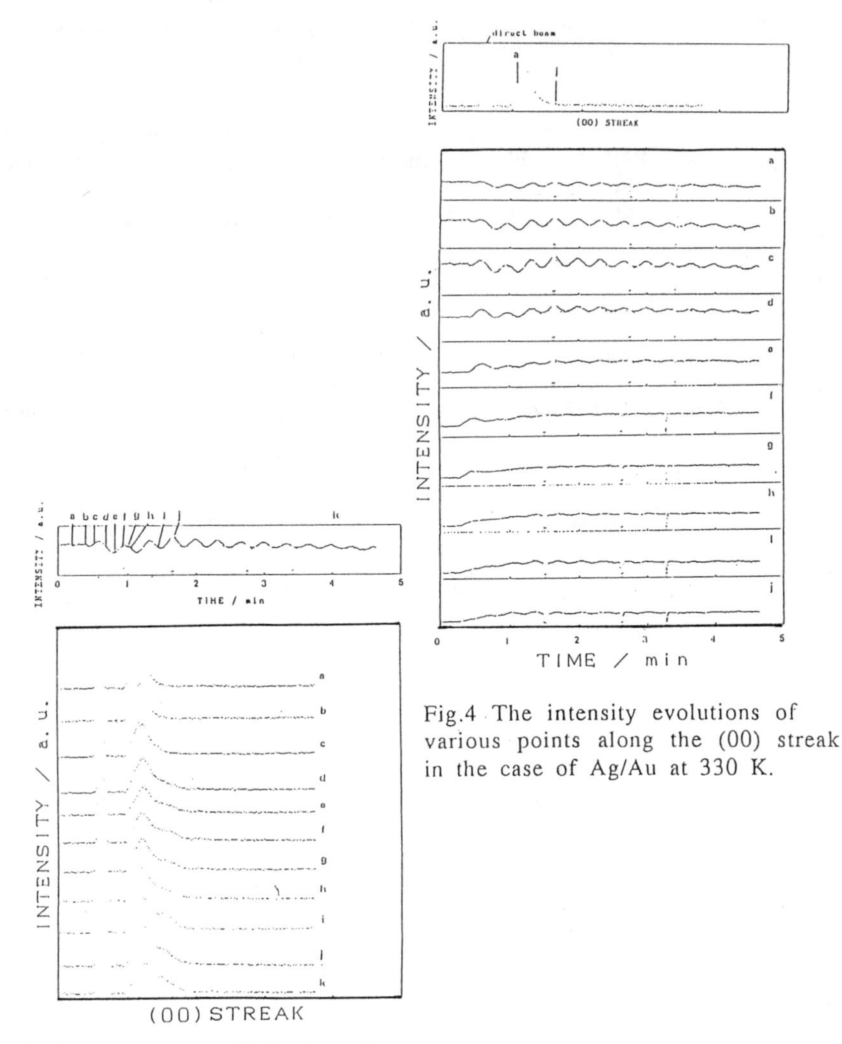

Fig.4 The intensity evolutions of various points along the (00) streak in the case of Ag/Au at 330 K.

Fig.5 The changes of the intensity distribution on the (00) streak in the case of Ag/Au at 330 K.

ACKNOWLEDGMENT

We are grateful to Yasushi Inoue and Shinzo Ogu for assisting the experiments. This work was supported by a Grant-in-Aid for Scientific Researh on Priority Areas from the Ministry of Education, Science and Culture, Japan, No.02254103.

REFERENCES

[1] G.Lilienkamp, C.Koziol, and E.Bauer: 'Reflection High-Energylectron Diffraction and Reflection Electron Imaging of Surfaces', ed. by P.K.Larsen and P.J.Dobson (Plenum Press, New York, 1988), pp489-499.

[2] C.A.Lang, M.M.Dovek, J.Nogami, and C.F.Quate: Surface Sci. 224(1989)L947.

[3] M.M.Dovek, C.A.Lang, J.Nogami, and C.F.Quate: Phys.Rev.B 40(1989)11973.

[4] Y.Suzuki, H.Kikuchi, and N.Koshizuka: J.J.Appl.Phys. 27(1988)L1175.

[5] H.Kikuchi, Y.Suzuki, and T.Katayama: J.Appl.Phys. 67(1990)5403.

[6] N.Takeuchi, C.T.Chan, and K.M.Ho: Phys.Rev.Lett. 63(1989)1273.

[7] P.J.Dobson, B.A.Joyce, J.H.Neave and J.Zhang: J.Crystal Growth 81(1987)1.

[8] P.A.Doyle and P.S.Turner: Acta Cryst. A 24(1968)390.

[9] C.P.Flynn: J.Phys. 18(1988)L195.

[10] G.Meyer-Ehmsen, B.Bolger, and P.K.Larsen: surface sci. 224(1989)591.

IN SITU RHEED AND HREM STUDY OF INITIAL STAGES OF INTERFACIAL REACTIONS OF UHV DEPOSITED TITANIUM THIN FILMS ON SILICON

M.H. WANG and L.J. CHEN
Department of Materials Science and Engineering, National Tsing Hua University, Hsinchu, Taiwan, Republic of China.

ABSTRACT

The initial stages of interfacial reactions of ultrahigh vacuum (UHV) deposited Ti thin films on silicon have been studied by in-situ reflected high energy electron diffraction (RHEED) and transmission electron microscopy (TEM).

An amorphous interlayer was found to form during the deposition of the first 1.7-nm-thick Ti layer. In samples annealed at 450 °C for 30-120 min, Ti_5Si_3, located at the Ti/a-interlayer interface, was identified to be the first nucleated phase. Ti_5Si_3, Ti_5Si_4, TiSi and C49-$TiSi_2$ were observed in samples annealed at 475 °C for 30 and 60 min as well as at 500 °C for 10 and 20 min. Fundamental issues in silicide formation are discussed in light of the discovery of the formation of the amorphous interlayer and as many as four different silicide phases in the initial stages of interfacial reactions of UHV deposited Ti thin films on silicon.

INTRODUCTION

Among metal/Si systems, Ti/Si stands out since its stable phase, C54-$TiSi_2$, is the most attractive candidate material as the gate electrode and ohmic contacts in the submicron silicon device technology. Although extensive investigations have been conducted to study the interfacial reactions of titanium thin films on silicon, the close proximity and/or overlapping of diffraction rings of titanium silicides rendered the unambiguous identification of phases rather difficult [1-3]. In addition, the reactions of Ti thin films on single-crystal Si were known to be strongly influenced by the presence of impurities in the surface region of Si [4]. As a result, although the final stable phase is known to be C54-$TiSi_2$, the phase formation sequence is still unclarified. Ti_5Si_3, TiSi, and C49-$TiSi_2$ were variously reported to be the first nucleated phase [4-9]. In addition, diversified results on whether there is an intermixing between Ti and Si during Ti deposition were reported [10-14]. In this paper, we report the results of an in situ reflected high energy electron diffraction (RHEED) and high resolution transmission electron microscopy (HREM) study of initial stages of interfacial reactions in ultrahigh vacuum (UHV) deposited Ti thin films on (111)Si.

EXPERIMENTAL PROCEDURES

Single crystal, 3-8 Ω cm, 2 inches in diameter, phosphorus doped (111) oriented silicon wafers were used in the present study. The wafers were cleaned chemically by a standard procedure. The samples were then dipped in a diluted HF solution (HF:H_2O = 1:50) immediately before loading into an UHV electron beam evaporation chamber. The base pressure of the UHV chamber was better than 1 x 10^{-10} Torr. The wafers were first heated to 850 °C for 10 min (Shiraki-clean) to evaporate away the native oxide layer so that an atomically clean surface was obtained. 4N5 purity Ti pellets were used as the evaporation source. Thin titanium films, 30 nm in thickness, were then deposited on silicon at room temperature. During the Ti deposition, RHEED was used to monitor the growth of Ti on Si surface. An a-Si layer, about 10 nm in thickness, was subsequently deposited onto the metal layers to protect the metal thin films from oxidation during annealing. The vacuum during deposition was maintained to be better than 1 x 10^{-9} Torr. The deposition rate was about 0.1 nm/s.

The heat treatments were carried out in a three-zone diffusion furnace from 350-500 °C for various periods of time in N_2 ambient. High-purity N_2 gas was first passed through a titanium-getter tube maintained at 800 °C to reduce the O_2 content. The accuracy in the temperature measurement was estimated to be within ± 2 °C. Detailed procedures for polishing of planview specimens were reported previously [15]. XTEM samples were prepared following the procedures outlined by Sheng and Chang [16]. A JEOL 200CX scanning transmission electron microscope operating at 200 kV was used for conventional TEM observation. For HREM observation, a JEOL 4000EX operating at 400 kV with a point-to-point resolution of 0.18 nm was used. No objective aperture was inserted into the beam path during imaging. Most of the XTEM micrographs were taken under a systematic [110] diffraction condition. An optical diffractometer was also used to analyze the lattice fringes from the negative films. A Perkin-Elmer 595 scanning Auger electron spectrometer (AES) was utilized for concentration-depth profile analysis.

RESULTS AND DISCUSSION

RHEED patterns revealed that after 0.2-nm-thick Ti layer was deposited, the diffraction peaks corresponding to the (7x7) Si surface reconstruction disappeared. The patterns became very dim following 0.2- to 0.6-nm-thick Ti deposition. As the thickness of the deposited Ti film was increased from 0.6 to 1.7 nm, only the primary beam was detected. The characteristic diffraction pattern of Ti showed up following 2-nm-thick Ti deposition. Examples are shown in Fig. 1. Furthermore, as the sample was rotated, the RHEED pattern did not change. The results indicated that the film was not epitaxially related to the Si. Instead, the Ti film is strongly textured. In addition, the growth mode of the surface film was found to be layered rather than islanded since the pattern did not become spotty even after 30-nm-thick Ti film was deposited. From the absence of diffraction spots, it may be inferred that an intermixed amorphous layer was formed for the deposition of the first 1.7-nm-thick Ti layer. XTEM examination of as-deposited samples revealed that a 2-nm-thick amorphous interlayer was indeed formed. An example is shown in Fig. 2. The Ti thin films were also found to exhibit an (010) texture. AES depth profile of the as-deposited sample showed that the oxygen and carbon levels are below the detection limit of the AES as shown in Fig. 3.

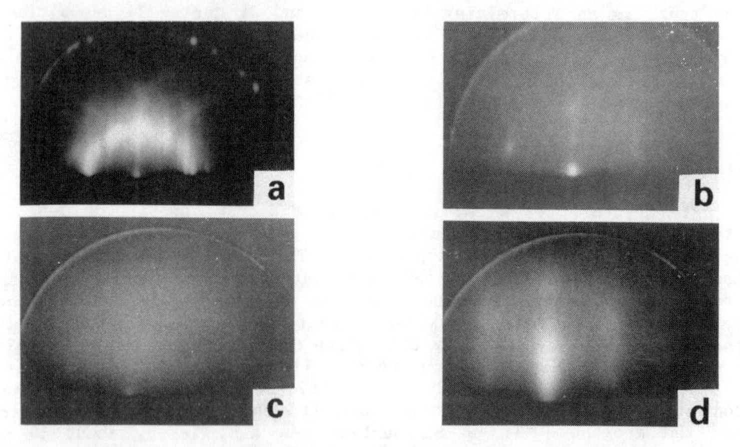

Fig. 1. RHEED patterns, (a) as cleaned, (b) 0.4 nm, (c) 1.7 nm, (d) 2.8 nm.

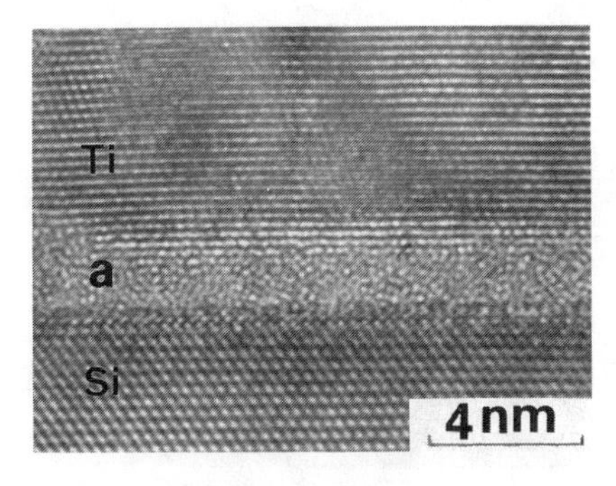

Fig. 2. HREM image of an as deposited sample.

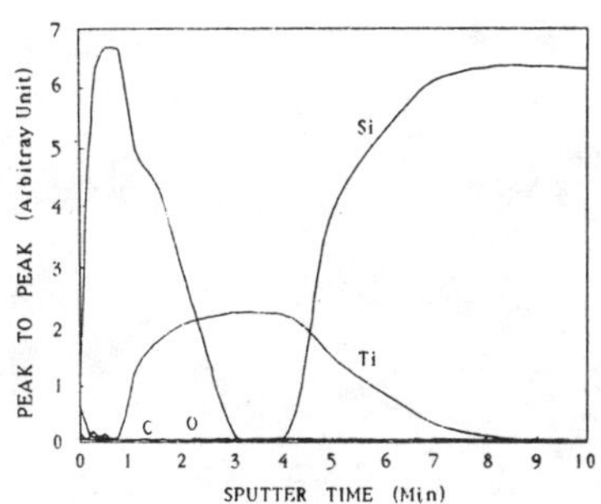

Fig. 3. AES profile of an as deposited sample.

In samples annealed at 350-425 °C for 1 h, the thickness of the amorphous interlayer was found to increase with annealing temperature and time. The maximum thickness was measured to be 10 nm. No crystalline silicides were detected to form from either planview or XTEM examinations. For samples annealed at 450 °C for 2, 5, and 10 min, no diffraction rings corresponding to crystalline silicides were evident in the diffraction patterns. For specimens of samples annealed at 450 °C for 30-120 min, Ti_5Si_3 crystallites, about 5 nm in average size, were found to distribute at both upper a-interlayer/Ti and lower Ti/a-interlayer interfaces as shown Fig. 4. Ti_5Si_3, Ti_5Si_4, TiSi and C49-$TiSi_2$ were observed in samples annealed at 475 °C for 30 and 60 min as well as at 500 °C for 10 and 20 min. The Ti_5Si_4 phase, located at the a-interlayer/Si interface, was determined from diffraction patterns along [001], $[3\bar{1}8]$ and $[3\bar{1}10]$ directions. The phase was found to be epitaxially related to the silicon with [001]Ti_5Si_4//[111]Si and (010)Ti_5Si_4//$(2\bar{2}0)$Si. Examples are shown in Fig. 5. The C49-$TiSi_2$ phase was identified from DPs along [001], $[0\bar{1}3]$ and $[0\bar{1}6]$ directions. The phase was also found to form at the a-interlayer/c-Si interface. Examples are shown in Fig. 6. The TiSi phase was determined from DPs along [010] and $[2\bar{4}1]$ directions as shown in Fig. 7.

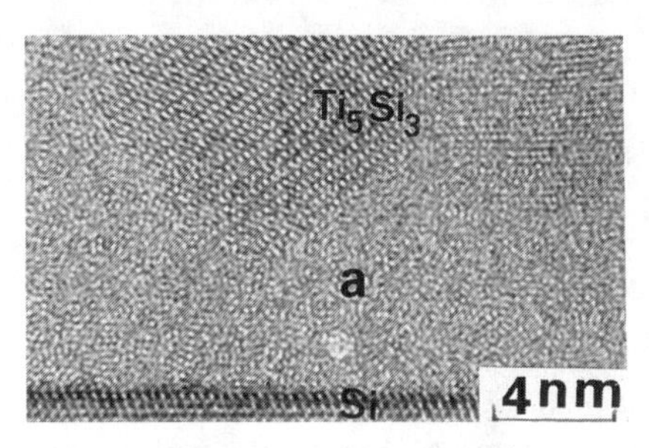

Fig. 4. HREM image of Ti_5Si_3, 450°C, 1 h.

C49-$TiSi_2$ was found to form at a-interlayer/Si interface in samples annealed at 500 °C for 10 min. The phase was found to grow rapidly in both grain size and extent with annealing time. In samples annealed at 500 °C for 1 h, C49-$TiSi_2$ was found to be the dominant phase with a small amount of TiSi present. An example is shown in Fig. 8. The carbon and oxygen contents were also found to be below the detection limit of the AES. The result indicated that the a-Si capping layer was very effective in preventing the indiffusion of C and O to participate in the interfacial reactions of Ti/Si system.

Fig. 5. (a) HREM image of Ti_5Si_4, 475°C, 1 h, (b) dark field micrograph, 500°C, 10 min.

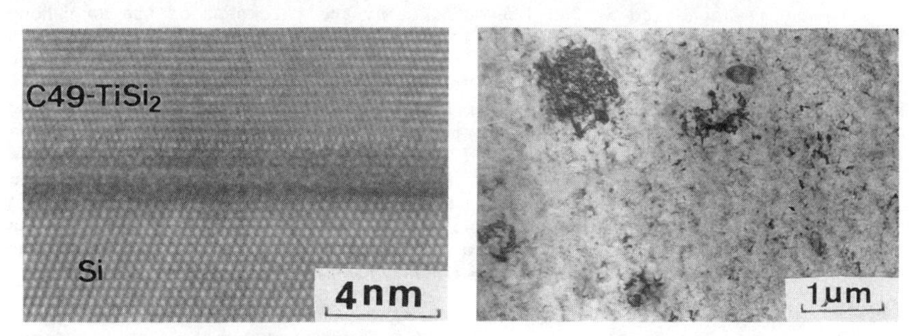

Fig. 6. (a) HREM image of C49-$TiSi_2$, 475°C, 1 h, (b) bright field (BF) micrograph, 475, 1 h.

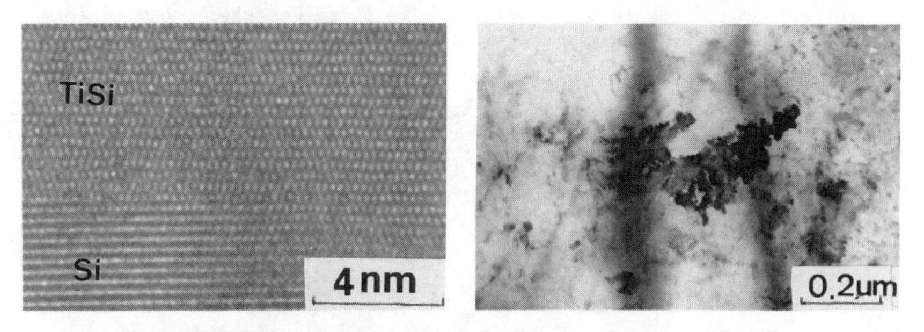

Fig. 7. (a) HREM image of TiSi, 475°C, 1 h, (b) BF micrograph, 500°C, 1 h.

Walser and Bene proposed a phenomenological theory to predict the first nucleated phase in the interfacial reactions of transition metal thin films on single-crystal Si. They assumed that a thin "glassy membrane" would form between the metal and Si substrate with a composition close to that of the deepest eutectic in the binary phase diagram before the formation of the first crystalline phase. The first nucleated phase is then close in composition to the glassy membrane [17]. The rule is about 80-90% successful in predicting the first nucleated phase. Bene further proposed that the reactions would follow the path of maximum in decreasing rate of the free energy change (maximum rate of energy degradation) [18]. Bene also predicted that th

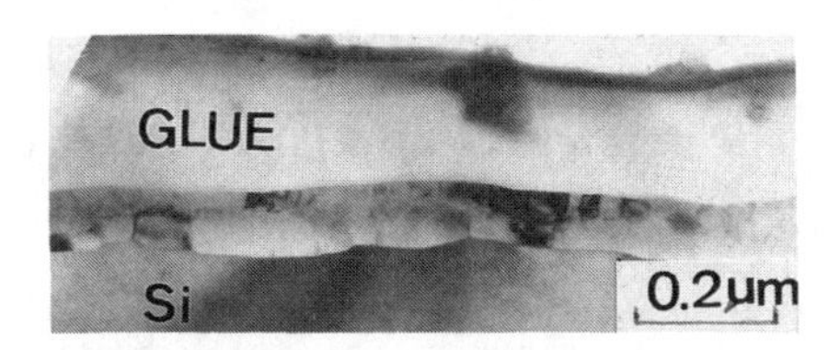

Fig. 8. BF micrograph, cross-sectional, 500°C, 1 h.

crystalline phase with composition and structure nearest to that of the amorphous alloy will be first nucleated from the glass interlayer by diffusionless transformation. The formation of an amorphous interlayer may serve as a kinetic constraint until the composition near the interface is adjusted to favor the nucleation of the crystalline phase. However, the validity of the reasoning had been in doubt owing to the lack of experimental evidence of the presence of a glassy membrane at the metal/Si interface. Recently, the formation of an amorphous interlayer in metal and single-crystal Si systems during low temperature annealing was reported to occur in Pt/Si, Ti/Si, Zr/Si, Hf/Si and a number of other refractory metal/Si systems [19-22]. Ti_5Si_3 was found to be the first nucleated phase. In view of the recent advance in the experimental techniques, a reassessment and refinement of the phenomenological theory for the first phase nucleation appear to be in order.

For a Ti overlayer on c-Si, the composition of the amorphous interlayer was found to range from 28 to 72 at.% Ti measured from Si to Ti overlayer after annealing [23, 24]. Ti_5Si_3 with a molar fraction of 62.5 at.% Ti is close to that of an amorphous interlayer at the side of the Ti overlayer. An examination of the published Ti-Si phase diagram showed that the Ti_5Si_3 has the highest congruent melting point, i.e. it has the largest driving force to nucleate from a supercooled melt which is close in composition to the Ti_5Si_3. In addition, the grain boundaries of a Ti overlayer served as nucleation sites which might serve to lower the energy barrier of the nucleation process. Previous measurements indicated that that the heats of formation of Ti_5Si_3, TiSi and $TiSi_2$ are 17.3, 15.5 and 10.7 kcal/g-atom, respectively [3]. The driving force for the formation of Ti_5Si_3 is therefore higher than those of other silicides. Accordingly, Ti_5Si_3 was found to form first at the interface of the Ti/a-interlayer.

Holloway et al. used multilayer samples with an overall composition of 40 at.% Si and found that the amorphous layer had chemical short range ordering (CSRO) near to the structure of Ti_5Si_4 [25]. The composition and local structure of the amorphous layer are often dependent on the way an sample was prepared, which in turn determines the first nucleated crystalline phase from the amorphous layer [23, 25]. After higher temperature (475 and 500 °C) annealing the composition of the amorphous interlayer is supposed to become more uniform and rich in Si. The nucleation of Ti_5Si_4 and TiSi is expected to be more favorable if the average composition of the amorphous interlayer is close to TiSi and the local chemical structure is similar to Ti_5Si_4. The formation of the Ti_5Si_4 was further aided by its epitaxial growth since the epitaxial silicide/Si interface energy is lower than that between polycrystalline silicide and silicon. The process window for the formation of the Ti_5Si_4 phase was found to be rather narrow making its detection difficult.

SUMMARY AND CONCLUSIONS

The initial stages of interfacial reactions of UHV deposited Ti thin films on silicon have been studied by in-situ RHEED and TEM.

RHEED and HREM analysis revealed that an intermixed amorphous layer was formed after deposition of the first 1.7-nm-thick Ti layer. In samples annealed at 350-425 °C for 1 h, the thickness of the amorphous interlayer was found to increase with annealing temperature and time. In samples annealed at 450 °C for 30-120 min, Ti_5Si_3 located at the Ti/a-interlayer interface was identified as the first nucleated phase. Ti_5Si_3, Ti_5Si_4, TiSi and C49-$TiSi_2$ were observed in samples annealed at 475 °C for 30 and 60 min as well as at 500 °C for 10 and 20 min. After annealing at 500 °C for 1 h, C49-$TiSi_2$ was the only silicide phase detected. Fundamental issues in silicide formation were discussed in light of the discovery of the formation of the amorphous interlayer and as many as four different silicide phases in the initial stages of interfacial reactions of UHV deposited Ti thin films on silicon.

Acknowledgment

The research was supported by the Republic of China National Science Council.

References

[1]M.E. Alperin, T.C. Hollaway, R.A. Haken, C.D. Gosmeyer, R.V. Karnaugh, and W.A. Parmantie, IEEE Trans. ED-32, 141 (1985).
[2]K.N. Tu and J.W. Mayer, in Thin Films Interdiffusions and Reactions, edited by J.M. Poate, K.N. Tu, and J.W. Mayer (Academic, New York, 1978), P. 329 and p. 399.
[3]M.A. Nicolet and S.S. Lau, in Materials and Process Characterization, edited by N.G. Einspruch and G.R. Larrabee (Academic, New York, 1983), P. 453.
[4]S.P. Murarka and D.B. Fraser, J. Appl. Phys. 51, 342 (1980).
[5]L.S. Hung, J. Gyulai, J.W. Mayer, S.S. Lau, and M.A. Nicolet, J. Appl. Phys. 54, 5076 (1983).
[6]G.G. Bentini, R. Nipoti, A. Armigliato, M. Berti, A.V. Drigo, and C. Cohen, J. Appl. Phys. 57, 270 (1985).
[7]R. Beyers and R. Sinclair, J. Appl. Phys. 57, 5240 (1985).
[8]M. Nathan, J. Appl. Phys. 63, 5534 (1988).
[9]I.J.M.M. Raaijmakers and K.B. Kim, J. Appl. Phys. 67, 6255 (1990).
[10]E.J. van Loenen, A.E.M.J. Fischer, and J.F. van der Veen, Surf. Sci. 155, 65 (1985).
[11]M. del Giudice, J.J. Joyce, M.W. Ruckman, and J.H. Weaver, Phys. Rev. B 35, 6213 (1987).
[12]J.M.M. de Nijs and A. van Silfhout, Appl. Surf. Sci. 40, 333 (1990).
[13]R. Butz, G.W. Rubloff, T.Y. Tan, and P.S. Ho, Phys. Rev. B 30, 5421 (1984).
[14]J. Vahakangas, Y.U. Idzerda, E.D. Williams, and R.L. Park, Phys. Rev. B 33, 8716 (1986).
[15]L.J. Chen and I.W. Wu, J. Appl. Phys. 52, 3310 (1981).
[16]T.T. Sheng and C.C. Chang, IEEE Trans, Electron Devices ED-23, 531 (1976).
[17]R.W. Walser and R.W. Bene, Appl. Phys. Lett. 28, 624 (1976).
[18]R.W. Bene, J. Appl. Phys. 61, 1826 (1987).
[19]J.R. Abelson, K.B. Kim, D.E. Mercer, C.R. Helms, R. Sinclair and T.W. Sigmon, J. Appl. Phys. 63, 689 (1988).
[20]A.E. Morgan, E.K. Broadbent, K.N. Ritz, D.K. Sadana, and B.J. Burrow, J. Appl. Phys. 64, 344 (1988).
[21]W. Lur and L.J. Chen, Appl. Phys. Lett. 54, 1217 (1989).
[22]J.Y. Cheng and L.J. Chen, Appl. Phys. Lett. 56, 457 (1990).
[23]E. Ma, L.A. Clevenger, C.V. Thompson, and K.N. Tu, Mat. Res. Soc. Symp. Proc. 187, 83 (1990).
[24]I.J.M.M. Raaijmakers, P.H. Oosting and A.H. Reader, Mat. Res. Soc. Symp. Proc. 103, 229 (1990).
[25]K. Holloway, P. Moine, J. Delage, R. Bormann, L. Capuano, and R. Sinclair, Mat. Res. Soc. Symp. Proc. 187, 71 (1990).

MORPHOLOGY OF VAPOR EVAPORATED Mo THIN FILMS

Y. CHENG AND M.B. STEARNS
Department of Physics, Arizona State University, Tempe, AZ 85287

ABSTRACT

Studies were made of the dependence of the morphology of Mo films, prepared by e-beam evaporation in an UHV system, on the substrate temperature and deposition angle. The main characterization techniques used were large angle x-ray scattering and cross-sectional high resolution electron microscopy.

INTRODUCTION

Multilayered thin films containing Mo are being used in many technical applications such as monochromators and other "optical" elements for soft x-rays. As part of a program investigating the structural dependence of Mo/Si multilayers on the deposition conditions, we have studied the variations in morphology of vapor evaporated Mo films with substrate temperature, T_s, and deposition angle, α. Here α is defined as the angle between the substrate normal and the incident vapor beam.

EXPERIMENTAL CONDITIONS

The Mo films were prepared by e-beam evaporation on oxidized single crystal (001) Si wafers. The growth pressure was $\leq 10^{-8}$ torr. The deposition rate used was 1 Å/sec. The films were deposited at 0°, 30° and 60° for $T_s = 300K$ and at 0° for $T_s = 550K$. The films are denoted by α and T_s; e.g. 30°-300K means the film was deposited at 30° from the normal at $T_s = 300K$. The thicknesses of Mo films were measured using a Sloan Dektak Surface Profilimeter. They were found to be between 3000Å and 3500Å.

The x-ray scattering spectra were measured using a Rigaku D/Max-IIB diffractometer with K_α radiation of a fixed Cu anode. The angular resolution of the diffractometer is 0.1°. Cross-sectional TEM specimens were prepared using a method involving dimpling and ion-mill thinning. These specimens were then imaged by a JEOM-4000EX high-resolution electron microscope operated at 400kV with a point-to-point resolution of better than 1.7Å. The specimens were tilted by a double-tilt top entry holder so that the single crystal Si substrate could be imaged along a [110] projection, with the (001) surface plane parallel to the electron beam direction. The structural characterizations of the films were made on the HREM images and selected area electron diffraction (SAED) patterns using (111) Si planes as an internal calibration.

RESULTS AND DISCUSSION

Figure 1 shows the large angle x-ray scattering (LAXS) spectra of these films. The maximum 2Θ of the diffractometer is 144°. Thus the angle of scattering (2Θ) was varied from 16° to 140°. The shutter was closed between 66° and 71°, in order to avoid damage to the x-ray counter by the extremely strong scattering from the (400) Si substrate (located at ~69°). It is seen that for all the Mo samples, the predominant diffraction peak occurs at $2\Theta \approx 40°$. This corresponds to the crystallite growth of (110) Mo which is the orientation having the smallest in-plane area per atom. Weak Bragg peaks from other orientations are also visible. These are usually less than a few percent of the (110) peaks. Thus the Mo crystallites grow predominantly in the bcc (110) orientation. This result agrees with the well-known general rule of growth of single component films that the preferred growth orientation is that which minimizes the energy by maximizing the number of bonds present between atoms in the plane of the films[1]. This corresponds to the orientation which has the smallest in-plane area per atom or the largest in-plane density of atoms. Note that there are few or no (200) Mo crystallites at 0°-300K in contrast with Mo films prepared at 30°-300K and 60°-300K, or the film prepared at 0°-550K, in which the

(200) Mo orientation is clearly visible. The increased (200) orientation growth has been observed for fcc Ni[2] and Pd[3], and bcc Cr[4] e-beam evaporated thin films at substrate temperatures higher than 300K. This is likely due to the enhanced coalescence of crystallites by surface as well as volume diffusion at higher T_S. Increased (200) crystallite growth at higher deposition angle, as seen in these Mo films, indicates that the α has effects similar to that of T_S on crystallite growth of thin films.

It is found that for (110) Mo crystallites, the values of 2Θ decrease for higher deposition angle α at T_S = 300K corresponding to a larger d-spacing at higher α for the (110) Mo crystallites. The peak widths also become slightly broader as α increases. Since the peak widths vary inversely with the crystallite lengths, this shows that the Mo crystallites have a smaller size in the direction of growth for the Mo films prepared at higher α. This limitation of crystal size is due to the shadowing effects at higher angles.

It is also seen that the Mo film prepared at 0°-550K has peak widths which are much narrower than those prepared at T_S = 300K. In particular for the (110) Mo crystallites, the peak width was found to decrease from 0.83° for 0°-300K to 0.28° for 0°-550K. This corresponds to crystallite lengths of 120Å and 520Å, respectively, in the direction of growth. This increase in crystallite size at higher T_S is primarily due to the increased mobility of the adatoms at higher T_S[5],[6] and has been generally seen in many thin film systems.

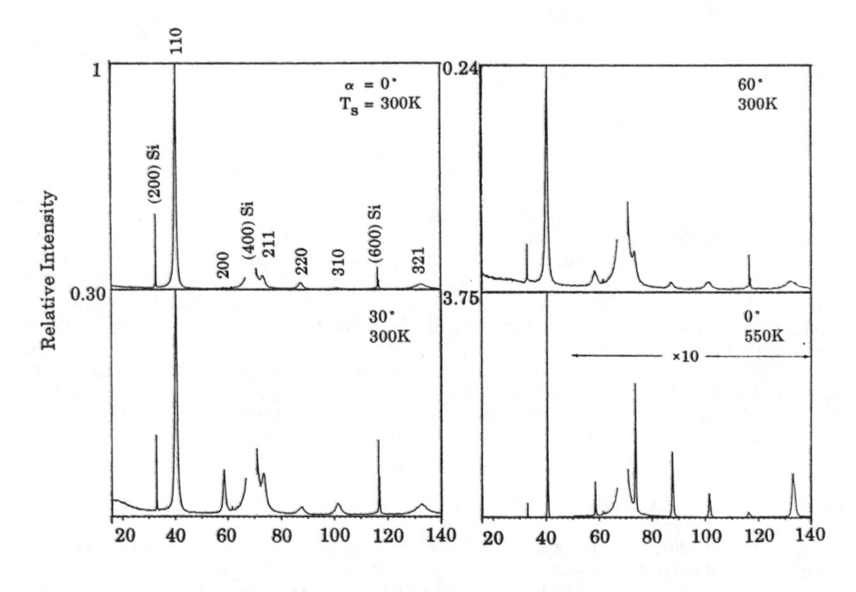

Fig. 1 Variation of the intensity of the LAXS spectra of Mo films with deposition angle and substrate temperature. The peaks are identified by their scattering planes (h,k,l). The peaks at 33°, 69° and 116° are due to the nominal (100) single crystal Si substrate. The (110) orientation of Mo is seem to dominate with small amounts of other orientations also visible.

Table I The average d-spacings and sizes of (110) Mo crystallites in the direction of growth obtained by fitting the measured and calculated LAXS spectra for the (110) peaks.

α (°)	T_s (K)	d-spacing ±0.002Å	Crystallite Length ±3Å
0	300	2.225	120
30	300	2.228	95
60	300	2.232	110
0	550	2.222	520
Bulk Mo: [8]		2.225	

Table I lists the d-spacings and average crystallite lengths in the direction of growth of Mo films obtained by fitting the measured LAXS spectra of the (110) peaks with the calculated spectra[7]. For films prepared at $\alpha = 0°$, the d-spacing of (110) crystallites is found to be about the same as the bulk Mo and there is a slight increase in the d-spacing with angle α. Moreover, the (110) Mo crystallite lengths in the direction of growth for the films grown at $T_S = 300K$ and α of 30° and 60° appear to be somewhat smaller than that prepared at $\alpha = 0°$. Both of these effects are reasonable due to the columnar growth at higher α.

Figure 2 shows the selected area electron diffraction (SAED) pattern for the pure Mo film grown at 30°-300K. The strong spotty annular diffraction ring is due to the (110) Mo crystallites. Weak diffraction rings due to other Mo crystallite orientations are also visible. This agrees with the results of the LAXS, where the dominant x-ray diffraction peaks were observed to be due to (110) Mo crystallites. Using the (111) Si diffraction spots as calibration, we calculated the d-spacings of the Mo crystallites. They were in good agreement with that obtained from the LAXS.

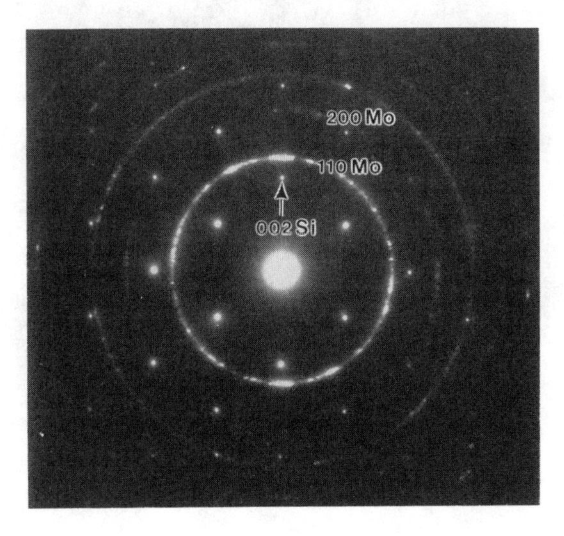

Fig. 2 Selected area electron diffraction pattern for the Mo film prepared at 30°-300K.

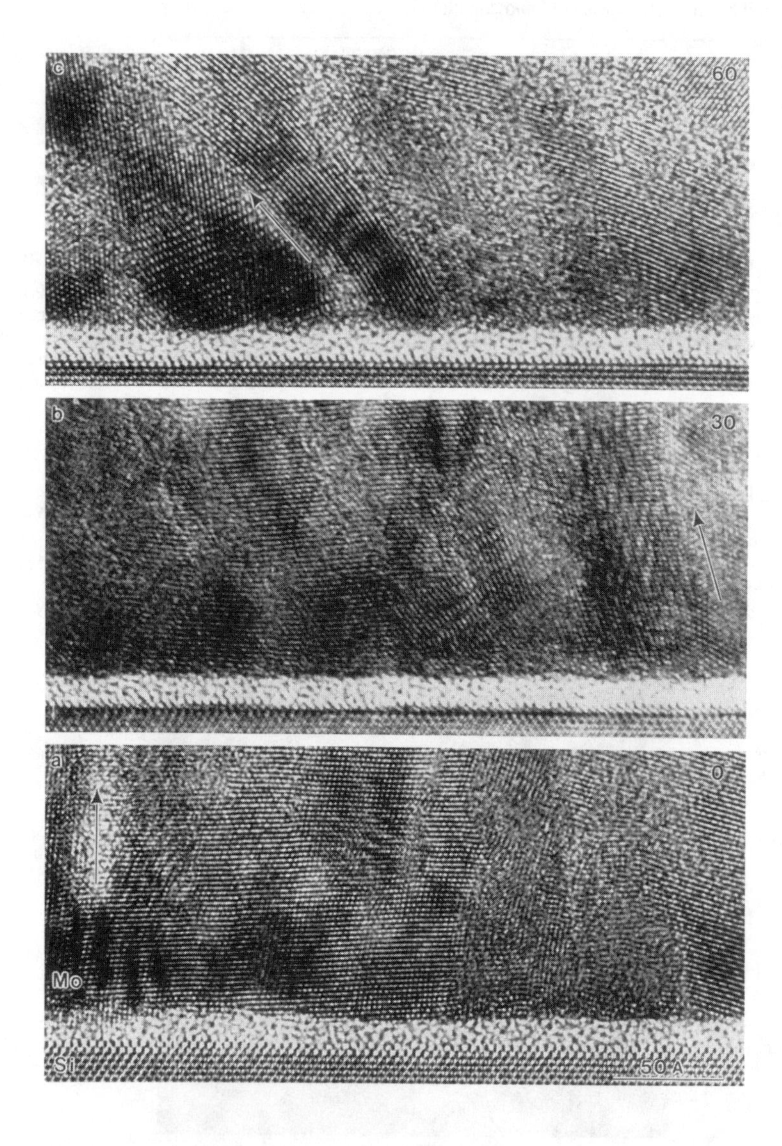

Fig. 3 High resolution micrographs of Mo films prepared at T_s = 300K and α = 0°, 30° and 60°. Mo lattice fringes and the columnar structures of Mo crystallites are clearly visible. The lattice fringes of the Si substrate and the amorphous SiO_2 can be also seem at the bottom of each micrograph. The directions of columnar growth are indicated by the arrows.

Figure 3 shows some typical regions of the cross-sectional micrographs for Mo films prepared at T_S = 300K and $\alpha = 0°$, 30° and 60°, respectively. They were taken in transmission with a magnification of 500k. Columns of Mo crystallites surrounded by lower density, amorphous Mo, are clearly visible for all the samples. Their directions of growth are indicated by arrows. These directions could also be determined from the tilt of the spots in the SAED patterns. For the Mo film prepared at $\alpha = 0°$, shown in Fig. 3(a), the widths of the Mo crystallites were found to vary from ~30Å to ~180Å in the plane of the film, with a typical width of ~35Å - 60Å.

As seen in Fig. 3b and 3c for the Mo films prepared at $\alpha = 30°$ and 60°, the Mo crystallites have a columnar structure which grows at an angle away from the film normal. Many reports of columnar microstructures have appeared in the thin film literatures[9],[10]. This microstructure consists of a network of low density material that surrounds tilted columns of higher density material. The columns arise from the shadowing of the adatoms by atoms that have been previously deposited. Empirically, both measurements and computer simulations have often found that the angle of inclination on the columns from the film normal, β, obeys a tangent rule given by[11]

$$2\tan\beta = \tan\alpha$$

The angle β measured on the micrographs for Mo films prepared at $\alpha = 30°$ and 60° were ~16° and ~41°, respectively. These angles are in good agreement with those calculated by the tangent rule. It was also found that the typical lateral widths of the Mo crystallite in the direction perpendicular to the column growth were ~60Å for $\alpha = 30°$ and ~30Å for $\alpha = 60°$.

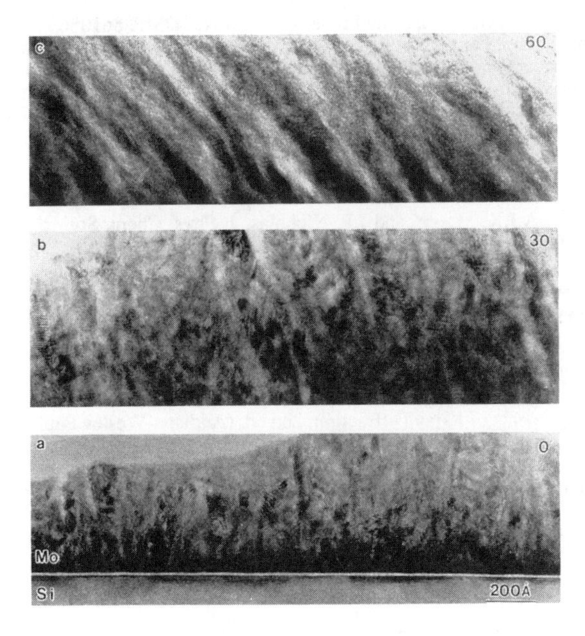

Fig. 4 Low Magnification TEM micrographs of Mo films prepared at T_S = 300K and $\alpha = 0°$, 30°, and 60°.

Low magnification micrographs for films prepared at 0°-300K, 30°-300K and 60°-300K are shown in Fig. 4. The light and dark stripes clearly show the columnar structures. It is seen in Figs. 3 and 4 that the columnar structure is much more noticeable in the films prepared at higher α. In particular, note the larger areas of light streaks corresponding to more low density amorphous Mo in the film prepared at $\alpha = 60°$. Thus the mass density of the films is seen to change with α. We directly measured this change by using a digital balance with an accuracy of ±10μg to obtain the weight of the Si substrate and the weight of the Si substrate plus the Mo film after the deposition. The mass density of the Mo film was then determined using the film geometry as measured by a surface profilimeter. The average densities of Mo films deposited at 300K were found to be 8.1, 7.9 and 5.8 g/cm^3 for $\alpha = 0°$, 30° and 60°, respectively. These are all smaller than the density for bulk Mo, 9.0 g/cm^3. The density is seen to decrease strongly as α increases.

CONCLUSIONS

Both the LAXS and TEM studies showed that the microstructures of the e-beam evaporated Mo films depend strongly on the growth parameters, T_s and α. The (110) orientation was found to be the preferred growth direction for the deposition conditions we investigated. The (110) Mo crystallite length in the direction of growth increased for higher T_s due to the increased adatom mobility. TEM micrographs showed strong columnar growth in these films. The columnar growth direction was found to follow the tangent rule which is derived from the self-shadowing of the adatoms.

ACKNOWLEDGMENTS

This work was partially supported by NSF Grant No. DMR-8610863 and by Lawrence Livermore National Laboratory through the Department of Energy contract No. W-7405-Eng-48. The x-ray data were obtained on equipment purchased under NSF Grant No. DMR-8406823. The HREM work was done on the facilities partially supported by NSF Grant No. DMR-8611609. The authors wish to thank Dr. David J. Smith for taking the HREM micrographs.

REFERENCES

1. J.K. Mackenzie, A.J.K. Moore, and J.F. Nicholas, J. Phys. Chem. Solids, 23, 185 (1962).
2. C.-H. Chang, Ph.D. thesis, Arizona State University, 1989.
3. G.D. Lewen, MS thesis, Arizona State University, 1990.
4. Y. Cheng, M.B. Stearns and D.J. Smith in Thin Film Structures and Phase Stability, edited by B.M. Clemens and W.L. Johnson (Mater. Res. Soc. Proc. 187, Pittsburgh, PA 1990) pp.151-156.
5. S.G. Fleet, Mullard Res. Rept. 466 (1963).
6. K.L. Chopra, Thin Film Phenomena, (McGraw-Hill, 1969) p.185.
7. M.B. Stearns, Phys. Rev. B38, 8109 (1988).
8. B.D. Cullity, Elements of X-ray Diffraction, 2nd ed. (Addison-Wesley Publishing Company, Inc., 1978) p.506.
9. C. Kooy and J.M. Nieuwenhuizen, Basic Problems in Thin Film Physics, edited by R. Niedermayer and H. Mayer, (Vandenhoeck & Ruprecht, Gottingggen, 1966), p.181.
10. J.M. Nieuwenhuizen and H.B. Haanstra, Philips Techn. Rev., 27, 87 (1968).
11. H.J. Leamy and G.H. Gilmer, Current Topics in Material Science, Vol.6, edited by E. Kaldis, (North-Holland Publishing Company, 1980), p.309.

Author Index

Subject Index